Climate Adaptation Through Environmental Justice

Climate Adaptation Through Environmental Justice

Comparative Approaches to Enhance the Resilience of Indigenous Peoples and Minority Groups

Jeremy Julian Sarkin

Distinguished Research Professor of Law, NOVA University of Lisbon, Portugal

EE Edward Elgar PUBLISHING

Cheltenham, UK • Northampton, MA, USA

Published by

Edward Elgar Publishing Limited
The Lypiatts
15 Lansdown Road
Cheltenham
Glos GL50 2JA
UK

Edward Elgar Publishing, Inc.
William Pratt House
9 Dewey Court
Northampton
Massachusetts 01060
USA

Authorised representative in the EU for GPSR queries only: Easy Access System Europe – Mustamäe tee 50, 10621 Tallinn, Estonia, gpsr.requests@easproject.com

A catalogue record for this book is available from the British Library

Library of Congress Control Number: 2025947041

This book is available electronically in the Elgaronline
Law subject collection
https://doi.org/10.4337/9781035364008

MIX
Paper | Supporting responsible forestry
FSC® C013056

ISBN 978 1 0353 6399 5 (cased)
ISBN 978 1 0353 6400 8 (eBook)
ISBN 978 1 0353 8258 3 (ePub)

Printed and bound in Great Britain by
TJ Books, Padstow, Cornwall

Contents

Acknowledgments

I have been working on issues relating to the protection of people from human rights abuses for my whole career. I have worked on these issues in many places. I have written on these issues extensively including on many places. Many of my publications are empirical studies in nature, or a combination of theory and practice. While I work in the law discipline, I have become very interdisciplinary and multidisciplinary in my research. Approaching legal issues from an array of vantage points is a goal of mine. This book is a testament to that. This multidisciplinary and interdisciplinary approach has allowed me to publish in a variety of impact journals, almost all in SCOPUS or Web of Science indices. Various methodologies, many implicit in my work, inform my research work in general. This type of legal research is essential for many reasons, including assisting in the law reform process, in which I am very heavily involved in various settings. I believe that legal research is, at times, about finding the law, engaging with it, determining its characteristics, and determining whether the role it is playing is useful or not. It is often, although not always, about making suggestions about how it ought to be changed or reformed.

One area that has been a key focus of mine, is how vulnerable groups are particularly affected by massive human rights abuses. My work examines broader issues such as the role of reparations, reconciliation, amnesties, and truth commissions. It has focused on many countries as well as various mechanisms. These themes are the role and function of international courts, internationalized courts, domestic courts, traditional courts dispensing justice, giving amnesty for truth, reparations claims for human rights violations, achieving reconciliation, as well as the operation of a number of truth commissions.

I have worked on various matters such as bringing to the fore the issue of reparations for colonialism. I served for many years as the Legal Advisor to the Paramount Chief of the Herero of Namibia for their claims for reparations from Germany for the genocide committed between 1904 and 1908. This led to the instituting of a number of court cases in the United States for claims for reparations. Other countries then followed suit to also institute claims. These types of cases are now growing all over the world.

As a result of this work, as well as my work in South Africa on the truth and reconciliation process, I have worked in the field of transitional justice for many years. I have worked on such processes in Uganda, Zimbabwe, the Maldives, Nepal, Burundi, Morocco, Bosnia and Herzegovina, Armenia, Azerbaijan, Georgia, St Helena Island, Timor-Leste, the DRC, Iraq, Libya, Tunisia, Indonesia, Philippines, Bangladesh, Syria, and Lebanon.

An interesting project that I was involved in was to begin a process in Iraq in 2017 to write a new constitution. In this regard, the United Nations Development Programme (UNDP) asked me to help develop a constitutional process that delivers a constitution that is respected and could help to resolve the crises that exist in many parts of the country. I thus met with the prime minister and other senior officials. I also drafted a law "Towards a More Secure, Unified, and Peaceful Iraq Law."

Another project was when in 2019 I was asked by the UK Government to review the governance system on St Helena Island. I went there a few times in 2019 and 2020 to facilitate a debate and gather ideas on the Committee-based system and whether it ought to change. I was tasked with exploring the understanding and appetite for alternative systems, for example, ministerial, executive councillors, how to redefine the role of the Legislative Assembly as well as the Executive Committee, the role of a Chief Islander, whether there ought to be a restructuring of the government's directorates, etc. This was done to have more clear political leadership and responsibility. The task was to shape an accountable system of governance. I wrote two reports, both of which were accepted. A process then for adopting such change was then embarked on. The UK Privy Council then endorsed those changes. That Order in Council was laid before Parliament in London, which was the final step in the process in the UK. St Helena Governor Phillip Rushbrook commented in a Press Release:

> Back in 2019, Professor Sarkin identified the public wanted clearer individual responsibility for making political decisions, greater political accountability for delivering services and reforms, and a governance system that progressed its delivering services and reforms, and a governance system that progressed its business more swiftly. The change to a Ministerial system provides the opportunity for the next elected government to achieve all three.

Another process I was involved in was to get the Human Rights Council (HRC) of the country Georgia revitalized and operating again after many years. In 2018, I was requested by UNDP to evaluate the role and function of the Council and the Secretariat. My work evaluated why the HRC ought to be reinvigorated, and why and how it ought to be transformed into a National Mechanisms for Reporting and Follow-up (NMRF) institution. My work also was to meet with the prime minister of the country and other senior officials and persuade them of the necessity of doing so. Shortly after my visit the HRC was revised. I received written praise from the UN Resident Coordinator for my work. I have written a number of articles about these institutions.

A recent project for vulnerable people, was my work on the need for a mechanism to search for those who were arbitrarily detained, or suffered an enforced disappearance in Syria. This led to the UN General Assembly creating such a process in 2023. It was with the study released in June 2021 that I conducted for five Syrian NGOs (the Truth and Justice Charter Group), and then my book and various journal publications thereafter, that the matter was taken up by the United Nations. My work reviewed the possibility that one million victims have been arbitrarily detained or enforcedly disappeared in Syria. It discusses the general processes that have been set up to deal with Syria. It then reviews which mechanisms deal with disappearances and detentions in Syria that are available for families to report to and the limited role those institutions have played so far, before offering options for the creation of a new mechanism dedicated to finding out what happened to Syria's disappeared and detained. What occurred is that the General Assembly adopted Resolution 76/228 on 24 December 2021. The Resolution requested the Secretary-General to perform a study on what could be done to clarify the fate and whereabouts of missing people in Syria and requested a report by mid 2022. That report, issued on 2 August 2022, recommended a new international mechanism to deal with the missing. The General Assembly then adopted a Resolution in 2023 setting up such a process.

The difficulties in getting the process established can be seen in the comment of David J. Scheffer, a former US Ambassador at Large for War Crimes Issues, who noted that the:

> proposed creation of a new mechanism to conduct searches for disappeared and detained people in Syria and find information for their families ... are ambitious proposals, most of which would confront political firestorms by powerful nations and a stubborn U.N. bureaucracy. But Sarkin puts his case on the table forthrightly and with significant evidence and his views are worthy of serious consideration in both the policy and academic worlds.

This book, on climate change and environmental justice, emerged out of a Foreword I wrote entitled "The Interrelationship of Climate Change, Climate Justice but also Environmental Justice" that was contained in the book edited by Subhram Rajkhowa and Sriparna Rajkhowa entitled *Climate Change and Climate Justice* (Satyam Law International 2024). As I wrote that I realized that little was written on these matters.

I was able to do much research work whilst on sabbatical in 2023–2024 at various places. These included as a Senior Research Fellow at the Institute of Commonwealth Studies, School of Advanced Studies, at the University of London. I was also able to use the libraries at the University of Cincinnati, KU Leuven, in Belgium, Wroclaw University, Poland, American University in Washington DC, and the University of Illinois in Champaign, Illinois. I thank them all for the opportunities that the visits to do research in their institutions accorded me.

My sabbatical year was an incredibly busy, productive, and enriching experience, both academically and personally. It provided me with the opportunity to travel extensively, engage with scholars from different disciplines, and contribute significantly to my field. The topics covered range from human rights and international law to climate justice and artificial intelligence. These publications reflect my diverse research interests and the interdisciplinary and empirical nature of my work. The sabbatical also allowed me to make substantial progress on long-term research projects and establish new lines of inquiry. The publications produced will hopefully contribute meaningfully to the fields of human rights, transitional justice, international security, and environmental law and environmental justice besides others. The time allowed me to focus deeply on my research, collaborate on various international projects, and ultimately produce a number of high-impact publications.

There are many to thank for their assistance in the process to write this book specifically. At NOVA University Lisbon Law School, the research center CEDIS were very helpful in various ways. The funding agency FCT in Portugal assisted with funding. Various people assisted along the way in various ways: Benjamin Dugdale, John McManus, Emilie Weingarden, Maya Huffman, Chelsea Szilagyi, Nair Luana Cardoso, Catarina Vaz Carvalho, Maria Joao Marques, and Raquel Relvas. A lot of time was spent on sorting the formatting and checking the accuracy of the references, a very thankless task. I thank my family (Rosanne, Eryn and Hannah) for assisting this project in various ways, including reading, editing, checking references etc. and for their love and support.

List of abbreviations

Term	Definition
ACECs	Areas of Critical Environmental Concern
BLM	Bureau of Land Management of the United States
CCE	Climate change education
CCS	Carbon capture and storage
CEE	Climate, energy and environmental
DAC	Disadvantaged Communities
DEFF	Department of Environment, Forestry and Fisheries of South Africa
FEMA	Federal Emergency Management Agency of the United States
GHG	Greenhouse gases
IAA	Impact Assessment Act of Canada
IEA	International Energy Agency
IMO	International Maritime Organization
IPBES	Intergovernmental Science-Policy Platform on Biodiversity and Ecosystem Services
IPCC	Intergovernmental Panel on Climate Change
LGBTQ+	lesbian, gay, bisexual, transgender, queer and others
LIHEAP	Low Income Home Energy Assistance Program in the United States
MDGs	Millennium Development Goals
NAP	National Action Plan of South Africa
NAS	National Adaptation Strategy of Canada
NCCR-WP	National Climate Change Response White Paper of South Africa
NDCs	Nationally determined contributions
NEA	National Energy Act of South Africa

Term	Definition
NGOs	Nongovernmental organizations
NHPA	National Historic Preservation Act of 1966 of the United States
PCF	Pan-Canadian Framework on Clean Growth and Climate Change of Canada
ppm	Parts per million
SDGs	Sustainable Development Goals
SEC	Securities Exchange Commission of the United States
UFET	United Front for Energy Transition of Canada
UN	United Nations
UNEP	United Nations Environment Programme
UNFCCC	United Nations Framework Convention on Climate Change – Paris Agreement
UPR	Universal Peer Review
WHO	World Health Organization
WMO	World Metrological Organization
2SLGBTQIA+	This term is used in Canada. Its component parts refer to 2S which denotes two-spirit people, L denotes Lesbian; G for Gay; B for Bisexual; T for Transgender; Q for Queer; I for Intersex; and + is meant to signify inclusivity for others who use further terms.

1. Introduction to *Climate Adaptation Through Environmental Justice*

INTRODUCTION

Severe climatic and environmental shifts mark the modern era (Caesar et al. 2024). However, as the 21st century unfolds, we hear really only the term "climate crisis" which is used with increasing frequency and vehemence. This term, although all-encompassing, can distract from the reality that the world is, in fact, enduring a range of environmental catastrophes simultaneously (Rodríguez-Garavito 2022). This is besides a range of other problems including increasing conflict and human rights violations. Only one of these environmental catastrophes is climate change. There are other interrelated issues affecting the planet including nature and biodiversity loss, as well as those of pollution and waste and a range of other matters. These are, however, interconnected. However, some issues, such as the matter of the loss of biodiversity, get far less attention, especially by the media (Legagneux et al. 2018).

These issues cannot be seen in isolation though. They have deep effects on each other. Other issues too, occurring concurrently, are rising conflicts and increasing violence and an increase in human rights abuses around the world. As was noted by The Elders in October 2024:

> All of humanity is confronted with existential threats – the climate and nature crisis, pandemics, nuclear weapons and unregulated Artificial Intelligence – that can only be addressed by countries defusing these geopolitical tensions and working together. (The Elders 2024a)

A range of problems are likely to surge in the future as the effects of climate change are felt. As people need to relocate and as more minerals and other types of resources are sourced for the green transition, so too will people's environment be negatively affected by mining and other processes, thus exacerbating the likelihood of conflict and violence. All these issues act to exacerbate the problems that exist especially for vulnerable groups.

As far as climate change specifically is concerned, the World Metrological Organization (WMO) has predicted that there is a high likelihood that by 2027 the global temperature will increase to about 1.5 degrees Celsius above pre-industrial levels. Already, 2023 was the hottest year on record seemingly for 2000 years (Büntgen et al. 2024), and the 1.5 degrees above the pre-industrial revolution target is very close to being breached (World Meteorological Organization 2024). The WMO states that there is a high likelihood that by 2027 the global temperature will increase to above that predicated level. It is also likely that at least one of the years before 2027 will be the hottest on record (UN News 2023).

What is clear, however, is that the negative environmental impact of climate change will continue to accelerate over time unless more action is taken to address this pressing issue. In fact, some predict that keeping within the 1.5 degrees is already not possible (Hansen et al. 2023). While some argue that breaching this level, and examining "overshoot pathways," is possible and ought not to be such a problem if it is later reversed, research in 2024 indicated that such a later reversal may not be possible and may not have the desired positive effects (Schleussner et al. 2024).

Nevertheless, the target of 1.5 degrees may be academic, as some argue that the planet is moving towards an average global temperature (Warren et al. 2024) of between three and five degrees higher than that of the pre-industrial era (Urbancic 2020). More than three-quarters of the hundreds of climate scientists surveyed in 2024 believed that 2.5 degrees was more likely (Carrington 2024a). In 2024, the United Nations Environmental Programme (UNEP) found that cuts in greenhouse gas emissions of 42 percent need to be obtained by 2030 and a 57 percent cut in such emissions is needed by 2035 to keep within the 1.5 target (UNEP 2024). The UNEP found in 2024 that little progress was being made to arrest emissions around the world and the amount of gases in the environment was actually increasing. Thus, it was reported, according to new data from the Global Carbon Project, carbon dioxide emissions from fossil fuels were expected to rise by 0.8 percent over 2023 levels to a record 37.4 billion metric tons in 2024 (Global Carbon Project 2024). As will be noted below, achieving the goals set seems difficult unless drastic steps are taken around the world.

If changing course does not occur, it is believed that the effects of climate change on the world's population will be severe. It is estimated that possibly 1.4 billion people may need to relocate by 2060 (Geisler and Currens 2017). This is because of the concern about the ability of humans to adapt to the changing environment. The major issues affecting the globe are the "stability, resilience, and life-support functions of our planet" (Caesar et al. 2024). These three core dimensions of the earth system are vital for a thriving earth. Thus, climate change is a major concern for all of humankind in one way or another.

One of the major problems is that the world's population is using more than 1.6 times the resources than the earth can rejuvenate. This is having a massive negative effect on sustainability, and is placing huge burdens on the seas, forests, biodiversity, and the whole environmental system. This in turn further feeds into climate change issues.

Some have postulated that the effect of unaddressed climate change will be the sixth mass extinction on earth (Cowie et al. 2022); that is, the extinction of a large percentage of existing species in a brief period of time (Kaiho 2022). However, while the previous mass extinctions were the result of natural events, the projected mass extinction is being triggered by the activity of humans (Ceballos and Ehrlich 2018): by the problematic and unsustainable use of the earth's soil and water resources; the overexploitation of various natural resources, particularly those used for energy production; and the misuse and abuse of various plant and animal species as human food sources (Etieyibo 2022).

Climate change is also being caused by rampant pollution, unchecked deforestation, the emission of enormous quantities of waste (United Nations Environment Programme 2024b), and other similarly destructive human activities (Fowler 2023). This new geological epoch has been called the Anthropocene, by some, as it depicts a world altered by the impacts of industrialized humankind (Crutzen 2016). This is the case because it is seen to be an epoch in which humankind has itself become a geological force with a level of technological and scientific power that renders it capable of transforming the entire planet (Saito 2023). However, that official designation is controversial and has not yet occurred (Witze 2024). Thus, the earth remains in the Holocene Era (Fletcher et al. 2024; Wallenhorst 2023), which commenced 11,700 years ago (Zhong 2024).

It is, however, the effect of humanity on the environment of the planet that has been enormous. The effect of it will be to see:

> land surface changes and alterations in regional climates and weather extremes … large-scale shifts in atmospheric and oceanic circulations, altering global warming rates and sea level rise … widespread shifts in temperature and precipitation patterns … marine ice cliff collapse, [that] have the potential to accelerate sea level rise, affecting flooding hazards like coastal inundation. Biosphere tipping points, such as Amazon dieback, intensify greenhouse gas concentrations, hastening global warming and its associated extreme weather events, regional climate shifts and sea level rise. (Betts et al. 2024)

Some estimate that climate change effects could see more than 230,000 people dying each year in poorer countries by 2030 (Lawson and Jacobs 2022). The problems with the biosphere integrity of the planet is causing losses to the planet's adaptability and resilience (Lawson and Jacobs 2022). Importantly, it means that the biosphere is increasingly becoming incapable of absorbing the effect of climate change. As a result, the extinction rate of various species is up to 100 per million species years. This means that there is a constantly increasing global biodiversity loss. It further signals that biosphere integrity of the planet is accruing losses to adaptability and resilience. Importantly, it means that the biosphere is increasingly becoming incapable of absorbing changes to planetary processes (Caesar et al. 2024). This can be seen in the fact that in 2024, it was found that one in five of the 1,189 species on the land, sea, and skies covered by the UN Convention on the Conservation of Migratory Species of Wild Animals (CMS) are threatened with extinction, with 44 percent of them experiencing population declines, according to the first-ever State of the World's Migratory Species assessment (Convention on the Conservation of Migratory Species of Wild Animals 2024).

One massive example of a die-off of a species was the effect of a marine heatwave caused by climate change, which saw 4 million birds, the common murres, being killed between 2014 and 2016 (this was half the population) (Renner et al. 2024). Also affected, but not to the same effect, were Humpback whales as well as Alaska Pacific cod.

To try and limit the extinction trend, an important nature protection roadmap was established in 2022 at the COP15 UN Biodiversity Summit, as part of the Kunming-Montreal global biodiversity framework. The "30x30" agenda, agreed to conserve 30 percent of the world's land and oceans by 2030. However, its goals are not being met, specifically as far as the oceans are concerned, but also generally (Fox et al. 2024). This is alarming because of its enormous potential effects on the planet. Again, as is noted throughout this book, all these issues are interlinked and need to be tackled simultaneously, and in an integrated fashion.

Climate change will have other dire effects on the planet. It is believed, for example, that climate change will bring more infectious diseases. As far as noncommunicable diseases are concerned, already 41 million persons die each year from these, making up 75 percent of all deaths. This number will increase over time, at least partly because of climate change, unless it is arrested.

In this context, research by the World Health Organization (WHO) indicates that identifiable preventable environmental hazards cause about 13 million deaths annually (Prüss-Üstün et al. 2016). This will increase, as by the 2050s eight times as many children will be affected by extreme heatwaves, twice as many by wildfires, and three times as many by floods (UNICEF 2024). The increasing number of floods is going to have enormous effects, including the release of contaminants from landfills and other hazardous waste sites. The groundwater and rivers are very negatively affected by this (Edirisinghe and Suchet-Pearson 2024), but there are also harmful chemical releases into the air. Already, the various types of pollution have an enormous combined negative effect on humans. Studies have shown that 2.3 billion people around the world are already harmfully affected by soil degradation (Gupta and Vegelin

2023). It is estimated that another four million people die every year because of outdoor pollution (Li et al. 2019). Pollution caused about 8 million people to die in 2021 (Health Effects Institute 2022). A million children die annually, two-thirds of whom are under five years of age (Landrigan et al. 2019).

While these problems will have enormous effects on humans, they also will affect other flora and fauna more so than occurred before (Mahon et al. 2022). One example, reported in 2024, was that 350 elephants that died in Botswana in 2020 were poisoned by toxic algae that grew in waterholes as a result of an abrupt change from drought to heavy rainfall because of climate change (Lomeo et al. 2024).

Air pollution is one of the problems causing massive health implications. Air pollution affects many more people, as it has an extremely negative impact on people's respiratory and cardiovascular systems. It places stress on, and damages, people's central nervous systems, cognitive functions, and skin, interfering with normal human development, which, combined, results in seven million avoidable deaths each year (WHO 2019a). Air pollution is also causing an increased resistance to antibiotics, resulting in millions of otherwise preventable deaths each year (Zhou et al. 2023). In fact, the Air Quality Life Index report for 2021 stated that air pollution stands as the "most significant external menace to human life expectancy" and is responsible for a decrease in the average lifespan globally by two years and three months (Air Quality Life Index 2021). However, that reduction in life expectancy was more than six years for people living in some countries, most notably in the Global South. Thus, of the 7.3 billion people who suffer high pollution concentrations, 80 percent live in poor and middle-income nations (Rentschler and Leonova 2023). A critical issue is that the quantity and toxicity of air pollution is not the same everywhere, and often the people most affected are historically marginalized groups (Wang et al. 2023).

The reality is that many millions of people suffer the consequences of the combined effects of soil, water, and air pollution (Gupta and Vegelin 2023). However, as has been noted:

> Unevenness is a defining feature of pollution … Unevenness is not only a description of pollution and its harms, not only a side effect, and not remotely an accident: unevenness is an accomplishment of pollution. It is its goal. (Liboiron 2021)

Thus, the problems are not evenly spread. Pollution affects people differently in the sense that poorer people, usually members of minority groups, are more likely to have pollution on their doorsteps. Interestingly, studies indicate that communities characterized by social and economic disparities may have a higher propensity to cause environmental degradation or pollution (Morello-Frosch and Obasogie 2023). Nevertheless, the effects on the health of people are disproportionately felt by the most vulnerable (Dialesandro et al. 2021). One glaring example concerns access to clean water (Petersen et al. 2022). It is minorities who have the least access to, and the worst quality of, water. Unsafe unclean water, limited and poor sanitation, and hygiene problems cause nearly two million deaths each year worldwide (WHO 2019b).

Pollution also has a devastating effect on the soil, air, and water of the community where it occurs. One source of a particularly dangerous and insidious form of pollution is the problematic use of various types of chemicals that have been released into the air, soil, and water, which have been on the rise since 2000, with a steady increase in their presence in the 2000–2017 period. These facilitate antibiotic resistance, bioaccumulation, cancerous and hormonal health effects, as well as "ecosystem disruption" and general pollution. Their safety risks are extremely pronounced (Caesar et al. 2024). Specific examples are pharmaceutical drugs and chemicals in pesticides that cause extensive health problems (Daughton and Ternes 1999). Also problematic are a range of other chemicals such as per- and polyfluorinated substances or PFAS (known as forever chemicals) (Benesh 2020), which are having widespread harmful effects on people, as well as the flora and fauna (Pelch et al. 2019). These synthetic chemical products are negatively affecting the environment, including the soil (Ruppel 2022a), and the diversity of the organisms therein, but also in the air, and water across the globe (Hagstrom et al. 2021). They are causing contamination generally but in some areas are so pervasive that they have created what appear to be dead zones, where little grows (Tachovsky 2021). A range of different chemicals with harmful effects are found in the plants that are harvested by local communities causing health problems. The effect of chemicals is also found in food generally, especially in the cheaper types of products that are eaten by poorer communities including Indigenous Peoples and minorities (Marya and Patel 2021). These issues combined with climate change (Müller et al. 2024) are also having a negative effect on various species. One example is on the insect population, which is critically important for reasons that include their role in decomposing dead matter to release nutrients into the soil, aiding in the management of pest species, and, some that fly (Srivathsan et al. 2023), assisting with pollination (Outhwaite et al. 2022). Insects therefore play critical roles in the earth's sustainability (Harvey et al. 2023). However, insect populations have been affected, with butterflies in Europe having declined between 30 and 50 percent (Warren et al. 2021). Bees have also seen their numbers declining. This is having a huge effect everywhere and affects a range of issues including agriculture.

All these issues are all massive problems, with long term dire consequences for the sustainability of the planet (Klein 2020a). What needs to be focused on is what are the mitigation and adaptation strategies to overcome these or at least reduce the decline. In this regard, various groups, especially Indigenous

Peoples, are being more affected by these changes (Reyes-García et al. 2023). This is a key focus of this book, as they have a large role in ensuring the sustainability of the earth. They have vital roles to play, for a variety of reasons, including to protect large swathes of land where many species live (Estrada et al. 2022).

The problem, however, is that humanity has almost universally failed to use and care for the natural resources and the environment in a sustainable manner (Atapattu et al. 2021). As a result, issues concerning the environment are taking center stage everywhere. While much of the focus is on climate change, other interrelated environmental issues include a lack of access to clean water by billions of people, but also widespread contamination of potable water, ineffective waste management (Ajibo 2016), (including e-waste, which like other such unwanted supplies are dumped either in poor areas or in poor countries) (Ogunseitan 2023), air pollution, and contaminated soil, to name a few. Many of these are purposeful and some are already criminal acts (Agnew 2012; Bergenas and Knight 2015). Besides the harm to the environment and causing major problems for the people and other species living in a variety of places (Seiyefa and Oyosoro 2024).

As far as the tremendous problems caused globally by waste are concerned, the World Bank has found that the amount of waste generated around the world each year is expected to rise from around 2 billion tons in 2016 to about 3.5 billion tons by 2050. High-income countries, despite having only 16 percent of the Earth's population, generate more than one-third of all waste (Heidegger and Wiese 2020).

This is a massive problem for a variety of reasons, including that waste "is a big source [of methane] and in countries like India, Pakistan and Bangladesh it's not only a huge source of greenhouse gas emissions but it's also a lost opportunity to tap a fuel resource that could help meet the country's energy needs" (Carrington and Clarke 2024). The World Bank notes that "those most affected by the negative impacts of poorly managed waste are largely society's most vulnerable – losing their lives and homes from landslides of waste dumps, working in unsafe waste-picking conditions, and suffering profound health repercussions" (Kaza et al. 2018).

Another huge problem is the use of plastics. The extent to which this is problematic is only now being understood. A report in 2024 found that about 2.24 gigatons of carbon dioxide equivalent was produced worldwide due to the manufacture of primary plastics in 2019. This amounts to 5.3 percent of greenhouse gas emissions worldwide, but these numbers exclude emissions from forestry, land usage, and agriculture (Karali et al. 2024). Therefore, the extent of the problem is not really understood.

What is known is that recycling and other measures are not doing what they are supposed to be doing to reduce the effect of plastic (Allen et al. 2024).

The problem is that "[p]lastic waste is ubiquitous – from our rivers, lakes, and oceans to roadways and coastlines. It is in the air we breathe, the food we eat, and the water we drink" (Allen et al. 2024). Therefore, a critical problem is the ingesting of plastics (WHO 2022) and the health effects of that (Senathirajah et al. 2021). In this regard, according to recent research, people consume and breathe microplastic on a regular basis when they eat various foods including fish and shellfish. Microplastics have also been discovered in bottled water, tap water, and even widely consumed foods like beer and salt (UNDP 2023; Kim et al. 2018). The process to draft a treaty to deal with this continues without an end in sight at least in part because of the participation of those representing the industries who do not wish to have a robust outcome.

What is happening is that air pollution, water contamination, a lack of waste management, massive problems with degraded rivers (Mehta and Jade 2024), and others, specifically caused by humanity in the natural environment, are causing dangerous situations (Rockström et al. 2023). A range of other intentional and unintentional human activities, individually and in combination, are causing the treacherous condition we live in today (Rockström et al. 2023). What is clear is that our current environmental challenges are socioenvironmental in nature, caused by unrestrained human activity. It is not by chance that the world is seeing more disasters (Wen et al. 2023) in the form of floods, drought, wildfires, heat waves, more and more intense hurricanes and other similar weather patterns, destruction of arable land, coral bleaching (Datta et al. 2024a), and reduction in water supplies (Sloan and Schmitz 2019).

Droughts around the world, for example, increased 29 percent from 2000 to 2020. This is causing major problems, which will be further exacerbated by climate change over time causing further human displacement (United Nations Convention to Combat Desertification 2022). To deal better with the consequences of the increasing number and intensity of droughts, the World Drought Atlas illuminates the systemic nature of drought threats. It shows how drought risks are linked to industries, including energy, agriculture, river transportation, and international trade, using dozens of maps, infographics, and case studies. It also shows how these risks can have a domino effect, escalating inequality and conflict (European Commission Joint Research Centre 2024).

Despite that, the role of conflict and specifically the role, for example of militaries, in contributing to climate change is rarely focused on, even though it is estimated that they are responsible for 5.5 percent of greenhouse gas emissions (Parkinson and Cottrell 2022). Climate change is having a massive effect and that will only see more problems in the future (UNICEF 2024). Rising prices across the board caused by climate change will have enormous effects on political stability in states (Kotz et al. 2024a). These are expected to increase significantly each year in the future because of climate change (Kotz et al. 2024a). Critically, however, all these issues result in various groups being more susceptible to environmental injustices and ensures that the effects of those harms are particularly severe. These groups might suffer more discrimination and targeting because of this.

Wildfires, too, have doubled in number and in intensity over the last 20 years (Cunningham et al. 2024). The human and other costs of these events have been dramatic. A study in 2024 found that the effect of wildfires was to increase the amount of carbon dioxide being released (Jones et al. 2024).

The various weather-induced problems that are occurring are mostly because of humanity's decisive role in climate change. One of the biggest problems relates to the emissions of greenhouse gases, which is causing the planet to warm up. What has been found is that, while there is a varied pattern of energy balance changes in the first two decades of the 21st century, the general and evident trend is that of warming. This is in large part because of the increasing concentration of greenhouse gases in the atmosphere. The concentration of these gases in the atmosphere in 2024 was sitting at 419 parts per million (ppm). This is well above the 350 ppm that accords with the Paris Agreement's goals. The warning is that any concentration above 350 ppm can contribute to "significant global ice sheet loss," and a concentration over 450 ppm can likely result in a planet that is predominantly "ice-free" in the coming millennia (Caesar et al. 2024). The changes that are happening are measured by what is called "radiative forcing," sometimes known as "climate forcing." This is used in climate research to describe how much the Earth's atmosphere's energy balance has changed. What was found in 2024 is that radiative forcing has been rising steeply (Caesar et al. 2024).

The effect of these changes is significant. For example, research blames deforestation for less precipitation around the world (Smith et al. 2023; Duku and Hein 2023). This is because at the global level, potential forest cover is at 59 percent – well below safety levels. This is highly problematic as, as will be discussed in more detail below, forests play an essential role of removing carbon from the atmosphere and storing it in their roots and leaves. Thus, forests are able to keep carbon dioxide out of the atmosphere. However, that role that forests play has been reducing over time, as both regional and global forests have been declining over decades, with losses to "primary forest" and "climate change impacts" being the main accelerators of this decline (Caesar et al. 2024). Hence, changing or halting the human activity causing environmental degradation would help ameliorate global warming (Jaureguiberry et al. 2022). If not, much more focus needs to be placed on carbon capture and storage, to reduce the effects of global warming. However, care needs to be taken as some seek to use carbon capture as a means to continue using fossil fuels at high levels. Thus, carbon capture has its place but for some it is controversial in that it is being touted to avoid dealing with the problems of fossil fuels.

The cost also of climate change weather-induced events, such as floods, cyclones, droughts, a variety of more intense storms, such as hurricanes and tornados, wildfires, etc., have not been the subject of much attention until recently (Lenton et al. 2023). The effect of these events was estimated by the insurance company Aon for just the year 2019 as being $232 billion, with more than half of those costs occurring in the United States alone. The protection gap between losses and what was covered by insurance was 69 percent (Klein 2021).

As far as the future costs of climate change, it has been estimated that it will cost $35 trillion per year by 2049 to address the overall global costs of climate change, while the costs of mitigating climate change have been estimated as being only one-sixth of those costs (Kotz et al. 2024b). Another research paper published in 2024 noted that the costs that will be incurred because of climate change will be six times the amount previously thought and each one percent degree Celsius rise in temperature will cause a reduction in the global world product (GWP) of 12 percent (Bilal and Känzig 2024; Nath et al. 2023).

THE USE (OR NOT) OF LAW TO DEAL WITH CLIMATE CHANGE

Around the world, the law is being used at times to mitigate the effects of climate change (Kuh and Leach 2022). In some places, it is being used also to deal with the effects of environmental injustice. However, one of the problems that various groups that need the legacy of the past dealt with as far as environmental justice is concerned, or to ensure that their problems which are being exacerbated by climate change are dealt with, is that the law is not created in a neutral environment. Most modern legal systems at best are not designed to sufficiently protect the rights of Indigenous Peoples (Westra 2012), minorities, or otherwise marginalized populations (Vieira et al. 2024). At worst these system are designed to actively protect the privileges of the wealthy, and particularly landowners – that is, the so-called elites of the society. Too often, the law is used as a means to oppress or suppress. As has been observed in the United States, but which is relevant elsewhere as well: "for one segment of the population of the United States, the Indigenous Peoples, law has been a weapon of oppression" (Diamond 2022).

This is certainly the case when it comes to resource ownership and usage, including of the land itself. Problematically, law enforcement often results in, or supports, environmental degradation. In this regard, it has been argued that the state plays a key role when it uses the law to permit environmentally damaging conduct or does not use the law to halt abuses that occur. While it ought to be the case that where the law is deficient, the appropriate governing body has a duty directly or through different institutions to amend the law to address the deficiencies. However, it is not unusual for these bodies to fail to act in that regard, often intentionally.

Further, even if the law is not always the problem, the law often does not assist in remediating the various challenges that marginalized groups face. In many places, too little is done to understand how identity factors impact people, particularly minority and Indigenous groups. Further, even as the legal framework to deal with environmental issues has developed, the degradation of the environment by both corporations and governments has continued and even increased.

In this regard the UN Special Rapporteur on human rights and the environment, David Boyd, noted in his report to the UN Human Rights Council in 2024 that it is mainly large corporations that are principally to blame because they produce and market unhealthy and unsustainable food, contaminate the air, water, and land, cause and exacerbate the climate crisis, destroy biodiversity and ecosystems, and release toxic substances that endanger people, animals, and ecosystems. Changing their role is one of the most important obstacles to moving towards a fair and sustainable future (Boyd 2024). In this regard, when he spoke at the Human Rights Council on 6 March 2024, he noted that a fundamental problem is that:

> States have failed to adequately regulate, monitor, prevent and punish businesses for their abuses of the climate, environment and human rights … The situation is further exacerbated as States often encourage, enable and subsidise destructive business activities. (OHCHR 2024a)

It is dealing with the role of business, especially large ones, which is one of the keys to environmental justice, but also dealing with climate change.

A further common problem worldwide is the taxation system. In many places it remains a major problem with respect to climate change issues but also environmental justice ones. Still, many large enterprises are not taxed on their problematic business practices, and, in fact, many get tax breaks and tax incentives to continue to use massive amounts of environmentally unfriendly fossil fuels. Many industries continue to have massive climate change effects. However, nearly 40 countries have introduced some form of carbon tax and there are many other initiatives to reduce the world's carbon footprint. Ways need to be found to further dramatically reduce emissions. This is because of the realization of not dealing with the issues connected to carbon (Barrage and Nordhaus 2024).

Despite this, one of the major problems still with a range of industries, is emissions. The impact of increasing carbon dioxide concentrations and radiative forcing is that elevated amounts of energy are being captured and trapped on land, in the atmosphere and in the oceans. This leads to rising temperatures which, in turn, provoke extreme modified weather events – impacting on land and underwater life simultaneously (Caesar et al. 2024).

However, it is not only carbon dioxide that is a problem. There are many other types of gases that are causing major problems, including, nitrous oxide and methane emissions. While most focus on carbon dioxide, the other two are also causing much harm to the planet. Thus, the 2024 Global Nitrous Oxide Assessment indicates that nitrous oxide is currently the most significant ozone-depleting material being emitted, putting a large portion of the world's population at risk for increased UV exposure and a rise in skin cancer and cataracts. The report finds that taking aggressive steps to cut nitrous oxide emissions might prevent up to 20 million premature deaths worldwide by 2050 as a result of poor air quality (UNEP 2024a).

The other major emission is methane, which also is having a major effect on the globe. It is critical that it is also focused on, as methane is seen to be second after carbon dioxide in adding to greenhouse gas emissions (Gauci et al. 2024). However, it has a warming power 80 times that of carbon dioxide. Only in 2024 was it realized how fast methane emissions are rising and causing the trapping of heat in the atmosphere (Shindell et al. 2024). As will be noted below, fossil fuel operations are a major cause of methane emissions, but methane emissions is also a major problem from farming. However, the extent of it being emitted from the oil and gas industry has only really become known recently (Sherwin et al. 2024). The levels of emissions of various gases such as carbon dioxide, methane, and nitrous oxide (Global Monitoring Laboratory nd), rather than being radically reduced, continue to climb at record pace (NOAA Research 2024). However, some believe that global emissions will peak in 2024 or 2025 (Fyson et al. 2023). Whether or not this occurs, many more ways of dealing with methane need to be found (Malley et al. 2023). One new finding is that forests have an important effect in absorbing methane (Gauci et al. 2024). Thus, protecting forests is important because of the role they play. Thus, protecting those who protect forests is critically important for mitigating climate change. This is even more important as climate change is affecting the ability of forests to absorb carbon and thus protecting them is even more important (Ke at al. 2024).

Indeed, large inroads can be found by focusing on the largest emitters. One database indicates that over the last ten years it is only 57 cement, coal, gas, and oil companies that have caused 80 percent of all fossil fuel carbon emissions globally (Carbon Majors Database 2024). This is important, as another study in 2024 found that none of the 25 largest oil and gas producers have remediation plans for their future production accorded to the goals of the 2015 Paris Agreement (O'Connor 2024). These sectors could, however, play a large role in reducing climate change. To some degree, this is being done in various countries such as in Norway (Dienemann 2023) and the United States until the end of 2024, as will be discussed. However, other economic imperatives are getting in the way of making more progress. Research in 2024 found that rather than reduce reliance on coal, coal-fired power stations are not being shuttered to the same extent in Europe, while China is opening more of such plants (Global Energy Monitor 2024). It has, however, been reported that coal usage is supposed to fall in 2024 (International Energy Agency 2023). This is crucial as the International Energy Agency (IEA) has stated:

> Achieving the goal adopted at COP28 of net zero emissions of greenhouse gases from the energy sector by 2050 hinges critically on the rapid transition away from the unabated use of coal for generating electricity. The scale of the task cannot be overstated: coal accounts for over one-third of global power supply, in many cases from recently built plants. As the most carbon-intensive fuel, coal's role in emissions is even bigger: globally, coal is responsible for over 40% of all energy sector CO2 emissions. If existing coal power plants and industries were to continue to operate as they do today, they would "lock in" emissions pushing the world well beyond the 1.5 °C limit. (International Energy Agency 2024a)

One of the problems as to why coal and other fossil fuels continue to be used to such an extent is because, at least in part, their sustained use is aided by international financial institutions who still provide financial support to these types of industry. A report in April 2024 noted that between 2020 and 2022 multilateral development banks supplied at least $142 billion in foreign public money for coal, oil, and gas projects (O'Manique et al. 2024). Thus, rather than reducing production, and therefore consumption, of these fuels, in the United States for example, the output of these fuels is increasing. Rather than taxing these industries for doing this, the producers continue to get tax breaks and incentives to produce and use these products, rather than being incentivized to switch to sustainable energy. The effect of this is to exacerbate the climate emergency (which some argue is the wrong terminology and that climate-change-epic ought to be used) (Maloy 2024), rather than trying to arrest it. What is being done elsewhere, such as in Europe, is a focus on those who are the biggest polluters, including the fossil fuel industry to get them to contribute sufficiently to arrest the problems and pay to deal with the problems more.

It is useful that other industries are trying to find ways of mitigating their impact on climate change. Thus, for example, the International Maritime Organization (IMO) agreed in March 2024 to an "IMO net-zero framework" to significantly cut greenhouse gas emissions from international shipping (International Maritime Organization 2024). This would be important because of the major effect ships have on climate change (Li et al. 2023). This process would see ships pay a tax (Luo et al. 2024) for every ton of fossil fuel they use. This would be income generating to use for various purposes but would also be a disincentive to continue to use large amounts of fossil fuels (Ye et al. 2024), and rather use other types of energy (Balci et al. 2024). This is an important development considering that ships cause three percent of global emissions while aviation causes 2.5 percent (Tavakoli et al. 2024). However, the issue of funding climate change projects is far from sufficient, even though a global climate fund has been established (Nhamo and Chapungu 2024). While carbon pricing is being rolled out also to fund climate change problems, these need to be better understood by consumers, as will be discussed later. What is important as well is the notion of voluntary commitments to comply with climate change standards. For example, various banks agreed to ensure greater compliance with climate change processes and only lend to those that complied with these. However, a study by the European Central Bank found that the impact of these commitments was not what it was supposed to be and that voluntary commitments were not having the effect they were expected to (Sastry et al. 2024). Further, much more needs to be done to make electricity available to the many people who do not have it, particularly to those in the Global South (Cantoni et al. 2022). This will ensure a just energy transition and move towards universal access to energy, particularly in places like Africa (Sterl et al. 2024).

A range of technologies can be helpful in this regard (Vijayan et al. 2023). However, as has been noted:

> Quite ironically, while the realisation of the climate change agenda implies redressing a series of injustices, dispossessionary modalities of solar development, rapidly elaborated for climate mitigation, may reproduce climate vulnerability for marginalised groups. (Sareen et al. 2023)

Thus, moving to renewables, which is happening quickly, with 30 percent of energy being produced by them in 2023 (Wiatros-Motyka et al. 2024), has downsides. Thus, for example, as rain levels change so some places will not be able to use their hydroelectric power stations to the same extent. Thus, Canada in 2024 had to import electricity from the United States for this reason. Other countries, such as China, Colombia, and India, have reverted to using more fossil fuels (International Energy Agency 2024b).

Thus, moving to use renewables brings its own difficulties. A problem, for example, in the transition to using renewables is what to do with solar panels at the end of their life span. This is due to be a million tons by 2035 in Australia alone (Australian Centre of Advanced Photovoltaics 2024). To ensure that this waste is not dumped in the Global South and where minority and other groups live this issue needs more attention.

Another problem is that the rush to find and extract the minerals necessary for renewables (including aluminium, cobalt, copper, graphite, lithium, manganese, and nickel) will have many effects. This is because large amounts of the minerals needed for the transition are found on Indigenous Peoples' land (Owen et al. 2023a). Thus, mining will come up against people's livelihoods, especially that of Indigenous People, such as the Sami in Europe, but also in the rest of the world (Holroyd 2023; Heikka-Huber 2023). This has already, and will in the future, seen further injustices, particularly in the Global South where these minerals are being sourced. Thus, the UN has established a Panel on Critical Energy Transition Minerals (Allan et al. 2024). Its task is to create a voluntary, shared set of guidelines to safeguard social and environmental norms with the goal of integrating justice into the energy transition. Thus, justice must be a critical component of dealing with climate change and therefore a reason that environmental justice and climate change issues need to be integrated so as to ensure that solutions for climate change have broader effects and do not rather exacerbate the problems that certain groups already face. This is because these issues intersect and cause double negative effects for the people who already suffer the harmful effects of environmental harms (Anyanwu et al. 2024). This will be taken up below on the need to integrate climate change and environmental justice.

One of the problems, when looking at the role of law in dealing with these matters, is that those who draft laws and those who interpret and apply them, often the courts, often do not understand or apply the nascent environmental protection law fully or properly, leading to problematic outcomes for the people affected. If they did and they were more able to play a more central role, then real and meaningful arresting of climate change but also reversing environmental injustices would occur. The critical issue is that, for a long time, many environmental injustices have occurred specifically to Indigenous Peoples and minority groups. They are now also the worst affected often by climate change but also by the broader issues relating to environmental injustices.

ENVIRONMENTAL JUSTICE AND CLIMATE CHANGE

One way to view the challenges of the environmental and climate challenges that exist, is through the lens of environmental justice, a discipline that seeks strategies and tools to deal with the disproportionate negative effects of environmental destruction and other influences. While this will be pursued in greater detail in the next chapter, briefly, simply, and broadly, environmental justice is both a concept and a movement, with the aim of ensuring that all people, regardless of their identity, are treated the same, and are protected equally from environmental hazards. This is because the environment (with all its component parts and different forms of life) that people live in, everywhere on earth, is critical for life in general, and specifically for humanity's survival. Unless humanity sustains our current environment, most forms of life on earth will perish. However, access to environmental resources is not equitable, and the harms resulting from environmental degradation fall to a much greater extent upon the poor and marginalized peoples. As a result, environmental justice advocates agitate for the fairer allocation of environmental benefits and burdens and for equitable access to clean water, air, and soil, as well as for the safeguarding of people from injurious contaminants and other threats. This is not the case for the majority of people around the world, often because of choices made by various authorities but also because of environmental injustices which allowed the degradation of these resources in particular places. These problems are more of an acute problem with climate change. Environmental justice is therefore also about providing equitable access to necessary resources such as uncontaminated land, clean air, and potable water, for Indigenous Peoples and other marginalized groups, as they often are discriminated against in the allocation of services.

However, some governments and private sector actors attempt to deprive Indigenous Peoples and minorities of their assets and resources, as well as subject their lands to environmentally dangerous practices. These actors benefit from placing facilities, such as hazardous waste sites and pollution emitting factories, and conducting environmentally damaging mineral extraction activities, where Indigenous Peoples and minorities live. These communities are easy targets for such activities because, unlike other communities, they often do not have the ability and resources to halt the intrusive actions. These environmental injustices are having deleterious effects on the soil, water, air, and crops that these groups depend on for their survival and pose a threat to their lives and lifestyles. However, the polluting effects are felt way beyond the location of the activities.

Environmental justice also demands the participation of all people – including, in particular, Indigenous Peoples and minority groups – in the conceptualization, advancement, implementation, and administration of environmental rules (Jodoin et al. 2023; Bullard 1996). This has seldom occurred. Moreover, environmental justice requires that these groups be meaningfully consulted and genuinely heard when consultation takes place. Yet this has often not been the case, even where formal consultation has occurred (Fjellheim 2023). Such groups must be included in decision-making processes, especially in matters that directly affect them or their land (Gustafsson and Schilling-Vacaflor 2022).

This position, and the conceptualization and use of environmental justice (Mels 2023) as it is interrelated with climate change and climate justice more generally, will be addressed throughout this book, as what environmental justice is, what its contours are, how it ought to be applied, and a range of other matters connected to it. These issues remain controversial.

Despite this, there needs to be a link between environmental justice and issues of equity and justice (Rivadeneira and Carton 2022) in dealing with issues concerning the environment and climate change (Biermann and

Kalfagianni 2020). This is to ensure that problems connected to environmental issues are not overlooked, as they are a critical part of the puzzle to deal with them (Evans and Phelan 2016). Thus, the term "just transition" is used by some (McCauley and Heffron 2018).

What is necessary to realize, however, is that environmental justice needs to be seen as something deeply interrelated to dealing with climate issues. Not dealing with environmental justice, and rather dealing with climate change in isolation, will not have the full positive effect that is needed. This is because environmental injustices that linger will ensure that climate mitigation or adaptation strategies will be less effective and will have less than desirable outcomes. This will certainly be the case where resistance to such strategies occur. This is happening in many places. As has been noted:

> Climate adaptation is not a neutral or apolitical process, but one that ignites social resistance. Government responses to risks of floods, droughts, or hurricanes – even those using a language of participation – might follow historical development pathways, strive to maintain the status quo, and directly or indirectly serve elite interests. (Brink et al. 2023)

These are issues that need to be considered much more, and the effect of past environmental injustices taken into account so as better deal with climate change but also do so in fairer and more equable ways. More work is also needed to consult and educate communities on what is being done, where and why to ensure understanding and support for the measures that are to be taken. Without that, greater resistance and misunderstanding will result (Brink et al. 2024).

While many states have adopted National Adaption Plans (NAP) (Garschagen et al. 2021) and various laws (Verschuuren 2022), rarely has environmental justice been included in any substantial way. Some countries do not even mention environmental justice (although in places it could be read to include it indirectly) in their NAPs. Thus, it is argued that without such a focus, the goals set to deal with climate change are less likely to be met. Issues of adaptation will be returned to below.

Certainly, around the world, there is much focus on climate change issues but less so on environmental justice issues. They are, however, deeply interrelated and interconnected. This is the key focus of this book. Therefore, at a minimum, environmental justice ought to be directly referenced if only to educate the wider community. However, environmental justice needs to be much more incorporated directly and be much more part of the process of dealing with climate change. It must move beyond policy and academic inquiry towards action that is taken by legislatures, by various authorities, and by the courts. However, while beyond the scope of this book, it is important to note that there is an interrelationship to a range of issues beyond these. As was noted by Naomi Klein:

> The most important lesson to take from all this is that there is no way to confront the climate crisis as a technocratic problem, in isolation. It must be seen in the context of austerity and privatization, of colonialism and militarism, and of the various systems of othering needed to sustain them all. The connections and intersections between them are glaring, and yet so often, resistance to them is highly compartmentalized. … Climate change acts as an accelerant to many of our social ills (inequality, wars, racism, sexual violence). (Klein 2020b)

In her 2016 Edward Said lecture "Let them Drown: The Violence of Othering in a Warming World," Klein operationalizes Said's "othering" and applies it to the climate crisis (Klein 2016). Elaborating on how "othering" insidiously operates in environmental policy and discourse, Klein lays bare the justifying ideologies behind the human and ecological cost of growth, noting how environmental "othering" sustains dispossession, violence, and culturicide (Klein 2016).

While climate change issues have been taken up by the public broadly around the world in growing numbers (Fisher et al. 2023), this has not occurred, generally speaking, for environmental justice issues. This has occurred in specific places where local communities have taken up specific issues. However, what has occurred is a growing willingness, even in mature democracies, to restrict climate protests (Gayle et al. 2023). Certainly, dealing with climate issues in democracies seems to be easier than elsewhere. As there has been an overall regression in the democratic situation around the world, with more countries moving away from being classified as free and now being rated partly free or not free, it is issues of climate and environmental justice that have suffered most. As has been noted "the world seemed to have swerved sharply away from democracy and toward autocracy – and in the process dramatically limited our ability to fight the climate crisis" (McKibben 2022). It is in less democratic states that moving away from fossil fuels is seemingly more difficult because of vested interests that are more able to secure their own interests. Thus, democratic backsliding is likely to limit progress being made in dealing with climate change. Thus, democratic advances are needed as a vital part of the process of dealing with climate issues. Thus, the interrelationship between democracies and environmental and climate justice needs to be more understood and be more on the agenda of the international community, but also regionally and locally. It needs to become part of the work of governments in dealing with climate and environmental issues.

As noted already, the issues of climate change and environmental justice are very interrelated. What is also clear, is that to deal with climate change much more effort is needed. Some of the interesting ways that are being examined include using diverse technologies to reduce the quantity of problematic gases in the atmosphere. While the effect of these processes at present is quite limited, more and more solutions are being sought, to reduce the carbon dioxide in the air and by capturing it and burying it underground (Soeherman et al. 2023). Ways to block the rays from the sun are also being pursued (Bromley et al. 2023), including sun shields (Szapudi 2023). Various geoengineering processes (Schubert 2023), however, bring various risks and problems. Thus, a variety of approaches are needed which limit the human impact (Mehmet and Yorucu 2024) on the environment, while mitigating the damage already caused. However, figures in 2024 show that rather than reducing the amount of gases in the atmosphere, those have gone up quicker than before. Thus, much needs to be done simply to arrest the increase, never mind removing it. A multitude of strategies are therefore needed, which include and are aimed at those most affected, especially Indigenous Peoples and minority groups.

Thus, more broadly, to assist in the attaining of the necessary climate goals there needs to be a focus on environmental justice as well. This is because Indigenous Peoples (Toki 2024), minorities, and other marginalized groups, around the world have suffered the most from environmental justice problems and will continue to suffer more from climate change. This is because of the ongoing massive climate change and environmental injustice problems resulting from both the actions and inaction of public entities such as state, regional, and local governments. Also responsible are other governing and oversight authorities, as well as private entities running the gamut from large multinational corporations to small, local companies.

INDIGENOUS PEOPLES AND MINORITIES

While climate change will affect everyone, the effects will not be same for everyone. A major problem today is that while climate change issues are being studied, they are not being studied to the same extent as far as its effects on Indigenous Peoples and minorities (Billiot and Parfait 2019). This is the major focus of this book. The problem is that the effects of climate change will not be felt by all people in all nations in the same way. In fact, there are great discrepancies. Even inside the same country, climate change is going to be felt differently. This is because some places in a country are more likely to be affected than other places. In the places where there are already environmental problems, the effects of climate change are to be felt in more problematic ways. These are the places where people live, not out of choice, but because of poverty and marginalization. Climate change is more felt by those who are situations in poverty. Thus, slums and other places where poor people live are more likely to feel the effects of floods, storms, droughts, and other climate change effects. As a result of where they live, and the deprivations in those places because of the effects of environmental injustices, it is the poor who are going to feel climate change even more. At the same time, they are less able to cope with these effects for a multitude of reasons, including lack of insurance (Herreros-Cantis et al. 2020; Bradt et al. 2021).

These issues therefore affect Indigenous and minority groups disproportionately. This is because the environment is central to their way of life in many ways (Reyes-Garcia et al. 2023). In fact, it is Indigenous Peoples specifically, in this regard, that are facing many challenges (World Bank Group 2023).

This is because:

> [t]heir traditional ways of life, livelihoods and practices are increasingly under threat owing to a range of factors, including lack of recognition and protection of their rights, exclusionary public policies, and the impacts of climate change. A combination of issues such as loss of access to traditional lands and natural resources, discrimination in the world of work, forced migration, and poor access to opportunities have rendered them still more vulnerable in social and economic terms. (ILO 2016)

Their situation is growing worse because of climate change in addition to the effects they have had to endure disconnected from climate change (Gerbatsch 2022). Thus, evidence from Indigenous People in the Torres Strait Islands, which are part of Australia, is that:

> climate change is putting our island home at risk. The Torres Strait has not contributed to climate change, but we are on the frontline of its impacts. I've seen the impacts myself very clearly as the years pass. Inundation and the erosion of the shorelines have washed away our roads, salinity is destroying our coconut trees and garden crops, and our drinking water wells have been contaminated. (Mosby 2020)

Because of the critical role the environment plays in their lives, safeguarding the environment for Indigenous Peoples especially, but also other minorities, is crucial. It is particularly the role of Indigenous Peoples in this regard that is important considering that around the world they have control over at least 38 million square kilometers of land, over 87 countries, which equals a total of about one-quarter of the planet's land (Fa et al. 2020). Their environmental role in this regard is crucial. As has been noted:

> Indigenous peoples are essential to promoting food and water security, and enhancing nature's ability to absorb harmful greenhouse gases. While representing a mere 5% of the global population, they manage approximately half of the world's land and forests, which are home to a staggering 80% of the globe's biodiversity. Furthermore, between 2001 and 2021, Indigenous peoples' forests were better net carbon sinks per hectare than most non-Indigenous forests. (Duarte and Al Mubarak 2024)

However, the position of these groups to deal with these issues is tenuous. As a result of the discrimination and segregation suffered by these groups, they experience significant inequalities which limit their access to the different socioeconomic, political, cultural, health, and public sectors of life (Shaheed 2020).

Indigenous Peoples and other marginalized groups have had to endure multiple injustices (World Bank Group 2023) that include losing their access to their natural resources through the allotment of designated reserve lands or resource rights to others through dubious processes ("land grabbing"), forest clearing, abusive use of their lands by the resource industry, and a host of other methods including "water grabbing" (Zoomers and Kaag 2014). These ongoing injustices have had, and continue to have, a domino effect on other social structure challenges faced by traditionally marginalized people, in that they contribute to an increase in overall inequality (Suradi et al. 2019), a decrease in their access to food and water, a reduction in social cohesiveness, and the loss of cultural integrity (Whyte 2018).

They are often excluded from public services because of where they live or receive very limited or poor services. One aspect of this discrimination is that these marginalized people are forced to live in degraded areas that are more likely to be the location of waste facility sites and similar industries (Williams 2021b), or to suffer greater negative effects from pollution, often because it is all they can afford (Hale et al. 2018). These places might be more low lying, and therefore are more likely to flood or face a range of other disasters. It is often the case that these sites are subjected to greater noise, air, water, soil, or other types of pollution (Casey et al. 2017).

The absence of measures to protect these places (Edirisinghe and Suchet-Pearson 2024) is only making the situation worse and will inevitably cause more harm in the future (UNEP 2020). The lack of remedial efforts by governments and industry is partly because these are often remote locations that are provided insufficient or no services because of the discrimination faced by the people who live there. As noted above, these challenges have limited the people's ability to gain anything other than low-paying, low-skilled employment, if any, and a range of other deprivations, which continue, and often exacerbate, the vicious cycle of poverty. They therefore live in places where land and houses are cheap or where rents are low, where the water, soil and air are of poor quality, where often the most toxic industries are placed, and where there is little or no access to environmental protection (Heidegger and Wiese 2020).

One area of significant concern is the effects of the resource industry on Indigenous Peoples and minorities in general. There are two aspects to the injustice caused by this industry: the first is the lack of equity, justice, and fairness in the distribution of resources (Schrock et al. 2015); the second is the danger resulting from the placement of hazardous facilities (Ajibo 2016) such as chemical plants or nuclear facilities, and the dumping of hazardous materials disproportionately in the areas where they live (Pearson 2022). This is often done specifically because these groups live there. These sites are chosen for the same reason that the people have been relegated to the sites; because they are remote from the general, more privileged population. The people living there lack the social, economic, and political power to object, and additionally because the lands are otherwise deemed less useful.

There is therefore a need to integrate dealing with these concepts especially when dealing Indigenous Peoples and minorities. This is because it this they who experience a disproportionate and significantly higher likelihood of being affected by environmental and climate degradation (Abate and Kronk 2013). For example, in the United States, it is estimated that people of color are exposed to air pollution at a rate of about 2.5 times that of others (Tessum et al. 2021). Environmental justice has become the primary tool to seek solutions for this inequality. It has paved the way for the protection of the lifestyles, culture, and social identity of vulnerable groups, including Indigenous Peoples and minorities. It tries to ensure these groups' well-being and provide them with a healthy and safe living environment (Shelton 2006). This is critical, as health issues have been major problems for groups in the underserved areas that they live in (Sunilra and Binu 2021).

Recognizing these groups, and valuing their roles on a variety of issues, including on dealing with climate change, can help to achieve justice for their past discrimination, marginalization, and prejudice that they have had to endure. It is not only about justice and human rights (Westra 2012), but also important as it is an appreciation of the diversity and richness of various aspects of our civilization. It could also help to achieve healing and reconciliation with these communities by acknowledging their past and by addressing the problems they face today (Krondorfer 2020).

WHAT IS THIS BOOK ABOUT AND THE METHODOLOGIES EMPLOYED

This book is a contemporary sociolegal examination of the many environmentally related challenges that Indigenous and minority groups face. This is done in order to learn lessons about what can and should be done for and about them. The book examines these matters from an interdisciplinary point of view. It reviews how an integrated approach to addressing climate change, incorporating environmental justice, can yield significantly better results. It does so by analyzing how these issues are understood and addressed – sometimes separately – while highlighting the more integrated responses seen in the United States, for example. The book explores why and how linking climate change and environmental justice can positively impact efforts to address climate change, particularly for marginalized and excluded groups, such as minorities and Indigenous Peoples.

It does need to be noted that the book deals with the issues until the end of 2024. This was because of the deadline of the publication, but also because it marks the end of President Joe Biden's term office, after which the issues of climate change and environmental justice were being dealt with very differently in the United States. Critically, the period until the end of 2024 saw a lot of positive steps being taken to mitigate and adapt to climate change in the United States but also in many other places around the world, but maybe not to the same extent and with so many resources being devoted to these issues. While there have been many problems there has been generally a commitment around the world to deal with these concerns.

This book in some ways is a testament to what is possible and what are the obstacles and pitfalls in dealing with climate change. The US chapter is most important as it shows the breadth of steps that can be taken and how dealing with climate change must also be done with an eye on environmental justice. That chapter also indicates how and why the most vulnerable need to be most assisted. It makes the point that if that does not occur they will be left behind as far as climate change is concerned. The recommendations made in the Conclusion try to pick up the general lessons that can be learnt from all the country examples, and make suggestions on what can be done in a host of different areas.

The book approaches these issues from the perspectives of various specializations to foster innovative viewpoints, uncover new methods for conducting research, and apply the findings to the practice of environmental justice. This integrated, interdisciplinary, comparative, and empirical approach enhances the understanding of the full effects of the issues under investigation (Ashton et al. 2007).

Using a variety of approaches, including an interdisciplinary one, is necessary as the difficulties that Indigenous Peoples and minorities have faced have been a long-term problem that needs to be worked on across the various disciplines both in the natural and social sciences. Critically, the problems these groups face began at least with colonialism and continue in many parts of the world today (Figueroa 2022). Thus, colonialism and its effects on environmental justice and climate change are important as they continue to play out today in a variety of ways (Singh 2023). Their lingering effects include structural racism and discrimination (Smith et al 2022).

This is important as far as Indigenous Peoples are concerned specifically, as colonialism had a devastating effect on them (Fuentes-George 2023). Their traditional ways of life, language, culture, natural resources, and local communities were all severely affected. As a result, they have in many instances struggled to maintain their identity (ILO 2007). While colonialism has formally ended in most places, new forms of colonialism persist (Lightfoot and Maddison 2024). In fact, in many ways, environmental injustice issues are vestiges of colonialism (Loyn 2006). The effects of colonialism are enormous and continuing (Lohmann 2022), affecting people in many parts of the world in many different ways (Arowolo 2022). That legacy continues in multiple ways including affecting environmental matters (Barclay et al. 2019). It affects climate issues specifically (Fuentes-George 2023). A vestige of these events is that many communities continue to be relegated to an inferior position in the societies where they live. This is done partly by labeling individuals and groups as different, which in turn creates social hierarchies, resulting in the marginalization of, and discrimination against, the "other" (Loppie et al. 2020). This marginalization has been a focus of environmental justice scholars for a while (Holifield et al. 2009). These same issues are relevant, and affect climate change matters as well. As has been noted:

> The extremely uneven and inequitable impacts of climate change mean that differently-located people experience, respond to, and cope with the climate crisis and related vulnerabilities in radically different ways. The coloniality of climate seeps through everyday life across space and time, weighing down and curtailing opportunities and possibilities through global racial capitalism, colonial dispossessions, and climate debts. Decolonizing climate needs to address the complexities of colonialism, imperialism, capitalism, international development, and geopolitics that contribute to the reproduction of ongoing colonialities through existing global governance structures, discursive framings, imagined solutions, and interventions. This requires addressing both epistemic violences and material outcomes. (Sultana 2022a)

Thus, the inequity and unevenness of approaches to deal with a range of different but related issues is clear. However, while it is useful that colonialism and its ills are attracting greater attention, including its ongoing effects, there is a need, academically, to democratize and decolonize the various studies that are being undertaken (Roy and Hanaček 2023; Cantzler and Huynh 2016). This is true concerning climate change as well (Sultana 2022a). Part of that, is the necessity of bringing more people into the discussions about dealing with all the issues (Safotso and Julius 2023), especially from the most affected communities (Colbourne et al. 2020). This will allow, for example, the Global South to have a larger voice on these matters. This is particularly important because much of the harm to the planet is caused by the Global North:

> [t]wenty percent of the world's population consumes approximately 85 percent of the planet's timber, 70 percent of its energy, and 60 percent of its food. This population is also responsible for more than 90 percent of the world's annual production

> of hazardous waste, some of which is exported to Southern countries and contributes to illness and widespread environmental harm. (Gonzalez 2012, references omitted)

Nonetheless, these interdisciplinary and multidisciplinary approaches are useful to broaden out the appeal and application of the field. However, they need to also be taken up practically, so as to apply these advances in understanding to the actual conflicts and disputes concerning environmental justice. While Pellow states that "we see EJ studies in law, education, and the environmental humanities" (Pulido 2017: 45), environmental justice needs to go further than being in the studies of these fields and actually be applied in policy and in domestic law, which is in many ways, deficient for many groups (Ahren 2024).

The book is also meant to be both applied research and as well as a sociolegal study. The aim is to examine the interrelationship between law and issues concerning climate change and environmental justice, and how the law impacts those concerns around the world (Webley 2019). This is because as William C. Clark noted in 2007, sustainability science is an "emerging interdisciplinary frontier" and a "use-inspired field" that "intersects environmental conservation, human health, and economic development, … natural, social, and engineering sciences, as well as insights from humanities and medical sciences; multiple sectors of human activity [and] knowledge and action, bridging fundamental and applied work" (Clark 2007 quoted in Klein 2020a).

The issues in the book are discussed within the frameworks of victimology, human rights, and comparative studies, as these disciplines can be helpful when searching for ways to deal with harm caused. They are also particularly helpful as a means to prevent future problems by applying a guarantee of nonrepetition approach to developing methods in addressing the challenges. Human rights remain the main avenue by which these issues are taken up and fought in the courts (Scott 2016). Although the role of the courts is relevant, it is not the focus here. To date, courts have played only a limited role. This is beginning to change especially at the international level. This is discussed is the contextual role of the US Supreme Court in the US chapter to understand their role generally including on issues concerning Indigenous and minority groups.

What is touched on briefly is how international human rights law provides a useful framework to deal with environmental concerns. Therefore, besides the natural and social sciences, this book also considers a range of legal issues (but as they merge into others), including human rights, transitional justice, and allied aspects including social and restorative justice (Nagy 2022). This is done as transitional justice and restorative justice, for example, can be helpful tools to address environmental issues, approaches that have not often been used in the past. As has been noted as far as transitional justice is concerned specifically:

> Conservation violence has not been the subject of transitional justice scrutiny. This is the case even though the traditional boundaries of the field are constantly expanding into areas far beyond what was initially contemplated, including in response to historical injustices suffered by Indigenous Peoples, as well as issues of environmental harm. (Luoma 2023: 90)

Although there has been some recent transitional justice research picking up on environmental issues (Clark 2021), especially as far as climate justice is concerned (Klinsky and Brankovic 2018), generally speaking, there has been little focus on environmental issues, especially environmental matters, in transitional justice (Ong 2017). This is the case even though there are ongoing discussions on the subject (Killean and Dempster 2022).

Transitional justice is, however, thought to be less useful when dealing with environmental justice issues because it is generally, wrongly, deemed to deal exclusively with the past, while environmental justice is typically forward-looking. However, it is nonsensical to draw a distinction between dealing with the past versus addressing present and future violations, as they are all interconnected. Transitional justice can, in fact, play an important role in resolving environmental justice matters (Forsyth et al. 2021). This is because issues related to the environment can divide societies and undermine trust in the legitimacy of the state (Bosi and Fazio 2017). Transitional and environmental justice advocates can work together to help prevent violence and conflict. This includes protecting groups that might otherwise resort to nondemocratic means to achieve some form of justice (Sarkin 2021).

The problem though is that even though ecological sciences, the social sciences, law, and environmental justice have much to collaborate on, they rarely do so (Miriti et al. 2021). Often the various disciplines remain within their own spheres, their insights seldom reaching other researchers and audiences. It does need to be understood, however, that environmental justice is not really only about the environment in the pure or natural sense. It is also about equal access to environmental resources by communities and ensuring that they are not harmed by the effects of things such as waste, pollution, and extraction, particularly in terms of how these things affect vulnerable and marginalized groups (Hernes et al. 2022). However, environmental justice does, of course, have a bearing on the environment *stricto sensu* and also the science involved. However, the attention given to dealing with climate change is not being mirrored in environmental justice matters. While some limited progress is being made in some places, advances have been extremely limited in many others, and there is a long way to go for a real impact to be felt.

The book's focus is global in nature (Gadson and Holm 2023). This is the case, as it is contended that what is happening in one part of the world is not divorced from what is happening elsewhere. It is also of interest to others elsewhere (Martinez-Alier et al. 2016). For example, the way that marginalized groups respond and their means of resilience and adaptation (Eriksen et al. 2021) (sometimes called transformations) (Moore et al. 2021) is just one area where a multitude of disciplines, including psychology, for example, or nursing studies, are relevant (LeClair et al. 2021). While there are many actors that have a role to play in the field of environmental justice, the emphasis of this book is on the role of national courts. However, a variety of other actors have played an important role, using a variety of strategic steps and different instruments, particularly in the Inter-American system (Garcia and Lixinski 2020; Cappuccio 2022; Fregoso 2022).

WHY THE FOCUS ON UNITED STATES, SOUTH AFRICA, AND CANADA?

A comparative methodology is also employed in the book. This is because comparative studies offer valuable insights and contribute to a deeper, more comprehensive understanding of the topics under scrutiny (Ali 2020). By comparing different entities – such as states, institutions, or other bodies – researchers can identify patterns that might not be apparent when studying a single case or country (Palmer 2024). Comparative analyses also illuminate how context influences outcomes and reveal the multifaceted nature of the issues being studied. This is particularly true when the study incorporates methods and theories from multiple disciplines, as this one does, providing a richer and more nuanced understanding. This approach avoids being one-dimensional or biased by viewing issues through a narrow lens. This book therefore examines practices in a number of countries as examples to demonstrate the necessity of linking and applying environmental justice and climate change together, and how it can be best achieved. This is discussed below further.

Regarding the specific countries chosen for this study, comparing South Africa, the United States and Canada is particularly valuable for several reasons. Their unique historical, political, and social contexts profoundly influence the issues under examination. Each of these states offers important perspectives on the problems of environmental justice and climate change, informed both by their historical legacies and current circumstances.

As is the case with a number of countries, Canada's efforts to address climate change and environmental justice are multifaceted. However, while some positive steps have been taken, as is the case mostly elsewhere, there are still major limitations with the approach taken. Like other nations, politics has a significant impact on the way the country deals with all the issues and especially as far as the way they affect specific groups. Crucially, one of the main reasons to look at this country is the enormous number of Indigenous Peoples who reside there. Indigenous Canadians make up a sizable portion of Canada's population, but other minority groups also face hardships. Importantly, as in other places, the majority of Indigenous Peoples in Canada reside in rural areas that have been and still are disproportionately impacted by climate change. Their experience of climate change differs from that of the general population due to their history, extreme social isolation, cultural traits and customs, and limited access to social services like work and education. Thus, Canada is a useful country to examine the issues of climate change but also the need to integrate environmental justice, as the legacy of those problems is having a major effect on the inability of groups to adapt to climate change. Thus, from the problems much can be learnt about ways forward to deal with some of these matters.

From a historical perspective, South Africa transitioned from Apartheid to a democratic society in 1994, making it a relatively young democracy. Its majority population endured centuries of oppression, characterized by severe human rights abuses and environmental injustices. This legacy remains stark, with widespread poverty, unemployment, and environmental challenges. South Africa's reconciliation process, epitomized by the Truth and Reconciliation Commission, is frequently studied for insights into transitional justice.

The United States, in contrast, is one of the world's wealthiest countries, yet it contends with significant economic inequality. As a state with over 250 years of history, its past includes racial segregation, discrimination, and systemic violations against minorities and Indigenous Peoples. The country has faced substantial environmental justice challenges. Efforts to address systemic racism and inequality in the United States can be meaningfully compared with South Africa's post-Apartheid initiatives, offering valuable insights into the shared and divergent pathways toward environmental social justice.

It must be noted, however, that the situation in the United States also reflects the uniqueness of the American condition, in the sense that environmental issues have become a political hot-button, with the Republican Party pushing back on climate issues and other matters that fall within the environmental justice sphere. The United States is a prime example of the challenges associated with having climate change contentions, including the long-term climate change-related damages suffered by marginalized groups, recognized. The challenges encountered while attempting to advance theories such as intersectionality in such circumstances are reflected on. Thus, the interwoven nature of politics in dealing with climate change and environmental justice in the United States cannot be underscored sufficiently.

The concern about the influence of politics on issues concerning climate change is far from an issue only of the United States. It occurs in many countries, but maybe not to the same extent. The issue of politics in the international climate change governance system has not received much attention, but the role of politics at that level is intense (Lidskog and Sundqvist 2015). In this context, and with regard to the role of states in addressing the science of climate change, the ongoing disputes over climate change reports, and the impact of such wrangling on international climate processes – particularly on how outcome documents are negotiated and finalized – there needs to be much more attention devoted to these issues. The intermingling of politics and science does not always produce the best result in addressing the challenges the world faces (Lidskog and Sundqvist 2011). This dynamic has had a significant impact on the Intergovernmental Panel on Climate Change (IPCC) (Hughes 2024). Thus, it has been stated that:

> Given that the IPCC was set up to serve a different purpose than the polycentric governance pattern we see today, it is unclear whether the IPCC processes and outputs are designed in ways that facilitate the creation, accumulation, diffusion and translation of the types of knowledge that we need to better cope with climate change. (Hermansen et al. 2023)

However, a focus in this book is on the role of politics at the domestic level. That is a focus of the US chapter. It is the nature of that country's political system and the way issues concerning minorities and climate issues are dealt with in that country that they are deeply affected by politics. It is which political party is in the presidency, and which of the two political parties control the two chambers of the Congress which is significant for both issues. In this context, as issues of populism (Moatti and Müller 2021) are deeply relevant (Webber 2023), in more and more political systems around the world (Norris 2020; Bergh and Kärnä 2024), so the lessons of the United States (Rodríguez-Pose et al. 2021), through the lens of dealing with climate change (Goldstein 2023), become more pertinent from this perspective (Gomes and Böhm 2023).

The United States is important, as will be indicated in the US chapter, as in recent years the Administration has recognized the interplay between climate change and environmental justice. Thus, especially during the Biden presidency, both issues have been dealt with in the law mostly by Executive Decree, but sometimes in legislation. From time to time, both issues have been dealt with collectively and in an integrated fashion. This has been done to overcome the legacy of past environmental injustices, and to pre-empt the greater effect that climate change is going to have on these same communities. Tremendous progress has been made in recent years in the United States to deal with both. However, that headway would be reversed if Joe Biden lost the election for the presidency in 2024. That is what occurred.

It is also crucial to focus on the United States since the United States, China, and Russia combined, account for 40 percent of the world's greenhouse gas emissions. Therefore, it is imperative that attention is paid to what occurs in the United States, as well as in the other two countries. This is because if the situation in the United States fails to meet the challenge of climate change, then the world will suffer the consequences of that failure. Therefore, the climate and environmental justice legal situation in the United States is used as a lens to examine the challenges encountered in such litigation, including the challenges in knowing and meeting the required legal standards. Environmental justice and the rights of Indigenous Peoples and minority groups are deeply political issues. Oftentimes the response to these issues is reflective of the politics of whichever political party is in power. This is an important consideration for anyone wishing to engage in environmental or human rights-related litigation.

Relying on research done by authors in the United States who contend that environmental justice litigation has not achieved many big wins, this book argues that more needs to be done to achieve greater success. While the climate emergency is spurring a lot more focus on climate change in the United States and around the world, it is argued that environmental justice is receiving insufficient attention. This is because most of the attention has been on climate change. While there has been some attentiveness to climate justice there is very little focus generally on environmental justice practically speaking outside the United States. While there is some environmental justice focus in the United States, in other countries environmental justice processes are less advanced, and there are significantly fewer matters being taken up. There are some court cases concerning broad climate and environmental issues being taken up but not too many environmental justice cases (Suedi and Fall 2024).

These cases are not focused on as the focus is on the laws and processes to deal with these two concerns. Thus, the United States, where there has been quite some focus on environmental justice, can provide useful comparative lessons. While the context and the laws are unique to each country, much can be learned from previous efforts. Critically, the focus here is the sociopolitical context and the political steps taken, or not taken, to deal with these matters. There is a limited focus on the role of the courts. The aim is to determine what are the positives and negatives of the steps taken to adapt to climate change and what steps can be taken to bolster or enhance such steps. Thus, lessons and recommendations are made, specifically in the expansive Conclusion to the book.

Critically, all three countries are valuable for study from a human rights perspective, as while they have rights protection framework and laws, each still faces significant challenges. South Africa, for example, has a strong post-Apartheid constitution renowned for its emphasis on human rights, but the practical implementation of these rights has been problematic. The United States and Canada with their constitutions protecting many human rights, face challenges in safeguarding individuals from various abuses, including discrimination.

In this context, understanding how each country treats its Indigenous and minority populations, especially concerning environmental justice, can shed light on broader issues of equity and human rights. The experiences of Indigenous Peoples and minorities – in the United States, and in Canada – can provide comparative perspectives on marginalization and advocacy. South Africa's situation is also instructive, as it reveals how the majority is addressing problems that were imposed by the White minority during the Apartheid years.

In terms of their economic situations, all three countries face economic challenges for their minority groups. South Africa, in particular, has significant economic challenges, but the high levels of inequality are a common feature across all three nations. Each country has strategies to address poverty and promote economic growth, but these strategies differ in many ways. What is clear is that the economic status of each country influences its capacity to address environmental issues. South Africa's developing economy, and Canada and the United States' highly industrialized economies, each present unique challenges and opportunities in dealing with climate change and environmental justice concerns.

As far as environmental justice is concerned, the three countries are useful to compare. This is because in South Africa, the legacy of Apartheid has left a significant impact on environmental justice, with marginalized communities often facing the worst environmental conditions. In the United States, environmental justice issues are deeply intertwined with the history of racial segregation and economic disparity. This is the case in Canada as well. Studying the environmental problems each country faces and how they are addressing environmental justice through their policies and governance can reveal best practices and areas needing improvement. Comparing policies and their effectiveness can provide valuable insights, and useful lessons can be applied elsewhere.

As far as climate change is concerned, a comparative approach is useful because climate change affects different regions in diverse ways. Such studies help identify regional differences, providing insights into specific vulnerabilities and adaptive capacities. By comparing the strategies and policies of different countries, it is possible to identify effective approaches to mitigate and adapt to climate change. This can help guide global efforts and inform policymaking. Comparing different climate policies and their results can help assess their effectiveness and efficiency. Policymakers can then refine existing policies and develop new ones that are more effective in combating climate change and better suited for adaptation. These comparisons can also provide useful information to educate the public and raise awareness about the impacts of climate change.

Further, taking a comparative approach and examining how each country is responding to global environmental agreements and frameworks, such as the Paris Agreement, is useful to understand how each country views its commitment to these issues and what actions they are taking in that regard. It is also useful to determine what the agenda should be to have a transformative effect (Newell et al. 2021). This is necessary as:

> effective climate justice policies will not likely result through deliberations in scientific journals or at academic conferences. While social scientists as well as climate, natural, and public health scientists have much to offer to inform and design future policies, particularly by providing an empirical basis for action. (Pastor et al. 2024)

Thus, much more needs to be done by governments at all levels to ensure that more integrated actions are taken to deal with climate change in ways that improve the lives of the most impoverished and downtrodden. These communities ought to be given greater access to funding and other resources to be better able to mitigate the effects of climate change.

SUSTAINABILITY

A major theme of this book is sustainability, a concept that has become one of the major buzzwords in the world today (Giovannoni and Fabietti 2013). There has been growing concern about climate change and the protection of the environment, together with social matters, such as poverty and inequality. International law concepts concerning the environment have been developed, specifically concerning the right to a healthy environment, as intergovernmental organizations attempt to lay down frameworks (which are typically nonbinding) to protect people from becoming victims of human-caused environmental degradation. However, these efforts, and the law, remain deficient in many respects.

Environmental justice proponents have been working to ensure that a clean and healthy environment is recognized as a fundamental human right and that everyone globally has access to clean air, clean water, uncontaminated soil, and a healthy natural environment (Tigre and Setzer 2023). They have sought to create and implement laws that enhance people's quality of life and contribute to more balanced and fair access to resources, especially for those usually most affected by environmental degradation, i.e., the poorest and least resilient members of our societies (Kameri-Mbote and Cullet 1996). At the heart of the concept of environmental justice, for some, is to effect changes so that Indigenous Peoples and minorities cease being the main groups affected by environmental harms. This is because, as Case notes, these "communities are typically poor and substantially lacking in the political acumen and power enjoyed by business and industry interests that create environmental impacts and risks for surrounding communities" (Case 2011: 701).

In this context, it is important to bear in mind that the World Bank has found that natural resources make up about 25 percent of the total global Gross Domestic Product (GDP) (Baba et al. 2021). Further, the world's economy has depended on relatively cheap natural resources, including fossil fuels, to meet its energy needs. Thus, their value to the economies of the nation-states around the world is enormous. However, until relatively recently, little attention was paid to the social and environmental costs of extracting, refining, and burning fossil fuels (Lehtonen 2004). That heedless approach has come to threaten the very foundation of continued economic growth. Indeed, the environmental pressures of climate change and its effects, in addition to other challenges, make economic progress difficult (Amprako 2016). This has been particularly challenging recently as the economic growth rate of many states has been declining. This slowdown, a continuation of what occurred at the end of first decade of the 21st century and then again during the COVID-19 pandemic, has brought about greater financial insecurity, widening inequalities, inflation, and unemployment. The question of what to do about these problems in the national context has divided societies, especially in terms of their environmental dimension. However, this situation has also raised questions about how the economy ought to be run, and what benefits should go to whom (The World Bank 2023). This time of economic uncertainty has triggered debate about how to deal with the demands of economic growth while ensuring environmental sustainability. As a result, many states are searching for alternative ways to drive economic growth (Fosu 2017). This will become a greater imperative as it is projected that the world economy will shrink about 20 percent within the next quarter century due to climate change (Kotz 2024b).

At the same time, the concept of progress has changed from being measured solely from an economic perspective to include other concerns, such as how to make development socially and environmentally sustainable (Sternfels et al. 2022). For example, the issue of mineral extraction has been the subject of much debate about how much profit it provides versus whether and to what extent it creates sustainable economic growth (McMahon and Moreira 2014). Mineral extraction activities and similar industries have a superficially large economic benefit, but at the same time have many negative effects on the societies in the region, something that has been called the resource curse (Sovacool 2013). The negative effects, and particularly the negative environmental effects, are felt to a greater extent by marginalized people.

The extractive industries alone have a massive effect on half the world's population, in most cases the people who live in the world's underdeveloped nations, where such resources are often found (Sovacool 2020). These are often the poorest, most marginalized, and dispossessed people, and include Indigenous and minority groups. In this regard, the High Commissioner for Human Rights, Volker Türk, noted in July 2023, that the devastating rights violations inflicted upon Indigenous Peoples had to end. He argued that what was occurring was a result of the "unprincipled and devastating impact of extractive industries on the environment and the rights of Indigenous Peoples. Their dispossession from ancestral lands, and the militarization of their territories" (OHCHR 2023).

Climate change and sustainability are major issues affecting Indigenous Peoples and minority groups. They experience a double impact from the environmental injustices they suffer. Firstly, they are affected by the harm caused by specific environmental injustices, and secondly, they are affected because the places where they live are underserved, which means they have a reduced ability to resist or recover from the effects.

What is also clear is that the detrimental effects of climate change are felt disproportionately by certain groups, usually the most vulnerable, including and specifically Indigenous Peoples and minorities (Dialesandro et al. 2021). This is because many of these groups live in environmentally depreciated areas, which are now being further depreciated by climate change (Minority Rights Group 2021). This is an issue that is being deeply felt by these groups, because of the intersection of climate change effects and poverty and other socioeconomic challenges that they face (Idiazabal Gorrotxategi and Pérez-Caurel 2019). Yet, these groups contribute the least to environmental degradation and climate change. They are also the least likely to receive assistance to deal with these challenges or to mitigate these effects. These groups are now suffering from the compounded effects of climate change and environmental injustice.

What is also clear is that even though there are efforts underway to bring about climate justice, progress is not being made quickly enough. For example, in the United States, greenhouse gas emissions, while not being reduced, has actually increased. It went up by 1.3 percent in 2022 as compared to 2021. This was, however, a reduction in the increase of the year before, when it was 6.5 percent. However, in 2020, there was a 10 percent reduction in the emissions levels in the United States, when the pandemic led to a significant contraction in the economy, resulting in reduced emissions from industry and automobiles (Rhodium Group 2025). There is, therefore, a need to find and implement ways of reducing harmful emissions, by using sustainable means of economic production and finding ways to mitigate the effects of the damage that is caused. Thus, sustainability issues are deeply relevant.

These issues are some of the matters that are supposed to be focused on by the Sustainable Development Goals (SDGs). These issues are interconnected and, as Gupta and Vegelin (2023) suggest, the concept of "inclusive development" highlights that there is a need to include most marginalized and excluded people (Pouw and Gupta 2017). For this reason, the SDGs seek not only to achieve social and economic sustainability, but also have the goal of achieving environmental sustainability. This is defined by John Morelli as:

> meeting the resource and services needs of current and future generations without compromising the health of ecosystems that provide them, and more specifically, as a condition of balance, resilience, and interconnectedness that allows human society to satisfy its needs while neither exceeding the capacity of its support ecosystems to continue to regenerate the services necessary to meet those nor by our actions diminishing biological diversity. (Morelli 2011: 23)

Morelli sees environmental sustainability as a "subset of ecological sustainability" (Morelli 2011: 23) and as an "intersection of human activities and ecological systems" (Morelli 2011: 23). Environmental sustainability emerges as a goal established in the SDGs based on the interdependence between nature and people, principally established in various of the SDGs, as discussed below. Without doubt, the goal of environmental sustainability is particularly important because the "entire world faces dire environmental threats from human-induced climate change and the loss of biodiversity" (Sustainable Development Solutions Network 2015). These issues are critically important from the standpoint of who is most likely to be affected by them and who can assist in achieving them. As will be discussed, it is those most vulnerable, such as Indigenous and minority groups, who fall within those parameters. Despite this, the limited reference is made to Indigenous Peoples and minorities are largely omitted, except by reference to the vulnerable.

Unsurprisingly, the very first SDG is to "[e]nd poverty in all its forms everywhere." Target 1.5 provides: "By 2030, build the resilience of the poor and those in vulnerable situations and reduce their exposure and vulnerability to climate-related extreme events and other economic, social and environmental shocks and disasters." However, as Knox has noted, there is a "particularly glaring omission" which "is the lack of attention to indigenous peoples, perhaps the group most vulnerable to environmental harms … which include only two passing references" (Knox 2015). This is surprising, since: "sustainability principles are fundamental to the cultures of many of the world's indigenous peoples and cannot be ignored in any consideration of desirable development paths for first peoples living on land that their communities have occupied for countless generations" (Throsby and Petetskaya 2016).

Nonetheless, it is argued that environmental justice efforts can benefit from the SDGs and that the SDGs are in many ways reflective of what environmental justice is about. It is contended that many of the SDGs are at one with the aims of environmental justice. It is also argued that the SDGs can be seen as a pathway for the achievement of environmental justice and that various actors, especially states, including their courts, have a role to play in this regard.

The effects of climate change must be reduced, not only because of the harm being done, but also because of the even greater effects it will have on the planet and communities in the future. Without addressing it sufficiently, for example, people's health, especially the vulnerable, including Indigenous Peoples and minorities, will be even more severely affected (Smith et al. 2015). This is because these effects are not being equally felt (WHO 2014). They are especially severe for marginalized groups (Alava and Singh 2022), and specifically Indigenous Peoples and minority groups (Crowley et al. 2022). Thus, places where there are forests, and other types of terrain that can mitigate or alleviate the effects of climate change, require greater protection (Gençay and Birben 2016). This is because, as noted earlier, forests are the source of many things, including food, clean air, and pharmaceuticals to name a few. Research indicates that forests are home to about 50,000 plants that are used for various therapeutic and medicinal purposes both by individuals and groups as well as by multinational companies (Shah and Bhat 2019). Additionally, huge amounts of products that are used as food comes from forests, feeding about one billion people (Chongtham and Bisht 2020). Wild food is an excellent source of essential nutrients required for sustaining people (Bélanger and Pilling 2019). Despite this, about two billion people have nutritional deficiencies, while nearly one billion people are malnourished (Bélanger and Pilling 2019). This will increase over time, as opportunities for gathering and growing food will decrease due to climate change. Forests are being dramatically reduced in size by the effects of climate change, including an increase in the number and intensity of forest fires, drought, flooding, the expansion of insect territories, and by relentless deforestation by humans, all of which cause about 150 million hectares of forests to vanish each year (Pongratz et al. 2021).

This book therefore touches on the role of the SDGs and how they interact with environmental justice issues that are of particular importance to Indigenous Peoples and minority groups. For example, SDG 11 calls for places where people reside to be "inclusive, safe, resilient, and sustainable." Therefore, this book examines environmental justice issues within the context of sustainability while also considering how "sustainability research and practice can benefit from a closer reading of environmental justice scholarship … [and how] vulnerability science can provide an important bridge between the two spheres of environmental justice and sustainability while also benefitting from the convergence of ideas, principles and practices from these fields of inquiry" (Boone 2010: 135).

This approach permits the examination of issues addressed by the SDGs, but also some that are not included. It is argued that the SDGs are lacking from the point of view of promoting issues concerning Indigenous Peoples and minority groups. In fact, it is argued that these groups are largely excluded from the SDGs. However, this is not meant to be an in-depth analysis, as that would be a book on its own. However, it is argued that the next version of the SDGs, which will be taken up when the present SDGs end in 2030, must address the need for real and meaningful sustainability for these groups. It is, however, argued that promoting and protecting the needs of Indigenous Peoples and minority groups will benefit humanity.

DEALING WITH CLIMATE MISINFORMATION AND THE NEED FOR ACCURATE DATA

An issue of concern when dealing with climate justice and environmental justice is the lack of accurate and complete information on such issues. This is even more pressing in the era of climate misinformation.

In this regard, UNESCO has claimed that this has a number of detrimental effects, including undermining scientific consensus, impeding the ability of authorities' to properly address these matters, and endangering the safety of frontline journalists and environmental defenders. The IPCC acknowledged the danger that misinformation poses to reaching climate goals when it said in 2022 that: "deliberate undermining of science" was causing "misperceptions of the scientific consensus, uncertainty, disregarded risk and urgency, and dissent." To deal with the problem at the international level in November 2024, the Global Initiative for Information Integrity on Climate Change was initiated by the United Nations, the Brazilian government, and UNESCO to improve research and devise policies to combat misinformation that might impede climate action (UN News 2024).

As far as scientific information on climate change is concerned, insufficient data can be an impediment in promoting climate science and to ensure the transparency of environmental justice concerns. Thus, institutions such as the National Oceanic and Atmospheric Administration which collect such data in the United States play critical roles in collecting a range of scientific information. However, this agency may be under threat in the future, either that it may be closed down or its independence severely curtailed.

However, the issue is not only about the formal scientific sector. This is a problems as not all information is being treated equally. As has been noted:

> Data is the currency of the 21st century. Data-driven insights have propelled multiple fields to new heights – from climate science to public health – improving our lives beyond measure. Too often, however, humanity overlooks another vital form

> of knowledge, one crucial in addressing the climate and biodiversity crises: the lived experiences, traditions, and wisdom of Indigenous peoples. (Duarte and Al Mubarak 2024)

As is noted, the issue of Indigenous Peoples and their knowledge is dealt with problematically. More broadly, and generally, around the world, the issue of date collection is a problem as governments:

> monitor the state of major environmental indicators, such as air and water quality, to analytical measures of major environmental aggregates. For the most part, however, measures of the economic contribution of natural resources and the environment have lagged behind physical measures … Natural resources such as petroleum, minerals, clean water, and fertile soils are assets of the economy … An important part of the economic picture is therefore missing if natural assets are omitted in creating the national balance sheet. (National Research Council 1999: 19)

This is contained in a 1999 report by a panel for the US National Academy of Sciences that was led by Yale economist William Nordhaus. He was later awarded a Sveriges Riksbank Prize in Economic Sciences in Memory of Alfred Nobel for his research on the economic effects of climate change (Barrage 2019). He was also involved in a study that developed a process to include environmental damages into the US national accounts system (Muller et al. 2011).

In January 2023, at the World Economic Forum at Davos in Switzerland, the United States announced a plan to monitor and quantify its natural assets, such as forests, wetlands, rivers, and deep snows, which have value, but which are not normally measured for their benefits to economic activity. This plan creates a method to assess the value of healthy ecosystems to humanity to create what is known as natural capital accounting (Nordhaus 1992). The information would then be used to help the state determine which business activity to support, which environmental assets to protect, and what directives to adopt. The special envoy for climate change, John Kerry, proclaimed that in this plan "the U.S. will put nature on the national balance sheet" (DePillis 2023).

In January 2023, the White House issued its "National Strategy to Develop Statistics for Environmental – Economic Decisions: A U.S. System of Natural Capital Accounting and Associated Environmental – Economic Statistics" (USA Office of Science and Technology Policy 2023). This is likely to be controversial in the United States, where Republicans and industry see attempts to gather data as a forerunner to greater regulation (Doyle et al. 2015). However, if issues concerning the environment and climate are to be dealt with then the data is absolutely crucial. This type of plan is one that other countries ought to take up as well in an effort to make good decisions and mitigate climate change at the national level. This is not an exercise that can be done quickly or that will give results rapidly. In the United States it is estimated that it will take 15 years to collect adequate data. Thus, the sooner this begins everywhere around the world, the better it will be able to track what is happening and what the effects of climate change and environmental issues are. This information is essential from an environmental justice point of view if, for example, the maldistribution of ecological resources, and how these affect various communities, is to be tracked and understood (Martinez-Alier 2003a).

Collecting information will also indicate how resources flow, often unfairly, and be helpful to try and remediate environmental injustice matters. In this context, there have been various attempts to create models to determine measures of such harms (Krieg and Faber 2004). These processes could help to assess the Net Present Value or to use other methods (Harrison et al. 2018) so as to value the various natural resources (Kallis et al. 2013), as well as the harm that is being caused to them. This is deeply contentious, as some believe it will not be helpful (Temper and Martinez-Alier 2013). However, some means to determine what the situation is, and how it changes over time, are essential, for example, mapping and accounting. As has been noted, valuation can be helpful because it "seeks to make environmental decision-making more inclusive of diverse human concerns and aspirations" (Tadaki et al. 2021: 190).

These impact issues should not be left only to governments: the other branches of the state also have roles to play in this regard. Legislatures ought to be taking up these issues and enacting laws requiring these measures to be taken, even if their government is also acting on these challenges. This is because governments might not set up these processes in the best way. Giving governments direction and parameters will hopefully see meaningful legislation. Parliaments ought to also monitor these processes and have hearings to determine their effectiveness. They can then make informed decisions to make the outcomes better in the future. Ongoing scrutiny is therefore of critical importance. Legislatures should also enact laws that demand that companies and the local communities they are in create good-neighbor agreements. These have been created in various places around the world, setting up the relationship between the parties in a positive way, giving a voice to the community concerned, and setting out the rights and duties of the parties (Illsley 2002). These agreements can go a long way in avoiding environmental justice issues and creating processes to deal with those problems that do arise (Walker 2010).

In addition to the executive and legislature, the courts also have a role to play in measuring what is happening in the environmental space. In general, when the courts review economic, social, and cultural rights cases, they ought to be demanding good data to make their assessments. If such data does not exist, then this must be obtained as a matter of urgency. When the courts must adjudicate on specific environmental justice cases, they should require, as a matter of course, impact assessments to be carried out by true independent experts, not those mandated by the wealthy parties in the case. True independence is essential. The courts also must ensure true assessments are obtained with regards to the impacts on the community as a whole and how issues of intersectionality play out regarding the specific issues that cases deal with.

TERMINOLOGY FOR MINORITIES AND INDIGENOUS PEOPLES

Various terms are used to describe the groups and communities discussed or referred to in this book. Despite the fact that minority and Indigenous communities are frequently grouped together, they are different even though some Indigenous Peoples might also be a minority group.

As far as minorities are concerned, who they are is subject to a great deal of controversy. Problematically, the numbers of people in minority groups around the world are not really collated. They are found only in some country statistics. This is because not all states collect or release this type of demographic data. In other cases, the data is vague, inconsistent, or unreliable. In many places, the information is scattered and the information available on minorities at times produces biased estimates. Another problem is that minority groups are often fluid rather than static. Gender, sexual orientation, race, ethnicity, religion, and disability are the main traits linked with minorities, though generalization is challenging. The emergence or identification of minority groups may also be attributed to language, nationality, and culture.

As far as Indigenous Peoples are concerned, while the generally accepted terms used internationally are "Indigenous," in some countries there are penchants for other terms, including "natives," "tribes," "first peoples/nations," "Aboriginals," "Indians," and, in the United States, "ethnic" groups (Corntassel 2003: 75). For example, in the United States, various terms are used to describe Indigenous Peoples, their communities and their lands, such as Indians, tribes, Indian tribes, reservations, Indian Country, and native nations (Missirian 2021). Some of these terms are considered derogatory, while not so in some other places (Browde 2022: 1). Thus, the term used in this book is either Indigenous group or Indigenous Peoples. In the American context, specifically, the term Native American is used. Interestingly, a United States Congressional Research Report notes that "In counting distinct groups, this report uses the term 'Indigenous People' or 'indigenous groups' rather than 'tribe,' 'nation,' 'ethnic minority,' or 'sociolinguistic group'" (Davis-Castro 2020). It should be noted that the Expert Mechanism on the Rights of Indigenous Peoples has called on the Human Rights Council to use capitalization to the term "Indigenous Peoples" in all United Nations documents.

2. Theorizing the interrelationship of climate change action with environmental justice

INTRODUCTION

Climate change issues are firmly on the agenda around the world. However, taking action in the narrow sense is still somewhat limited. While there are often lots of plans, action is not always being taken in many parts of the world. In this regard US Climate envoy John Kerry, appointed to the position by President Biden in 2021, noted as he left his position in 2024 that:

> If we did the things we could do – that we know how to do, and that we have the technologies for – we could actually do it. We're just not. We're not doing it on a global basis. Emissions are going up in too many countries. Oil and gas are still on a binge and their profits are obscene. I mean they're just shocking. And everybody seems to be locked into a place of indifference. (Wallace-Wells 2024)

A problem, however, is that while there is much focus on climate change, climate justice and ancillary issues, especially in academia (Giacomini 2022), what is lacking is a focus on environmental justice problems and how they interrelate to climate-related matters in policy work of governments, local authorities, and particularly by the courts. Part of the reason is the lack of understanding about their interconnections.

What can also be seen is that the public are very unfamiliar with some of these concepts. Thus, one survey in the United States found that while people knew about climate change, two-thirds of people had never heard of climate justice. The remaining few had heard of it but very few of them knew much about it (Leiserowitz et al. 2023). Many individuals, groups, and governments are unaware of what it is and what the issues entail. This is problematic, as much needs to be done (Kranz et al. 2022). Part of the reason for this is a lack of understanding about what climate justice is and how it can be used (Newell et al. 2021). Legally speaking, much needs to be cleared up to make room for requisite action. Thus, giving these issues the necessary focus may ensure greater awareness, and therefore lead to greater attention in dealing with the problems (Doucette et al. 2023).

While awareness about climate change issues is growing, there is still limited media attention on environmental justice problems. For example, one report in 2024 found that in 2023 there was little news coverage on environmental justice in the US. Networks passed on important environmental justice issues that would have emphasized the ways in which vulnerable groups are disproportionately harmed by racism, economic exploitation, and environmental degradation (Abi Deivanayagam et al. 2023). The report noted that although there were some news pieces that showed a greater comprehension of environmental justice issues, overall coverage remained patchy and uneven (Cooper 2024). Thus, there is a need to educate people about the wider environmental justice concepts and their relevance. This is crucial, everywhere in the world. However, there is also a need for focus on environmental justice internationally as well. This is not occurring in international bodies and international fora.

THE INTERRELATED CONCEPTS OF CLIMATE CHANGE ACTION AND ENVIRONMENTAL JUSTICE

While this book focuses on the broader issues of climate change and environmental justice, it is with the realization that issues of climate change and the interrelated concept of climate justice affect environmental justice concerns and vice versa (Schlosberg and Collins 2014).

While people often use these terms interchangeably, they are different. Climate justice and environmental justice are different concepts even though some see them as the same (Méndez 2020). What is the case, is that climate justice grew out of environmental justice (Pezzullo 2022). Therefore, they are deeply interrelated (Schlosberg and Collins 2014). They have similar goals (Siegfried 2024) but climate justice is more narrowly focused on achieving those objectives within the climate change sphere. However, justice is a critical component of them (Zimm et al. 2024). As has been noted:

> climate justice seeks to reframe the discourse around climate change from a purely environmental issue to a social, economic, and ethical one. It emphasizes the interconnectedness of environmental concerns with broader social justice and human rights considerations. It encompasses both the ethical and political dimensions of climate change, advocating for equitable and fair responses to the challenges posed by a changing climate. Accordingly, key aspects of climate justice include equity and fairness; vulnerability and marginalization; human rights; participation and inclusion; global solidarity; environmental racism and discrimination; just transition and accountability. (Totry-Jubran 2023, references omitted)

Therefore, both climate justice and environmental justice are also human rights concepts (Gilliam et al. 2023). They deal with how various rights including the right to life, dignity, a healthy environment, food, water, housing, health, etc. need to be protected and promoted. This is important as both environmental justice and climate change can have dramatic negative effects on the enjoyment of human rights (Atapattu et al. 2021). They are also human rights matters because climate justice accepts that the repercussions of climate change are not uniformly distributed, and that marginalized and vulnerable populations frequently suffer the most. They also both attempt to deal with discriminatory behavior and its consequences (Levy et al. 2024). They both involve dealing with the unequal distribution of climate change's effects and try to promote just and equitable responses to it (Alibudbud 2023). This is crucial for vulnerable groups. The law needs to deal with these issues and absorb them, so they are taken into account more often when dealing with a range of issues.

Critically, climate justice aims to address the disparities that exist in the distribution of the negative consequences of climate change (Maldonado et al. 2016) by ensuring that all people, regardless of class, race, ethnicity, or other identity issues, as well as where they are in the world, receive the necessary assistance and can equally benefit from laws, policies, and actions taken to deal with climate change. Therefore, climate justice imagines a global community where the realities of climate change are addressed for all, and the costs arising from climate change are distributed fairly and equitably. In so doing, all people regardless of their identity or where they live, can prosper in an environmentally safe setting. This is interrelated with environmental justice. Interestingly, the need to use an intersectional approach has also been advocated for climate justice (Amorim-Maia et al. 2022; Mikulewicz et al. 2023) further indicating the interrelationship between it and environmental justice. The law and the courts need to be more attuned to these issues and apply them.

Thus, while climate justice is about finding fair and equitable ways to address the unequal effects of climate change on certain places and particular groups, climate change also gives rise to environmental justice concerns. While climate justice is about taking significant steps to cut fossil fuel emissions and slow down global warming, so too is environmental justice. Thus, both need to be applied by the law.

Climate justice is centrally focused on putting measures in place that advance greening the globe whilst advancing fair and even-handed ways of tackling the problems caused by climate change. It is about taking steps to reduce the use of fossil fuels, using more clean energy, developing green jobs, and finding sustainable ways to ensure crops are more resilient and more resistant to climate change (Atapattu et al. 2021). This is because "Fossil fuel operations cause 40% of emissions, and *The Guardian* revealed there were more than

1,000 super-emitter events from oil, gas and coal sites in 2022 alone, many of which could be easily fixed" (Carrington and Clarke 2024). As noted above, one of these is methane. In fact, "methane emissions have accelerated since 2007 and cause a third of the global heating driving the climate crisis today. The acceleration has alarmed scientists" (Carrington and Clarke 2024). This is a major issue, as methane is seen to cause 30 percent of global warming (Scoones 2023). Therefore, the Global Methane Pledge (GMP) was crafted at COP26. By the beginning of 2024, it had been adopted by 155 countries who have agreed to cut 30 percent of their methane emissions by 2030 (Global Methane Pledge nd). There are, however, a multitude of other gases that are being emitted which cause problems (Nisbet et al. 2023; Climate Trace 2023). It is believed that there is an undercount of the amount of emissions that are occurring (Pearce 2024). A major reduction in the emission of various gases is essential for the earth's sustainability.

As can be seen, there is considerable overlap between climate justice and environmental justice. Therefore, to apply environmental justice concepts to climate change-related matters would produce very positive results (Malakar et al. 2023). Some do argue that: "Ideas of climate justice and environmental justice are separate but closely related" (Albertyn 2023a: 138). However, what is the case, is that both of their effects may engender:

> Anger, resentment, and frustration towards major polluting nations or corporations may affect geopolitical alignments, remaking alliances around climate considerations in addition to traditional security threats, or drive societal-level reactions such as boycotts and divestments of companies associated with emissions. It may even spur increased ecoterrorism. (Frederick and McCulloch 2024)

This sentiment has not, however, been sufficiently taken in domestic law and policy globally, with a few exceptions.

What can be seen is that the environmental justice movement inspired the development of the focus on climate justice (Schlosberg and Collins 2014; Tahir 2022). While climate justice and environmental justice, in fact, deal with similar issues (Evans-Agnew and Aguilera 2023) they are not exactly the same (Newell et al. 2021). These concepts do overlap substantially (Schlosberg and Collins 2014). While both are concerned with situations of control and power (Fernandes-Jesus et al. 2020) and who has been excluded from the benefits and disadvantages of the various environmental issues (Martín 2022) and remedies (Abi Deivanayagam et al. 2023), climate justice is narrower than environmental justice. Climate justice examines environment-related issues in the context of climate change alone. Climate justice is about recognizing the disparate, unequal, and disproportionate effects (Sultana 2022b) that climate change has on individuals and specific groups (Mattar et al. 2021). Climate justice endeavors to find just and equitable means to address the injustices that have arisen (Rice et al. 2022). Its objectives are to improve equity and justice while reducing marginalization, exploitation, and oppression (Biermann and Kalfagianni 2020). Using a climate justice strategy entails closely examining who is victimized, ostracized, or marginalized by processes related to climate change, and any mitigation or adaptation efforts that are made. It centers on identifying who gains, who loses, how, where, and why. It is an aid in reaching a clearer understanding of the connections between the many aspects that co-create and sustain injustices (Sultana 2022b). Thus, climate justice instructs us on how the ethical and social implications of climate change are to be addressed (Robinson 2018a). It acknowledges that not all people will be affected equally by climate change (Derman 2020). The effects of climate change are not the same everywhere, visiting more dramatic and devastating effects on some people's lives more than others (Giacomini 2022). Climate justice works towards addressing systemic and historic inequalities that affect various communities due to climate change. Thus, The Elders in 2024 published a report that noted that:

> As long as the needs and concerns of vulnerable and marginalised groups are not taken into account, a sense of fairness not restored and historical injustices not redressed, climate action will struggle to attract the political support needed in individual countries – and globally – to be effective. Climate justice is thus both a moral imperative and a political responsibility. Justice can take multiple forms, including distributive justice (sharing benefits and burdens fairly between people), procedural justice (ensuring representative decision-making) and restorative justice (encouraging accountability and repairing harm). All are essential to a just transition. (The Elders 2024b)

However, their report had no reference to environmental justice, although what they deal with has an implicit environmental justice dimension. These strands need to be brought together everywhere, to ensure that full scale of climate change, and its implications, are understood and dealt with. However, there has to be, as the Elders note, a reckoning with what has occurred in the past, and ways need to be found to deal with those problems, including the environmental injustices, to ensure that climate change can be adequately dealt with by everyone, including those transitionally left behind.

As far as recognition of the importance of these issues is concerned, generally, the confluence of environmental justice, climate change, and climate justice is being recognized. This is occurring only by some in academia and in civil society studies. Certainly, climate change research would assist (Magnan et al. 2022) in the development of environmental justice principles and vice versa (Kim 2021). In doing so, it would also help with solutions in praxis. This is not, however, the case, as countries, even developed ones such as the United

Kingdom, are not taking enough steps to mitigate the major effects that have already been caused but will be occurring to an even greater degree in the not-too-distant future. A report notes that the:

> evidence shows that the gap between the level of risk we face and the level of adaptation underway has widened. Adaptation action has failed to keep pace with the worsening reality of climate risk. (UK Climate Change Committee 2021)

Thus, in practice integrating the concepts and dealing with them in an integrated way would be helpful.

While there is an expanding multidisciplinary set of inquiries which underscore the exacerbated vulnerabilities of marginalized populations, those realities need to become much more integrated into state policy development and environment and climate damage amelioration practices. It is in praxis where the biggest gap in the integrated use of these concepts exists. For example, many countries treat them independently. One of the few countries with a more unified approach has been the United States over the last few years, as is discussed later. Thus, a combined approach incorporating all three of climate, energy and environmental (CEE) pillars is required across the board (da Silveira et al. 2018). This is because all three are intimately interlinked, intertwined (Natarajan 2021), and overlap (Porter et al. 2020). As has been stated:

> Enter climate change, for which injustice is both fuel and byproduct. On the one hand, the causes of climate change – fossil fuel extraction, industrial pollution, deforestation – are concentrated in marginalized communities. Arguably, exploitation has allowed those activities to continue longer, and more recklessly, than what is in humanity's best interest. The organizer Hop Hopkins puts it this way: "You can't have climate change without sacrifice zones, and you can't have sacrifice zones without disposable people." On the other hand, marginalized communities also suffer disproportionately from the consequences of climate change, including more frequent and more severe floods, storms, droughts, heat waves, and crop failures. Yet justice for communities is largely missing from the way we are attempting to solve the problem. (Maru 2023: 103)

Without this, actions related to climate change run the risk of maintaining, or worse, escalating, current disparities. To develop fair and long-lasting solutions, it is imperative that environmental justice and climate justice be addressed, as the effects of climate change continually worsen. That deterioration will continue to have greater dire consequences for some communities because of the environmental injustices they have suffered in the past, and continue to suffer today. To this end, there is a necessity for inclusive, justice-focused strategies for mitigating and adapting to climate change at various levels. In states, it must occur at both the national and local levels. Critically, adaptation strategies need to focus on the groups and areas most affected by past environmental injustices and how to improve resilience, influence policy, and encourage a more fair and just response to climate change issues.

A problem, however, is that even developed countries are not taking enough steps to deal with the current crisis, never mind the less developed ones. A more composite approach that examines issues and deals with them more holistically is needed. There needs to be a greater unification of a range of issues within the broader climate and environmental justice milieus. Some call for the uniting of:

> … climate, energy and environmental (CEE) justice to provide a more comprehensive framework for analysing and ultimately promoting fairness and equity throughout the transition away from fossil fuels. The urgent need to accelerate the transition could, and should, unite CEE justice scholarship … The transition towards a post-carbon world means that justice scholarship must unite to develop comprehensive frameworks of analysis. (McCauley and Heffron 2018, references omitted)

To take up a variety of these matters, the notion of a "just transition" has come to the fore (Bouzarovski 2022). In this regard, it has been suggested that CEE justice scholarship can be united by redefining the just transition notion beyond its initial strategic intent. Therefore, distributional and procedural justice – the two aspects of just transition that all three pillars of CEE research have in common – should be thoroughly addressed in the process of redefining the notion of a just transition (McCauley and Heffron 2018).

Overall, on the issues being dealt with here, what is needed is a united and integrated process that combines all elements of climate change, climate justice, and environmental justice in research and action. Already there have been claims that the environmental justice implications of climate change are insufficiently accounted for when dealing with climate issues (Kato-Huerta and Geneletti 2023). However, it is more than the implications that ought to be considered, as it is the effects of past environmental harms which are more important. This is not to argue that this does not occur. This does occur, from time to time, but implicitly or organically, rather than being specifically delineated as such, except in very specialized climate science-orientated research (Romero-Lankao et al. 2023; Anser et al. 2023). This subject has not been addressed much in the social sciences or legal studies. This is not surprising given that, in 2009, there was a recognition that the legal literature did not focus very much on the justice angle of dealing with climate change (Brunnée 2009). Since then, various authors have specifically identified the interrelationship between climate change, climate justice, and environmental justice (Ekhator and Okumagba 2024). For instance, it has been stated that:

> Environmental justice and climate justice were originally articulated in the dominant liberal discourse of distributive equity, given that some communities face disproportionally more environmental risks than others. This linking of environmental justice concerns to climate change has expanded the discourse on the environment, taking the analysis of the meaning of injustice beyond inequity, and creating opportunities to adopt more multiplicitous forms of social justice. (Lombard 2022, references omitted)

While this interrelationship has been taken up by some in academia (Schlosberg and Collins 2014) this has not occurred very often in the legal literature, and those occurrences remain infrequent (Tschakert et al. 2021). Thus, Schlosberg and Collins note that the majority of scholarly discourses on climate justice focus either on the pragmatic policies of more elite environmental nongovernmental organizations (NGOs) or idealistic ideas and normative arguments of justice theory (Schlosberg and Collins 2014). They demonstrate that the issues and tenets of environmental justice are evident and coherent in their articulations of climate justice.

As noted earlier, the argument often made is that the concept of climate justice and the developing climate justice movement emerged from the earlier environmental justice movement. However, what is not noted generally is that the two overlap, and as such, ought to be dealt with together. Their inherent connection and the critical need to deal with both fields of study collectively is not an issue that stands out in the academic literature, with some few exceptions. It has been argued, for example, that because racial minorities in the United States are frequently the most susceptible to environment-related dangers as well as the negative effects of climate change, the two issues ought to be examined together (Center for Health, Environment and Justice 2020). The case is well made that the communities that suffer the most from pollution and environmental degradation – an environmental justice issue – are also often those most at risk from climate change – a climate justice issue. They therefore should be used as a combined approach to addressing the multilayered issues resulting from our deteriorating ecosystem.

A similar position is taken by Robert Bullard (Bullard et al. 2016), who argues that both environmental justice and climate justice are about equity. As a result, it is necessary to tie environmental justice and climate change policies together to overcome systemic disparities in the allocation of environmental risks (Bullard et al. 2016). While establishing this link is essential to developing just climate policies addressing the evolution of climate change, recognizing and analyzing the links between the environment and the climate reveals a more subtle, but none-the-less important issue. While the interrelationship between environmental justice and climate change in that sense is clear, what also seems to be the case is that when combined their individual effects are magnified. While both clearly have disproportionate effects on certain groups, what is insufficiently considered is their compounding effects on each other and vulnerable groups.

For example, in the book *The Intersection of Environmental Justice, Climate Change, Community, and the Ecology of Life*, it is argued that communities are being impacted by these global environmental issues and urgent action is needed across the various fields to address the combined impact of environmental injustice, climate change, and other matters (Nesmith et al. 2021). However, it is also about how climate change inherently affects disadvantaged people more (Mohai et al.2009), and in fact intensifies the problems that these communities already have (Abate and Kronk 2013). This is because climate change exacerbates existing levels of poverty. For example, its negative impact on agriculture productivity leads to higher food prices, which hit poorer communities the most (Balasubramanian et al. 2021). The various communities which have been subjected to environmental harms in the past are also likely to be more affected by climate change for a number of reasons, including as a result of the limited amount of community resources that have been invested in those places. Critically, levels of poverty in these communities will inevitably increase because of climate volatility. Thus, there are a range of community, human rights, and race-related issues which intersect the environmental justice and climate change discourse (Robinson 2018). Therefore, targeting areas that already have severe environmental justice problems, to provide them with greater assistance in adapting to climate change, will bring greater climate change adaptation rewards (Thakrar et al. 2020). As is noted in the chapter on the United States, these areas were often "redlined" – provided with reduced assistance – by governing bodies. As a result, they continue to suffer massive environmental problems, such as poor to hazardous air quality (Lane et al. 2022). Without taking a more integrated approach when addressing environment- and climate-related issues by including an analysis of community, human rights, and race-related issues, it is likely that these underserved and poorly resourced areas will continue to suffer while other areas benefit from greater assistance. One result will be that these areas could become more depressed as house prices and rents in other places rise, making them even more unaffordable for minorities (Kim et al. 2024).

A critical issue is that while there is some degree of acceptance in the United States (Pastor et al. 2024), especially in the gray literature, about the link between climate change and environmental justice, this view has largely not been accepted or addressed outside the United States. This is often because the various multiple issues are addressed in fragmented ways, when they ought to be taken up with greater coherence (Amorim-Maia et al. 2022). There are many facets to the integration needs of climate change and environmental justice. As has been noted:

> In the realm of climate policy, issues of environmental justice (EJ) are often treated as second-order affairs compared to overarching sustainability goals … EJ is in fact critical to successfully addressing our national and global climate challenges; indeed, centering equity amplifies the voices of the diverse constituencies most impacted by climate change and that are needed to build successful coalitions that shape and advance climate change policy. (Pastor et al. 2024)

What is clear is that climate change will aggravate the differentials that exist between communities. Those who will continue to suffer the most are likely to be the poor and marginalized, notably groups consisting of Indigenous Peoples and minorities. These groups already suffer the consequences of various types of climatic catastrophes (Stein et al. 2023) to a greater extent than other groups within their societies (Renglet 2022). Already such events are increasing. 2023 saw a $66 billion loss to such events occurring, which was the highest monetary loss ever recorded (AON 2024). These groups will therefore have to endure even more environmental harm, proof that environmental justice issues influence climate change issues, and vice versa.

It is therefore clear that climate change overlaps with environmental justice. If climate change is to be mitigated fairly for everyone, both climate justice and environmental justice initiatives are required (Siegfried 2024). This needs to occur in the law and policy at the domestic level specifically so as to combine them in the way the legal system deals with all climate and environmental matters.

This is necessary, as both climate justice and environmental justice initiatives attempt to address climate change impacts that are unfair (Newell 2022) so they ought to be used together (including in the law and by the courts) where possible to strengthen their efforts. At present, different authors in different disciplines still refer to one or the other but rarely both. This is still often the case in the legal literature. For example, the term climate justice is used by some to deal with how restitution for land taken during colonialism is appropriate as a means to deal with climate change and that giving land back to those who lost it would be a form of climate justice (Racehorse 2023). However, those authors do not deal with these issues in terms of environmental justice. This is an example of when the two concepts are rarely discussed and combined.

This is problematic as climate justice and environmental justice have much in common. This is especially so, but not only, when they deal with the specific climate-related impact on groups, such as Indigenous Peoples and minority groups. This is because of the disparate spread of the impact of climate change globally (Langmack 2023). The negative consequences have a greater impact on these groups because of where they live and at times because of where they are forced to live because of economic and other hardships that they have had no control over. These issues need to be recognized and used by the courts to deal comprehensively and equitably with all the issues. This is a major theme of this book.

Equality and fairness must mean that the justice component (especially environmental justice) of climate change needs to be examined more often, and more thoroughly in the law and by the courts. This is because climate change affects some nations more than others. This means that poorer nations, particularly those in the Global South, and more so on specific vulnerable groups, suffer the effects more (Ngcamu 2023). These communities have already suffered the greatest damage and have the least ability to mitigate the effects of climate change and environmental injustice (Anser et al. 2023). They will also suffer more severe consequences in the future and again have less ability than other states to deal with those consequences. Part of the reason is that while Africa absorbed more carbon than it used in the past this is no longer the case (Ernst et al. 2024). This is deeply problematic, as from an historical point of view developed nations in the Global North have been the biggest contributors to greenhouse gas emissions. They have historically been the source of most of the industrial output which has caused the emissions that have led to climate change. These nations continue to be the largest polluters. As a result of their role in being the principal polluters, wealthier nations have ethical obligations, at least, and perhaps legal duties, depending on how this plays out legally over time, to assume the responsibility and make the larger contribution to the processes needed to lower emissions. This also means that they should assist other states with the economic and technical resources needed to reduce and soften the effects of climate change and find ways to adjust to life in the era of climate change (Sardo 2023). Climate justice thus sees the processes currently used to ameliorate the effects of climate change as unfair and seeks to bring ethics into the process of formulating climate change policy at the domestic level especially, and to encourage a more equitable and human rights-focused approach (Gilliam et al. 2023) when responding to climate change (Malakar et al. 2023).

This ought to be the basis for dealing with either issue in the law or by the courts. An agenda for climate justice in legal matters must include an awareness of the development disparities that are made manifest by climate change (International Bar Association 2014; Ruckstuhl et al. 2022). However, such an agenda, for many, is much wider. This is because:

> Actions taken in accordance with the current global order will most likely result in the same outcomes: environmental decay and humanitarian crises. Climate justice will not occur overnight, but it requires a reexamination of the global order to effectively deal with the crisis and injustice unleashed by climate change worldwide. (Deb and Genshaft-Volz 2023)

Thus, dealing with the global governance architecture is important, and reforming these processes is essential. What also must be addressed is the current state-centric process of international relations, which provides little scope for individuals or groups to play a role.

While there has been an emphasis on assisting states with the climate change burden that they already face, climate justice is not only about assisting states. It is also about assisting deprived or marginalized groups or communities (Ogunbode 2022), especially those in low-income countries, to deal with climate change and its impact (Williams 2021a). More meaningful involvement is required. It has to be about ensuring that vulnerable groups, including the wide swathe of minority groups and Indigenous Peoples (Giacomini 2022) that are disproportionately and unfairly affected by climate change, are better protected from, and are better able to deal with, the consequences of climate change (Barnwell and Wood 2022). These people, especially in low-income countries, are especially susceptible to the effects of climate change (Williams 2021a) because of their economic situation. This means that they often do have the means and ability to deal with climate change and adjust their lives accordingly. Thus, climate justice also means that vulnerable and disadvantaged people's interests and rights ought to be given top priority in the law, in climate policy and other related activities, especially in the domestic context (Malakar et al. 2023).

THE INSUFFICIENT FOCUS ON INDIGENOUS PEOPLES AND MINORITIES IN CLIMATE CHANGE ACTIONS

The lack of focus on Indigenous Peoples (Westra 2012), and particularly those in minority groups, is sadly far too common in efforts to address climate change and related issues, including environmental justice, except in certain countries, for certain groups, and at certain times. It is a strong indication of how little these groups are considered in general.

An example of how little these groups are considered within the context of climate change can be seen in the Paris Agreement, which does not prohibit discrimination, which is a key omission. It is also problematic that various groups are not mentioned. As one study notes, there is no mention of "ethnic, minorities, race, or ethnicity" (Grahn-Farley 2022: 85). In fact, a range of groups affected are not mentioned unless it is argued they are obliquely caught up under "local communities" or "people in vulnerable situations" (Fornalé 2023: 2). All the Paris Agreement does is to state in the Preamble:

> Acknowledging that climate change is a common concern of humankind, Parties should, when taking action to address climate change, respect, promote and consider their respective obligations on human rights, the right to health, the rights

> of Indigenous Peoples, local communities, migrants, children, persons with disabilities and people in vulnerable situations and the right to development, as well as gender equality, empowerment of women and intergenerational equity. (United Nations Framework Convention on Climate Change, Paris Agreement, 2015)

However, the fact these groups are only mentioned in the Preamble, not in the text of the treaty itself, and they are not provided for in the Agreement, is problematic. These groups are provided with no specific rights, and no duties towards them are demanded, meaning issues prevalent among groups facing challenges are not contained within the treaty.

While responses around the world to climate change have intensified since the Paris Agreement under the UN Framework Convention on Climate Change (United Nations Framework Convention on Climate Change 2022) much still needs to be done if targets for the maximum global temperature increase are to be met. The UN must act to ensure the 196 states that have signed the Paris Agreement meet their pledges made under it. State climate action plans, termed nationally determined contributions (NDCs), need to be scrutinized to ensure they are sufficient (Rodríguez-Garavito 2022). All segments of each state's population must be empowered to make an equal contribution, and efforts must be made to ensure that no one is left behind. This must include Indigenous and minority groups, who have major roles to play (Warren and Jackson 2002) but are in any event entitled to safeguards. Promoting environmental justice issues can assist in this regard. If these groups are protected and supported, they can continue to play their critical roles in ensuring greenhouse gas emissions are reduced and global warming is mitigated. This is urgent given that there will likely only be a limited amount of reductions in such emissions by 2030.

What can be seen is that while efforts to advance environmental justice since the Paris Agreement demonstrate an increased focus on Indigenous Peoples, very little additional work is being done on behalf of minorities. Evidence of this can be seen in the Synthesis Report for the Climate Change Global Stocktake meeting held in the UAE in November–December 2023. Minorities are only mentioned twice in the report, in both cases almost as an afterthought, for instance where it is stated that "It is also important to develop partnerships with traditionally marginalized groups, including women, youth, Indigenous Peoples and local communities *as well as ethnic and other minority groups*" (United Nations Framework Convention on Climate Change, Paris Agreement 2023: para 160, my emphasis). Interestingly, the focus is on "ethnic groups" although the wording denotes "other minority groups" as well. Indigenous Peoples again, as is usual, receive greater attention in the document. While the references to minorities (paras 40 and 160) are simply almost identical calls for partnerships to include them, there is much more of a focus on action for Indigenous Peoples. For instance, the report in para 5 states: "Climate action and support are *enhanced by catalysing action* by all Parties and non-Party stakeholders, including civil society, the private sector, financial institutions, cities and other subnational authorities, local communities and Indigenous Peoples" (my emphasis). While Indigenous Peoples are identified in para 5 as one of a number of stakeholders who should be brought together to enhance climate action, there is no mention of minorities in the same context. In other paragraphs, while local communities are noted, there is no mention of minorities. While the knowledge systems of Indigenous Peoples and local communities are acknowledged in para 148, it is stated in the same paragraph that "Non-Party stakeholders should endeavour to include and support stakeholders *who are often marginalized*, including women, youth, and Indigenous Peoples, so they can all effectively participate in and contribute to these initiatives" (my emphasis). Sadly, there is no mention of minorities here. This is surprising considering how marginalized minority groups are. While the word "including" denotes a nonclosed list, which could, in fact, include minorities, they ought to have been specifically mentioned, which would have given their issue the prominence they deserve in this context. Overall, while Indigenous Peoples and minorities are mentioned in the text of the Report, they are, in most part, pushed to the periphery.

The plight of minorities also appears to have been forgotten in the 2023 Report of the Special Rapporteur on the implications for human rights of the environmentally sound management and disposal of hazardous substances and wastes (UN General Assembly 2023). The topic of that mandate is obviously very relevant to environmental justice and the issues within that rubric. However, while there are numerous references, and four specific paragraphs, devoted to Indigenous Peoples, there is only one reference to minorities in para 86, and that is restricted to "ethnic minorities and persons of African descent." The paragraph continues by noting that "tribal communities of African descent have the same rights as Indigenous Peoples," seeming to equate the two groups, while leaving out all other minority groups. Clearly, greater attention should be devoted to the various and often dangerous situations faced by all minority groups.

This absence of dealing with minorities as far as climate change, climate justice, and environmental justice can again be seen in 2024 in the annual report of the United Nations High Commissioner for Human Rights on the rights of persons belonging to national or ethnic, religious, and linguistic minorities (OHCHR 2024b). While a whole of problems concerning minorities are discussed in the report, there is no mention of environmental or climate issues. This is surprising in an Annual Report of the UN agency devoted to human rights issues.

More generally, at the international level, more needs to be done to ensure states are responsibly and adequately addressing environmental justice issues that impact minority groups. Oversight processes, including Universal Peer Review (UPR) and human rights treaty bodies, need to perform due diligence checks from climate change and human rights perspectives that focus on these groups especially. Environmental justice issues should be a permanent fixture on their agendas. This hardly occurs at present.

INTEGRATING CLIMATE ISSUES AND ENVIRONMENTAL JUSTICE FOR MINORITIES AND INDIGENOUS PEOPLES

This section briefly reviews the effect of discrimination, marginalization, and ostracization on Indigenous Peoples and minorities. These issues are taken up in each country chapter. However, what is key here, is that those ills, particularly from the past, but not always, have had such an effect as to exacerbate the problems that climate change is bringing. For this reason, to deal better with climate change issues and vulnerable groups, there ought to be an integration in the way the challenges of climate change is dealt with. Climate change actions and environmental justice need to be dealt with concurrently. This is because there would be tremendous benefits to dealing with all the matters simultaneously and in an integrated fashion. This is because climate change, climate justice and environmental justice issues while somewhat different, overlap. Certainly, environmental justice includes issues of climate change, and climate change issues are concerns for environmental justice.

As was noted above, there is a deep relationship between climate change and environmental justice (Maldonado et al. 2016). This is because it is usually Indigenous Peoples and minority groups who live in poor, underserved areas that have not been adequately equipped or prepared to deal with the consequences of climate change. They have already been more severely affected by it, and will be even more affected by it as the climate crisis grows. This is the case, because of discrimination and domination that:

> people of color, minority groups, and indigenous people are often either pushed into areas affected by climate change because these are the only places they can afford to live, or they are unable to leave areas affected by climate change because they cannot afford to move to another place. (Grahn-Farley 2022: 85)

Thus, these groups are ill-equipped to deal with climate change.

These groups are not, however, treated equally. Critically, even amongst the various vulnerable groups, there has been disparate treatment. While environmental justice activists have focused on Indigenous Peoples, there has been much less effort to bring environmental justice to minority groups especially outside the United States. Even in the United States, the bulk of the work has been in support of racial minorities, while neglecting other vulnerable minorities.

The various minority groups are often lumped together despite the at times significantly different social, economic, and historical challenges they face. They are not, in fact, explicitly focused on in the Sustainable Development Goals (SDGs). While the 2030 Agenda involves a universal focus, it still therefore has significant gaps in that it fails to specifically target minority groups, including religious, linguistic, cultural, and ethnic groups (OHCHR nd). In fact, one of the most significant omissions is the lack of explicit acknowledgment of these groups (OHCHR nd). This is the case despite the fact that the goal of leaving no one behind was to emphasize sustainable development within the social, economic, and environmental pillars of development (Mac-Seing 2023). As a result of the failure to specifically include them, goals such as access to education, decreasing inequality, sustainable cities, climate action, and peace may fail to address their marginalization and exclusion from policies and programs. Therefore, slowing overall progress. This is particularly so when dealing with issues of "economic inequality" and poverty, which are fundamental objectives of the Sustainable Development Agenda (Winkler and Satterthwaite 2017). Former Independent Expert on Minority Issues Gay McDougall (McDougall 2016) has argued that if the broader strategy of tackling poverty does not target minorities, it may fall short in addressing the root causes of social and economic inequalities as well as the marginalization that these groups face (Winkler and Satterthwaite 2017). McDougall had already noted in 2007 that minorities "experience widespread discrimination or exclusion" (McDougall 2007). That is at least a reason to consider using intersectionality as a means to ensure that no one is left behind (Mac-Seing 2023).

Therefore, greater assistance should be afforded to people with disabilities, sexual minorities, and minority linguistic and religious groups, as the environment-based challenges they face are often overlooked by those on the outside. This is problematic as a variety of different groups, including religious, cultural, linguistic, and sexual minorities, as well as disabled people, are also very vulnerable to harm from environmental degradation (Chakraborty et al. 2019). This is especially the case as these groups are still discriminated against in multiple ways (Molero et al. 2019), across the board (Sabatello et al. 2020), and in all societies (Sargeant 2017), generally, and specifically as far as environmental issues are concerned.

As a result of discrepancy in treatment, many minority groups have not received the same level of assistance as Indigenous Peoples have, from those responsible for, or dealing with, environmental harms. In this regard, it is argued that the protection available for minority groups generally in international law, and as far as environmental injustice issues are concerned, has been far less. For example, while some work has been done to address environmental justice issues for one racial minority (African Americans) in the United States, very little has been done for other minorities there, or those living elsewhere in the world. In fact, while much of the existing work in this area focuses on Indigenous Peoples, there is much less focus on other minority groups, who will often bear the brunt of climate and environment-related degradation. Greater priority should be given to them being consulted, and they should be taking part in, or even play a leading role in, decision-making processes related to these issues. Their consent is vital when processes affecting them are being developed and implemented (Young 2019). These processes ought to be better legislated to ensure they occur properly. The court too must examine the reality of consultation and determine whether effective consultation occurred before finalizing any determinations.

As far as Indigenous People are concerned, at present, the discourse scrutinizing the effects of human-induced climate change on them particularly can be broadly divided into two discursive threads. On the one hand, there are those who emphasize the acute vulnerability of Indigenous Peoples to climate change (Hall and Crosby 2020). On the other hand, a distinct yet intersecting thread highlights the unique adaptive capacity of Indigenous Peoples to climate change (Howitt 2020). Indeed, the literature in this latter thread often argues that Indigenous knowledge, practice, and resilience represents a solution to the problems generated by climate change (Petzold et al. 2020). While both threads offer insight into the conditions confronted by particular Indigenous groups, each runs the risk of oversimplifying, misunderstanding, and further marginalizing the diversity of experiences and approaches of Indigenous Peoples (Howitt 2020). There is, however, little focus on minorities, which ought to be the case because of the adverse effects on them as well.

Sadly, the scope of abuse that many of these groups experience in their lives generally, but also more specifically related to the environment, is considerable. The damage resulting from these abuses is accentuated by the fact that these groups are often marginalized, invisible, and unnoticed (Zuni Cruz 2020). They are also silenced (Godsey 1995), and so have little ability to deal with the harm they suffer. While some believe that the problems of discrimination and marginalization of groups is a thing of the past, others argue that it is ongoing and in fact getting worse. Boaventura de Sousa Santos argues that:

> We live in a period in which the most morally repugnant forms of social inequality and social discrimination are becoming politically acceptable. The social and political forces that used to challenge this state of affairs in the name of possible social and political alternatives seem to be losing steam and, in general, appear to be everywhere on the defensive. (de Sousa Santos 2018)

In this regard, it must also be noted that it is often state activity and institutional violence that causes these problems. As has been indicated "state institutions that have enacted and continue to prosecute some of the most violent forms of colonisation, including assimilation, child removal, behavioral micromanagement, and land theft" (Strakosch 2024). It is also, however, sometimes state and institutional micromanagement which is mare harmful, but often there is indifference to the plight of Indigenous or local minority populations that results in "environmental injustice through interlocking systems of racial capitalism, settler colonialism, and enslavement" (Kojola and Pellow 2021).

A prime example of an instance where such injustice often, and perhaps inevitably, occurs is with the construction of large-scale projects, such as dams or irrigation systems, which alter or degrade the environment to the point that people are forced to flee their homes or are actively harmed by the environmental damage caused by these massive projects (Schwartz 1993). The reason their lands are targeted is that these groups often reside in places where there are significant natural resources or in physically isolated areas where developers can build facilities that would not be permitted in more developed or affluent areas. These physical conditions make their land attractive to state or private developers.

Further exacerbating these problems is the socially marginalized nature of these groups. Often, the very people who are meant to protect them from victimization – large intergovernmental organizations, as well as their own governments – are the ones selling off their land to such harmful effects. Their social marginalization makes these groups easy targets for exploitation.

POVERTY AND THE EFFECTS OF CLIMATE CHANGE AND ENVIRONMENTAL INJUSTICES

Critically, poverty is an issue that also connects climate change, climate justice, and environmental justice. It is the poor (Martinez-Alier 2003b) and marginalized who are most affected. Those who fall into these camps are more likely to be Indigenous Peoples and minority groups. The World Bank reports that while Indigenous Peoples constitute about 6 percent of the world's population, they make up about 19 percent of the world's extreme poor (The World Bank 2023), and live in dire poverty (United Nations Department of Economic and Social Affairs nd). While there have been some improvements, in some places Indigenous Peoples are still amongst the poorest of the poor. This is because they are more likely to be unemployed, have higher rates of illiteracy (Stavenhagen 2015), are more likely to be unhoused, have access to fewer health facilities, and have fewer educational opportunities (King et al. 2009).

Therefore, what is apparent is that environmental injustices have deeper impacts on particular vulnerable groups. They are also more profoundly impacted by climate change (Giacomini 2022). This is because the discrimination, prejudice, and social isolation that marginalized groups are subjected to, further exacerbates their ability to deal with the issues of climate change. For example, social discrimination against Indigenous Peoples contributes to low rates of academic achievement and early abandonment of education (O'Malley et al. 2018), which in turn leads to their continuing high rates of poverty compared to non-Indigenous people (Waller and McCullum 2021). Another result of inequality, marginalization, exclusion, and discrimination, which effects poverty levels, is that the life expectancy of Indigenous Peoples around the world is on average 20 years less than the rest of the population where they live.

Linked to these issues is the fact that Indigenous Peoples often lack access to health, water, and sanitation services (The World Bank 2023), often as a direct result of where they live (McGraw and Fox 2019). In fact, they endure tremendous environmental problems relating to sanitation and wastewater. Where wastewater from sewage is not treated properly (Hillary et al. 2020), as often occurs in places where Indigenous and minority groups live, it contaminates the soil, the air, and the water, and has massive health effects (Davis et al. 2022). This will be returned to below. However, all these effects are common-place phenomena around the world (Rajapakse et al. 2022; Waldron 2021; Brinkmann 2021). It has a particularly devastating effect on children (Contreras 2020). If the wastewater is not well treated, antimicrobials can enter the environment, resulting in enhanced resistance to treating various diseases (Mehta et al. 2019). This means that when future generations are ill there will not be the medication available to treat them because of rampant drug resistance. Health problems are already more advanced in some places (Chan 2019), with great disparities occurring in areas where minorities live (Baciu et al. 2017; Cullen et al. 2022). Thus, these are crucial issues for a variety of reasons, but mostly because they ensure that the poverty gap is not closed but is in fact extended.

While the SDGs identify these as issues which should be addressed under the mantra of leaving no one behind, this has not occurred to any large extent. More than halfway to the 2030 deadline for the completion of the SDGs, over half of the people of the world are being left behind (United Nations Food and Agriculture Organization 2023a), as far as extreme poverty, health, and a range of other SDG issues, are concerned. As the United Nations Secretary General Antonio Guterres noted in February 2023: "Halfway to 2030, we are far off track ... We need to go further and deeper. On climate, on conflict, on inequality, on food insecurity, ... we are closer to the edge than ever" (UN Secretary-General 2023).

These problems endure because high rates of marginalization and exclusion (Giordano 2020; Vorobjovas-Pinta and Hardy 2021). This is aggravated by a general governmental reluctance to focus on development initiatives for them (Masakure 2019). However, in reality, groups are frequently excluded from participation (Anguelovski and Martinez-Alier 2014) and representation in political decision-making and development plans and activities. This further erodes their exclusion from education, health care, employment, and access to land – in other words, their well-being. This has deep implications for long-term poverty levels.

As a result of all these problems, dealing with poverty is now recognized as a global imperative. The UN has stated that "Eradicating poverty is the greatest global challenge facing the world today and an indispensable requirement for sustainable development" (Li 2015: 305). Sustainability, in the context of Indigenous Peoples, can be seen in the "geographical overlap between biological richness and linguistic diversity and between indigenous territories and biologically high-value regions (actual and projected protected areas), recognized importance of indigenous peoples as main managers and dwellers of well-preserved habitats, and certification of a conservationist-oriented behavior among indigenous peoples derived from its pre-modem belief-knowledge-practices complex" (Toledo 2001). This issue is very relevant for various concerns, including poverty alleviation, which is usually a matter of deep relevance to Indigenous Peoples and minority groups.

While there was seemingly some progress initially on the SDGs, the reduction rates of poverty around the world seem to have slowed, and global hunger, for example, has increased (Lelechenko et al. 2020). More than two million people a year die from hunger already (Lawson and Jacobs 2022). This is taken up further below.

Critically, the situations of poverty that Indigenous Peoples and minorities find themselves in, is in much larger percentages than the rest of the population in the countries where they reside (UN Innocenti 2023).

Poverty therefore has an enormous effect on these groups. For example, children from Indigenous and other minority groups are often excluded from schooling, or the schooling they receive is substandard. The effect of being excluded from education is further compounded for some groups, such as minority group children with disabilities, who encounter additional barriers in the education system compared to persons without disabilities (UN 2019). Harassment and exclusions in the school system reduce morale and the desire to remain in the educational facility, and may bleed over into the work environment (International Labour Organization nd). As a result, Indigenous children and members of minority groups are less likely to complete primary or secondary education, and are therefore less likely to be literate, limiting employment possibilities (UN 2019). This ensures that they cannot break the poverty cycle. Their restricted educational and other opportunities reduce their employment possibilities as well as their ability to meaningfully engage in their societies.

HUNGER AND FOOD AND THE EFFECTS OF CLIMATE CHANGE AND ENVIRONMENTAL INJUSTICES

Despite the SDG pledge not to leave anyone behind, many people are left behind, and a large number of those are Indigenous Peoples and minorities. More of these people live in dire poverty today than was the case just a few years ago. In fact, it was reported by the UN that in 2023 there were 122 million more people in situations of hunger than in 2019; that is, a staggering 735 million people facing hunger in 2023, as compared to 613 million people in 2019 (FAO et al. 2023). This report also found that nearly a third of the world's population, or 2.4 billion people, were not able to constantly access food, and 900 million persons were subject to severe food insecurity (FAO et al. 2023). In 2024, FAO warned that a further 16 countries were facing acute food problems (FAO 2024). A high proportion of the people who face food insecurity are Indigenous Peoples or minorities. In this context, in July 2023 200 economists wrote to the UN Secretary-General and the World Bank President calling for these institutions to do more to ensure that the severest increase in global inequality in 75 years is reversed (Equalshope 2023). It is not surprising, therefore, that the 2023 SDG Progress Report noted that overall, only 12 percent of the SDG targets are on track to be met by 2030. The report also finds that for half the SDGs, progress is weak and insufficient, and that for 30 percent of the SDGs there has been stagnation or a slide backwards.

Related to this, is the effect of food production on the environment (Verkuijl et al. 2024). It is also about the amount of food that is wasted particularly in the developed world (United Nations Environment Programme 2024b). The amount of food that goes to waste rather than being used to feed poor people around the world is enormous. In the United States, it has been reported that a third of food is lost or goes to waste (United States Department of Agriculture nd). This has a massive impact on landfills and on the production of greenhouse gases. At the same time, millions of people go hungry each day.

Thus, the connection of hunger and food to both climate justice and environmental justice indicates how interrelated these concerns are. Food issues are deeply connected to climate change, as is taken up elsewhere, but moreover, food issues for certain groups have been deeply negatively affected by racism and discrimination. This has had disproportionate effects for certain groups. As they are more uninsured and underinsured (because of their status) compared to other groups (Becares 2015), they are even more vulnerable. Thus, there is a clear association to environmental justice. Thus, these issues need to be dealt with together. What emerges is that food production, distribution, and availability, need to better consider who people are, and where they live. Another related major problem is food wastage. Around 19 percent of food, which equates to a billion tons, is wasted (UNEP 2024c). This is a dire problem in the context of rising hunger and increasing difficulties in food production. This will get worse with further climate change (Raihan 2023).

One way to deal better with some of the issues concerning food is to use plant-based agriculture and avoid or limit the eating of meat and dairy products (Smetana et al. 2023). This is because of the problems this type of farming causes, including methane emissions. Already, Denmark has embarked on a plan to tax those emissions. This type of approach is necessary to help to reduce the environmental effect of food production on the earth while assisting in reducing levels of hunger (Scherer et al. 2023). This needs to be done in ways that balance producing agriculture with issues concerning biodiversity (Daum et al. 2023). It also needs to be done using smart agriculture and smart irrigation besides a range of other techniques to ensure sustainability (Pang et al. 2023). Water availability, particularly clean water, is, however, a problem. Additionally, the way it is used for agriculture is often unsustainable, often depleting groundwater for small amounts of agricultural output (UNEP 2022). However, what is occurring is that in many parts of the world, subsidies are given to farmers that continue to support animal-based farming, rather than plant-based ones. Thus, for example, the EU continues to provide overwhelming support for animal farming (82 percent) over other more sustainable types of agriculture (Kortleve et al. 2024). The same study found that economically those subsidies were problematic because animal products provided only 35 percent of EU calories and 65 percent of EU proteins but are linked to 84 percent of embodied greenhouse gas emissions that come from food production. Shifting the focus to more plants, and other alternatives, would have enormous positive effects on climate change. It will help reduce methane gas emitted by the animals farmed for food.

A focus on the food industry is also important to reduce methane, but also for the sustainability of the earth's purposes. While meat and dairy products offer 18 percent of calories and 37 percent of protein, they account for 83 percent of agricultural use and 60 percent of greenhouse gas emissions from agriculture (Elmore et al. 2023; Greenspoon et al. 2023). While the UN's Food and Agriculture Organization (FAO) acknowledged in a report to COP28 that human diets "absolutely" had to change, (United Nations, Food and Agriculture Organization 2023a) reducing the consumption of meat and dairy was not one of its 120 recommended actions (Carrington 2024b). This is surprising seeing the massive effects of these on the planet. Usefully, the United Nations Environmental Programme did issue a report around the same time calling for alternatives to these foods, arguing that it would positively affect the impact on the sustainability of the planet (UNEP 2023). What is clear is that better coordination is needed amongst UN bodies and agencies. However, it is necessary that food consumption patterns change, and the number of animals farmed is reduced (Harwatt et al. 2024).

However, as with other climate-related matters, the issue of food and food production has been politicized by some. Thus, climate skepticism has, for example, become part of the culture wars that are occurring on a whole range of fronts (Dobson 2022) in the United States (Sarathchandra and Haltinner 2021). However, conspiracy theories (Räikkä 2023) and suspicion (Tam and Chan 2023) around climate change are found in many other places (Forchtner and Lubarda 2023). The media (Sarathchandra and Haltinner 2023), and often specifically social media, has been used extensively to promote cynicism and distrust around these matters (Tyagi and Carley 2021). Thus, moving away from meat is attacked by those who see climate change as a myth and see creating meat alternatives as an attack on farming and farmers. This is especially true in the United States, where Florida has adopted a law that bans creating or selling meat alternatives. Thus, the focus on dealing with climate change, as noted earlier, must deal with the conspiracies and mythologies (Peters 2023) that are often used for political purposes (Douglas and Sutton 2015). This is a danger, as these same conspiracies have the potential to derail climate change mitigation or adaptation strategies in the rest of the world (Biddlestone et al. 2022). This allied to false or misleading advertising from industries such as that of fossil fuels causes many problems. Thus, the EU has adopted a directive that prohibits misleading advertising (European Parliament 2024). More like this needs to be done elsewhere to counter the various tactics that are being taken to foster an awareness of the dangers of climate change, in general, but for food production specifically (Schneider 2023).

However, more broadly, attitudes around the world support actions taken to counter climate change. In a 2024 survey by the United Nations Development Programme (UNDP) and the University of Oxford, 75,000 respondents in 77 countries, nearly 80 percent, supported taking steps to counter climate change, 72 percent wanted a move away from fossil fuels, and 78 percent wanted more protection for people at risk from extreme weather (UNDP 2024b). There are, however, differences by country. as will be indicated in the US chapter.

HEALTH AND THE EFFECTS OF CLIMATE CHANGE AND ENVIRONMENTAL INJUSTICES

As can be seen, there are a range of problems which specifically impact marginalized groups more. Besides the ones addressed above, they also face greater mental and physical health challenges (OHCHR 2012). Globally, Indigenous Peoples experience higher disease burdens than non-Indigenous people (Curtice and Choo 2020). They have lower life expectancy (Valeggia and Snodgrass 2015) and higher rates of infant mortality (Valeggia and

Snodgrass 2015). Indigenous Peoples also have increased mental health challenges (Stein et al. 2021), which sees elevated rates of depression, alcoholism, domestic violence, and suicide (King et al. 2009). Indigenous women have increased levels of teenage pregnancy (Valeggia and Snodgrass 2015).

Part of the reason for these is the effect of discrimination. Discrimination has, for example, been found to cause a variety of health problems (Emmer et al. 2024). This is worsened by the fact that there is a reduced likelihood that these groups would be able to access the healthcare system (Stein et al. 2021) for a variety of reasons (Al Shamsi et al. 2020). As a result, about five million people die each year in poor countries because of a lack of access to healthcare (Lawson and Jacobs 2022).

One of the problems is a lack of information available to these groups about health and healthcare in their own language. Healthcare workers often do not speak their language (Al Shamsi et al. 2020). In fact, it has been stated that "environmentally unjust state and corporate practices and policies are designed to maintain the health of highly valued populations – people and species who matter – at the expense of those whose lives matter less and are differentially valued by the state" (Kojola and Pellow 2021).

The result is that the health status of Indigenous Peoples and minorities is usually worse than the rest of the population, in part because of their severe disadvantage in regard to access to clean water and proper sanitation as well as other essential services (Winkler and Satterthwaite 2017). In fact, the situation of Indigenous Peoples globally is considerably worse than that of others living in the same location (Toki 2024). They are often worse off than the other marginalized groups in the same nation (Loyn 2006). This situation exists because they have suffered from historic injustices, such as the dispossession of their territories, resources, and particularly discrimination. The result of this has been dramatic, specifically regarding the environmental harms they have had to endure, which is taken up throughout this book.

These problems are being exacerbated in places subject to higher air pollution levels, environmental degradation, and degraded water (Sedlak 2024) resulting from climate change (Crowe 2021). For example, their communities are often subjected to higher levels of pollution than others, including poor air quality, soil contamination, the existence of toxic waste sites, and the release of toxic chemicals from local industrial activity, poor water quality, besides other environmental justice problems. In fact, examples of the existence of environmental problems in areas where minorities live seem to abound (Motairek et al. 2023). Statistically, minority groups everywhere live in greater proximity to industrial facilities (Shrader-Frechette 2022), and therefore are more exposed than the general population to the toxic waste these facilities produce (Stoett et al. 2021). While not much work has been done on these issues in the Global South, this seems to be the case in the North as well (Abubakar et al. 2022).

Exposure to industrial waste often leads to health problems such as cancer and cardiovascular disease, and higher mortality rates (Kaufman and Hajat 2021), which are exacerbated by fewer health facilities in the areas where they live, and less financial support for health care. Limited access to health care leads to additional deaths due to missed diagnoses, lack of treatment, and lack of disease prevention awareness (Zweig et al. 2021). In the United States, it has been noted that these effects will be further exacerbated for some vulnerable groups as a direct result of climate change (Nogueira and Yabroff 2024).

Piles of waste, free-flowing sewage, and inadequate drainage are just some of the environmental problems they may face in their communities (Heidegger and Wiese 2020). Climate change is going to exacerbate these problems (Etzel et al. 2024). In fact, there is a looming water availability problem as potable water is running out in many countries including France, China, and the United States by 2024, unless steps are taken to address this (Nikkhah et al. 2023). Already around the world, 2 billion people face water problems (United Nations Convention to Combat Desertification 2022). The ills caused by air pollution, environmental degradation, and degraded and less water supplies are being further affected by the loss of biodiversity (Suiseeya 2014) and deforestation, leading to large amounts of population displacement. Thus in 2021 natural disasters displaced 23.7 million people globally (O'Rourke 2024). This will increase rapidly in the future and therefore the issues raised by climate-induced displacement need more attention, specifically for groups such as minorities and Indigenous Peoples. Much needs to be done to mitigate the effects of climate change but also to allow adaptation. However, where that is not feasible, as will occur in many parts of the world, there is a need to ensure safe, regular migration pathways. This is going to be an enormous challenge in the context of already perceived "migrant crises" around the world.

CLIMATE MITIGATION AND ADAPTATION STRATEGIES FOR MINORITIES AND INDIGENOUS PEOPLES

Finding solutions is critical, for everyone who will be affected by climate change. This is especially so for the groups focused on in this book. This is because of their specific problems and the particular effects on them. The consequences of climate change on Indigenous Peoples, for example, include the loss of access to ancestral lands and the natural medicines found there, as well as reduced food security (United Nations 2003). In fact, food (and water) insecurity is worsening (Mirzabaev et al. 2023) partly because of climate change and rising conflict, amongst others. It is affecting communities such as Indigenous Peoples, more and more (United Nations Food and Agricultural Organization et al. 2023b).

Therefore what is being done is taken up in more detail in the country chapters is the situation in those countries to understand how a variety of countries are tackling these issues. The Conclusion to the book will then provide more detail about what can done to deal with these problems, particularly the way that dealing with environmental justice issues ought to be dealt with at the same time otherwise few gains will be made.

What can be seen already is that environmental justice proponents have sought to assist vulnerable communities in various ways, such as by working to ensure that the transition to a low-emission, carbon-neutral economy is achieved in ways that take into account the needs of Indigenous Peoples and minority groups (Patnaik et al. 2020). However, as the problems will worsen over time, there is a crucial need for the law to recognize how climate change and climate justice are deeply interrelated (Robinson 2018). Dealing with climate change needs to be done in ways that advance social justice (Damico et al. 2020). The inequity of the effects of the damage must be admitted, addressed, and arrested, and mitigation or adaptation strategies and laws put in place (Fayazi et al. 2020).

In fact, issues of mitigation and adaptation are intimately connected to climate change. While mitigation is about reducing the release of greenhouse gases to lessen the severity of future climate change, it is adaptation which can be done to deal with the effects of climate change.

Thus, adaptation is the strategy taken to plan for and counter climate change. The most accepted definition of adaptation by the IPCC is that it is the "process of adjustment to actual or expected climate and its effects. In human systems, adaptation seeks to moderate or avoid harm or exploit beneficial opportunities" (Pachauri et al. 2014). These are thus meant to be plans to help people, structures, and systems cope with climate change. Thus, part of it is about steps taken to limit the negative effects or detriments that are affecting people in specific places, and how localities adapt to those issues (Meerow and Mitchell 2017). This includes, critically, how people are assisted in resisting, reversing, and repairing the problems caused by climate change (Kinol et al. 2023). It is, at the minimum, about simply coping, or being resilient to climate change. It is about taking steps in this regard.

However, generally speaking, adaptation plans are limited in nature around the world. As has been argued:

> Global priorities focus on mitigation while local realities and challenges require adaptation. Proponents of environmental justice have been advocating for vulnerability and adaptation considerations, which are more pertinent in the sub-Saharan Africa. (Chanza et al. 2022: 144)

Thus, adaptation strategies have only in recent years become more focused and widespread. However, much more is needed, in a lot more places. A huge problem in this regard is that "[c]urrent adaptation measures are mostly reactive – that is, they are responses to disasters or other major events. To successfully respond to and prepare for the future of climate change, any land use plans must include a mix of proactive and reactive measures" (Fine 2023: 1868).

As far as who is affected the worst, it needs to be noted, as is a theme of this book, that if people have already suffered environmental justice harms, they are less likely to have the capacity (including the resources) to cope with the additional burden brought on by climate change. They usually do not have the ability to adapt and/or repair the harm caused by climate change. It is Indigenous Peoples and other minorities who are most likely not to be able to afford insurance to deal with these effects of climate change (Mann et al. 2024). Generally, however, there are going to be substantial problems concerning insuring property and a range of other items in the future as the effects of climate change grow. New ways of providing insurance will need to be found (Watts et al. 2023), especially for those who are more likely to be affected and less likely to have or afford insurance coverage (Sered and Fernandopulle 2023).

Such individuals, especially from various minorities (Makleff et al. 2022) and Indigenous groups, already have a range of problems, because of inequality, discrimination, and marginalization, such as health difficulties, which will be further exacerbated. Thus, these people and groups (Barankin et al. 2021) are therefore much more vulnerable to climate change (Benevolenza and DeRigne 2019). This is as a result of predating disparities, and a range of pre-existing environmental and other types of problems which make them less resilient to such changes (Smith et al. 2022). Some of these problems are environmental justice in nature, as is contended in this book.

As has been noted, a variety of groups are more susceptible to climate change, because of the legacy of past discrimination and the environmental problems they have already suffered, and thus need greater assistance to adapt (Lindsay et al. 2023). It has also been noted that "progressive climate adaptation planning can help bridge resource, capacity, and awareness deficits within cities and beyond" (Chu and Cannon 2021: 51). This is important, as many groups have already suffered enormously because of a variety of environmental injustices. Despite this, it was found in a 2021 study on the United States that:

> adaptation plans and projects to date not only focus on reducing the vulnerability of "the haves" over the "have nots," but also exacerbate the vulnerability of the most disadvantaged ... Many adaptation proposals embrace nature-based solutions such as dunes, wetlands, bioswales, greenways, and green roofs. These investments

> beautify neighborhoods, stabilize communities, and increase their desirability, which in turn attracts wealthier residents and contribute to "green gentrification." (Shi 2021: 1459)

While in some places, a justice approach is being used by some local authorities in their climate adaptation plans, and are a significant issue for some (Fiack et al. 2021), it seems that there are no clear broadly accepted principles for using justice (or how human rights issues ought to be considered) (Dana 2023) for planning purposes in climate adaptation processes in many places (Chu and Cannon 2021: 85). Further, adaptation strategies that aspire to be just and equitable remain theoretically fragmented, normatively scattered, and often not fully engaged with emerging socially transformative priorities on the ground (Amorim-Maia et al. 2022). This is problematic. Not surprisingly, not taking into account the situation of communities, and prior environmental injustices, and their continuing effects when designing adaptation processes has been called maladaptation (Forsyth and McDermott 2022). Thus, social equity and justice is seen to be a pivotal issue that is taken up in many local adaptation plans (Fiack et al. 2021). However, "while the language of equity has become increasingly popular, little is known about its inclusion, operationalization, or relationship to existing inequities in specific places, especially outside of 'best practice' cities" (Petersen and Ducros 2022: 61). One study in 2022 focused on the United States, found that:

> Almost 70% of cities in this study stressed the importance of addressing equity or justice, and 85% acknowledged the unequal climate risks faced by "vulnerable" residents. However, the rhetoric around climate vulnerability and justice did not translate into adaptation actions connected to social groups or inequalities. Less than 3% of all adaptation actions directed resources to low-income communities or people of color, and actions targeting those groups were concentrated in education, outreach, and coordination efforts. These findings align with previous research that documents a high degree of rhetorical commitments to equity or justice in climate or resilience plans. (Zoll 2022: 17, references omitted)

While this is unique to the United States it is probably not that uncommon globally. This is deeply problematic as "city-led adaptation interventions that ignore these realities are likely to maintain or exacerbate disparities and the systems that produce them" (Petersen and Ducros 2022: 3). These issues are taken up in the country chapters.

Adaptations must therefore incorporate both climate justice (Juhola et al. 2022) as well as an environmental justice component to allow the most vulnerable people to better cope with climate change. This is already happening in places, mostly in the United States, where justice, because of the strength of the environmental justice movement, is seen to be a key issue in dealing with climate change (Diezmartínez and Gianotti 2022). It is, however, not happening in many other places, but this should be encouraged (Williams et al. 2022). Already, though, the interconnection between environmental justice and climate justice has been recognized (Winker et al. 2024). So has the need to consider vulnerability in climate change and therefore the link with environmental justice. As has been argued, the implications for social vulnerability need to be considered (Woodruff 2022). These topics are also interconnected with the principles of sustainability (Fiack et al. 2021), bringing the notion of sustainable adaptation to the fore (Eriksen et al. 2011).

It also needs to be noted that the steps taken to mitigate or adapt to climate change must not worsen the effects of climate change, especially for those who are least able to deal with these effects. This is indeed the case, as again the potential for minorities and others to once again be marginalized, is present. In fact, what must be done is for the law and policy in nation-states to put more focus on those who are being more dramatically affected and who are the least likely to be able to mitigate the damages on their own. This is crucial for dealing with the problems brought on by a warming planet. Without taking up climate justice in the law, the problems of climate change will not be dealt with, and in fact, the problems will get worse. Thus, dealing with climate justice is in part about how the effects of climate change can be reduced.

Whatever processes are adopted, they must also consider local knowledge and preferences and satisfy the demands of the communities that are dealing with these problems. This does not always occur. These communities are not given the respect they deserve and are often excluded (Hedge et al. 2020), as will be indicated in the book. While at times lip-service is often given to consultation processes, real and ongoing involvement of communities, especially those more affected by climate change, is essential. This again needs to be legislated and enforced. This is because equity demands that local community empowerment takes place. It cannot be elites alone that make all the choices. This is largely the case at the moment because much of the published literature in scientific journals, concerning climate change, is in English. This limits the rest of the world from accessing such information.

There is also a need for the education of the wider public on these issues, and in so doing, to bring them to a better understanding of the risks associated with climate change and what they can do to mitigate these. It is about making minority communities more resilient to the short- and long-term effects of climate change by giving them the skills and resources to prepare and react to the effects of climate change. Raising awareness and improving readiness, both within minority groups and more broadly, within the general population, can enhance overall resilience to the effects of climate change. These are cross-cutting issues affecting both environmental justice and climate justice. They are extremely relevant to advance efforts that deal with all environmental matters including climate change.

CONCLUSION

Climate-related hazards are a major threat facing the world today. They pose a magnified risk for vulnerable groups and have the potential to further deepen social inequalities (Abate and Kronk 2013). Some of those problems are environmental justice in nature. These exist because of past discrimination and exclusion. This has occurred because these groups are often viewed as the "other," because they share an identity or identities which make them discernible and thus categorizable. Being categorized as Indigenous or as a minority has implicit sociopolitic or sociodemographic overtones that brands a group, usually in adverse ways (Seyranian et al. 2008). The perception of them as "other" can incite fear and hatred and lead to violence towards them (Shelley 2016) – or indeed by them towards others as a means to protect themselves.

Members of these groups face a range of problems including injustices: rejection, marginalization, exclusion, and inequality, to mention just a few. The link between relegation, ostracism, and disparity in general, on the one hand, and violence and conflict, on the other, is an important issue in a world where there is more conflict and violence all over the world now than at any time since World War II. Critically, a lack of focus on Indigenous Peoples and minorities, and specifically their environmental justice challenges means that climate change is less likely to be successfully addressed (Maldonado et al. 2016). If their issues, viewpoints, and laws, especially as far as Indigenous Peoples are concerned, were better recognized and incorporated it would make a world of difference in dealing with the many environmental challenges the world faces.

Practically speaking, there is a global realization of a need to switch to more sustainable and resilient systems, the need to cut emissions dramatically, and carry out adaptation strategies by putting into place laws and policies that deal with the effects of climate change (Verschuuren 2022). What must also be recognized is the need to ensure that steps taken to address the problems caused by global warming are taken in ways that do not make existing disparities even worse. What is important in this regard is the need to see the interrelation between sustainability and justice. They are rarely linked together. As Boone has noted:

> Justice is a core yet often ignored principle of sustainability. In efforts to find ways to reduce harm to the natural systems that support us while seeking ways to expand economies and improve livelihoods, many neglect or pay lip-service to justice as a central sustainability principle. Yet, sustainability for some at the expense of others undermines the principles and practice of sustainability as a force for positive change. (Boone 2010: 135)

Sustainability must therefore be linked to the concepts of climate change, climate justice, and environmental justice in the law. For this reason alone, these concepts need to be dealt with in a more integrated way. Without dealing with environmental justice, as well as climate justice, attempts to deal with climate change will be less successful.

What is clear is that as the climate emergency becomes more urgent, so does the need to find ways to limit its impact. Promoting environmental justice, in combination with other solutions, is one method to deal with the broader issues, while ensuring fairer, more equitable and more human rights-friendly outcomes for many people around the world (Giacomini 2022). This must mean that at all levels, including international and domestic, greater attention should be paid in the law and by the courts to environmental justice. This is steadily increasing at the international level but not so in the domestic context.

While environmental justice should guide efforts to address the effects of environmental degradation generally (Sant'Anna 2018), the concept of climate justice should also be used widely and by the law, and the courts specifically (de Jong 2024), to address the damage resulting from climate change. This is because of the interrelationship between climate change and environmental justice (Dietz et al. 2020). However, while climate change and environmental justice matters are interconnected, they are not treated the same. While much is being done to address climate change, the same discrimination towards Indigenous and other minority groups regarding environmental issues occurs when addressing climate change. This can be seen in the fact that there is often less investment to prepare for or ameliorate climate change damage in areas populated by marginalized groups, which means they have to bear a greater share of the problems caused by climate change, including the global warming effects of storms, flooding, or droughts (McGraw and Fox 2019), without the infrastructure or other resources to do so (Shapiro 2019).

3. Climate change actions and environmental justice in the United States

INTRODUCTION

This chapter examines the broad sociopolitical issues concerning environmental justice and climate change in the United States through the lens of Indigenous Peoples and minority groups. These groups are the focus because they have been disproportionately affected by environmental injustice and already by climate change generally but also in the United States. These groups have suffered and are likely to continue to suffer, the effects of climate change, and be the victims of environmental injustice, to a measurable extent greater than the general population. The purpose is to come to a better understanding of the difficulties these groups face in general, and specifically as far as climate change and environmental justice issues are concerned. This chapter can really only sketch the broad details and go into only a few of the many issues that exist. The focus is the federal level, as again examining the issues at the state level is outside the scope of this study.

The United States is a focus of this part of the study, partly because the environmental justice movement began in the United States 40 years ago, and much work has been done in the area both in academia and in praxis. There is therefore a lot to learn from the American experience from both a positive and negative viewpoint. Additionally, it is crucial to focus on these issues in the country as the United States, together with China and Russia, account for 40 percent of greenhouse gas emissions (Climate Watch 1990). Consequently, the failure to effectively deal with these issues in any of the three countries will have massive ramifications for the whole world. While the American political situation may be unique, the discrimination and adversity that various groups face is a worldwide phenomenon. The political and social response in the United States to these issues can provide guidance to other countries.

Climate change, climate justice, and environmental justice are, as noted earlier, still largely studied and addressed separately by those who take up the broad issues in academia and in praxis (Tahir 2022). This chapter reviews the issues distinctly because of this trend. However, it will note those instances where there is an important overlap of these issues, which is actually quite common.

It is clear that climate change is and will continue to have a wide variety of negative effects on the country, including a rise in crime rates (Harp and Karnauskas 2020; Hou et al. 2023). The effects, termed the climate gap (National Academies of Sciences, Engineering, and Medicine 2022), will be particularly marked for Indigenous People (O'Rourke 2024) and minorities (Mizutani 2018). These effects include increased incidents of drought, floods, fires, erosion, and rising sea levels (Lyons et al. 2023). This will trigger an increase in the demand for public services to respond to these crises (Williams et al. 2020). The cost of this will be enormous and it is questionable whether local authorities will be able to cope. The rising level of the sea will be particularly devastating (Carson 2020). As has been noted, "Owing to its geography and population distribution, the USA is a coastal nation, with more than 30% of its population residing in coastal cities, generating an estimated annual revenue of US$3.8 trillion. Consequently, socioeconomic losses from climate-induced SLR will represent a notable facet of climate-change consequences in the USA" (Ohenhen et al. 2024: 108). These issues will affect some communities more than others, particularly Indigenous Peoples and minorities (Carson 2020). These groups have already been suffering various environmental injustices, which are now being exacerbated by climate change problems (Mizutani 2018). Thus, these groups must be afforded environmental justice and climate justice to try and limit the devastating effects of climate change that are taking place today and that will increase over time.

According to a study conducted in 2024, 32 coastal US communities are sinking (Ohenhen et al. 2020). Rising sea-levels are expected to flood between 272 and 427 communities in the country by 2060. Flooding is becoming usual in certain parts because of excessive rainfall, and drought is affecting other parts. Minorities were being more affected by this than others (Lee 2024). To combat some of the problems that these groups face, already in 2016, $48 million were earmarked to move the Indigenous People inhabiting the Isle de Jean Charles in Louisiana (Crepelle 2018). However, the effects of climate change are being felt already in other parts of the United States, including Alaska, California, and Florida (Alarcon et al. 2024). In Alaska, climate change and development have caused the reindeer and caribou herds to decline at a rate of anywhere between 66 and 94 percent in recent years. This is having a dire effect on subsistence hunters (Rosen 2024).

Thus, as is occurring elsewhere, it is often Indigenous Peoples and various minorities who are bearing the brunt of climate change, together with a range of environmental injustices (Terrell et al. 2024). It is estimated that over the last 20 years ten million people have been displaced because of climate change in the United States (O'Rourke 2024) and these numbers are increasing (Carson 2020). Three million people were affected by disasters in 2022, many, in whole or in part, due to climate change (Frank 2023). There will be an increase in these disasters in the future, which will cause further problems for certain groups. It is believed that 200 million people around the world could be displaced by 2050 (Tahir 2022). It has also been estimated that 40 percent of Americans live in places that may be underwater by the year 2100 (Carson 2020). There are other effects though. For example, in Alaska, the loss of protective sea ice is exacerbating the permafrost and coastal erosion, which is having a serious effect on the Indigenous population. Water supplies, infrastructure, health, and safety are all under risk because of climate change (Cochran 2014). Thus, a key issue concerns adaptation and the costs thereof. As noted in the Introduction the costs of doing so are enormous. As far as health and environmental costs are concerned, in the United States alone, the cost to mitigate the damage caused just by using coal is estimated to be between $175 billion and $520 billion annually (Epstein et al. 2011).

The chapter therefore reviews the environmental justice and climate change situations of minorities and Indigenous Peoples in the United States. Their situation, part of the legacy of the past, still sees them being subjected to extensive racism and deprivation. They endure many environmental justice-related problems such as limited access to public services, as discussed elsewhere in this book.

To understand the country as far as Indigenous Peoples and minorities are concerned the demographics of the United States are laid out in the chapter. Then the effects of entrenched racism on Indigenous Peoples and minorities is examined to understand the long-lasting processes of racism in the country and its consequences. Then reviewed is the effect of climate change on minorities and Indigenous Peoples in the United States. Then reviewed is the effect of environmental injustices on minorities and Indigenous Peoples. How the United States has dealt with these issues is then examined to understand what has been done particularly during the Biden Administration to deal with environmental matters. Specifically reviewed is the Inflation Reduction Act, which is landmark legislation, but also other steps taken to deal with some of the issues. Also examined are some of the ongoing challenges in dealing with climate change and environmental justice. The role of the US Supreme Court is briefly examined to give some understanding of the role the Court has played on these issues. The focus in the chapter is on what has occurred at the federal level and not really on what has occurred in the states.

THE DEMOGRAPHICS OF THE UNITED STATES

This sections reviews who lives in the United States to understand the complexities of the makeup of the US population. The United States Census Bureau recognizes significant groupings of peoples within the United States, including White, Black or African American, Hispanic or Latino, Asian, Native American and Alaska Native, and Native Hawaiian and Pacific Islander (Creamer et al. 2022). There are, however, other groups as well. Of the 327 million people that were determined by the census to reside in the United States in 2021, 195 million identified as White, 61 million as Hispanic, 43 million as Black, 20 million as Asian, and 4 million were classified as "American Indian and Alaska Native" (Creamer et al. 2022: 60).

A critical issue is that the various Indigenous and minority groups often remain invisible. In this regard, in relation to Indigenous Peoples, it has been noted that "invisibility relates to the lack of respect, comprehension, and honoring of the political status of Indigenous peoples. ... Recognition, acknowledgment, and reconciliation are indeed healthy antidotes for invisibility, marginalization, and alienation" (Zuni Cruz 2020: 140, 142). It has been noted that "[t]he degree of invisibility of Indigenous peoples, in spite of the existence of 574 federally recognized tribes with political status, is a central issue in major cases and events of the decade" (Zuni Cruz 2020: 140). The same could be said of many of the minority groups in the United States as well.

So important is the issue of invisibility for such groups in the United States, that a bill was introduced in the US Congress in 2019 entitled the "Not Invisible Act of 2019" (Parsloe and Campbell 2021: 197). The issue of the invisibility of various groups such as those who identify as transgender or nonbinary led to the recognition in 2009 of 31 March of each year as International Transgender Visibility Day. The day became politicized when American Republicans criticized the recognition because, in 2024, 31 March fell by accident on Easter Sunday. This was somehow seen as a deliberate affront to Christians (Associated Press and *Guardian* Staff 2024). Thus, the issue of identity continues to divide. It is estimated that 1.6 million Americans identify as transgender (Herman et al. 2022).

As noted by Bullard in some of his pioneering work (Bullard 1993a), invisibility has for a long time been seen as a major environmental justice issue. It is a worldwide problem. In its Report on the State of the World's Indigenous Peoples in 2019, the United Nations detected a "persistent invisibility" of Indigenous Peoples in official data (Zuni Cruz 2020). In this regard, intersectionality can help to raise understanding about the difficulties they endure which render them less visible. This type of accommodation would make a huge difference to the way groups are seen and therefore treated. This will be taken up by this author in his next work.

THE EFFECTS OF ENTRENCHED RACISM ON INDIGENOUS PEOPLES AND MINORITIES

Entrenched racism in the United States has had a significant effect on environmental justice issues in the country. Various groups face a range of environmental-related injustices, including that they are more likely to have lead in their dwellings and neighborhoods, experience poor air quality (Sprainer 2022), live close to dangerous locations, incinerators, and factories that release pollutants into the air, and their water is more likely to be contaminated with dangerous chemicals (Bridges 2022). The rise in water prices is affecting such groups in disproportionate ways (ThummBorst 2023). This is true also in some places for energy costs, but also for rising insurance prices as well as others, which is disproportional for different race groups (Contat et al. 2024). As noted earlier, the cost of climate-induced weather events in 2019 globally was already $232 billion, with the United States accounting for more than half of that amount. There was a 69 percent protection gap between losses and insurance coverage (Klein 2021). There are going to be major effects further on insurance in the United States as the present situation of major losses to insurance companies is unsustainable. This is because in 2023, insurance companies in 18 states lost money on homeowners' insurance. Losses only occurred in 12 states in 2018 and in eight states in 2013. This is causing increased premiums, reducing coverage or the abandonment of some states completely (Flavelle 2024). The effect of this is that the number of people in the United States without home insurance soared to 12 percent in 2024 from 5 percent in 2019 (Schulz and Guynn 2024). This trend is likely to continue. This is supported by the Consumer Federation of America (CFA), who issued a report in early 2024 that found that more than six million homeowners already do not have insurance and that poorer people are twice as apt not to be covered by such insurance (Consumer Federation of America 2024). As is noted:

> Being uninsured can foster deeper economic precarity for millions of homeowners across the country, especially those with lower incomes, and it is an important contributor to racial inequality. Inequalities in who has homeowners insurance will likely widen the long-standing racial wealth gap, as uninsurance disproportionately impacts Hispanic, Black, and Native American homeowners. Over time, insurance access is likely to become a key decider of who can fully reap the benefits of homeownership, including maintaining their home and building wealth. (Consumer Federation of America 2024)

Thus, in the future, as insurance rates decline, again, it is the more vulnerable who will also suffer more from climate change and will be less covered by insurance. It has been argued, however, that the insurance industry can play a positive role in dealing with climate change by instituting various measures that ensure that sufficient mitigation and adaptation occurs (Fox 2024).

The disproportional effects on minorities and Indigenous Peoples, is partly at least because of environmental racism. Bullard has defined environmental racism as "any policy, practice or directive that differentially affects or disadvantages (whether intended or unintended) individuals, groups or communities based on race" (Bullard 1993b). Environmental racism is a very common occurrence. In addition to forcing African Americans and other minorities into distinct enclaves in the United States, this kind of racism leads to significant inequalities in housing, healthcare, education, and many other facets of daily life (Norton 2023: 85; Johnson-Jennings et al. 2020: 91). This is further exacerbated by climate change issues (Gerbatsch 2022), which means that the infrastructure in their areas cannot cope with the flooding and other effects of climate change (McGraw and Fox 2019). This is often because the areas these groups inhabit have not had the same level of investment in surface and subsurface infrastructure designed to protect the community from environmental extremes as other places (Shapiro 2019).

African Americans, as a group, but other minorities as well, have suffered environmental injustices out of proportion to the rest of the population. For example, in Mississippi, a hazardous waste facility was built in an area specifically because it was where minorities lived (Boerner and Lambert 1995). As early as 1983, a US government study found that 75 percent of hazardous waste plants were built in areas where minorities reside (Duncan 1999). North Dakota is another place where various groups are affected (Bridges 2022). In Louisiana (Mizutani 2018) there is a place called "cancer alley" (Castellón 2021) or "petrochemical America," where more than 150 oil refineries, petrochemical plants, and other dangerous industrial facilities exist (Castellón 2021; Blodgett 2006). A 2024 report found that coal-based steel and coke plants in the United States had startling human costs causing many preventable deaths, a quarter million cases of asthma, and $13 billion in medical expenses annually due to pollution from just 17 facilities in Indiana, Ohio, Pennsylvania, Illinois, Michigan, and Virginia. It found that persons who live close to coke factories have a 26 percent higher risk of developing cancer due to air pollution, while those who live close to steel mills have a 12 percent higher risk than the national average (Industriouslabs, 2024) This interrelationship is important, because as David Boyd, the Special Rapporteur on the issue of human rights obligations relating to the enjoyment of a safe, clean, healthy and sustainable environment, has stated generally, about where such toxic facilities are found: "And guess where they're located? In poor, predominantly Black communities. It's just, as I said, it's unconscionable" (Gayle 2022). Another commentator has noted:

> [I]n the U.S., the best predictor of whether you live near a hazardous waste site is the color of your skin. In Los Angeles, poisonous lead contamination at homes near the closed Exide battery recycling plant has yet to be fully cleaned up, hurting working-class Latino families. … Black, Latino and low-income California residents are especially likely to live near unplugged oil and gas wells that can spew pollution, a Los Angeles Times investigation found. People of color are more likely than white people to live alongside power plants, oil refineries and landfills … It's true in cities, where black, brown and low-income people experience hotter temperatures during deadly heat waves because their neighborhoods lack tree cover and parks, or because they're near freeways or industrial facilities. (Roth 2020)

As will be noted below, action has finally been taken in recent years to deal with some of these issues.

THE EFFECT OF CLIMATE CHANGE ON MINORITIES AND INDIGENOUS PEOPLES IN THE UNITED STATES

As far as climate change is concerned, one of the biggest challenges in the world and in the United States, is that extreme weather events cause significantly more devastation for poorer communities due to their limited ability to prepare for and recover from such events. This issue disproportionately affects minorities and Indigenous Peoples, who often face systemic economic disadvantages. They have fewer resources to take precautionary measures or cope with the aftermath of climate-relate disasters. Additionally, they are often more severely impacted simply because of where they live. Many of these communities reside in lower-income areas that are more vulnerable to major climatic events, such as flooding. When it comes to a variety of weather events, including hurricanes and severe storms, they are at greater risk because their housing is typically less weatherized and less equipped to withstand such occurrences. Furthermore, these communities tend to receive less assistance in recovery efforts. One study found that income inequality worsens as a result of climate-related damages. Alarmingly, this disparity appears to grow in correlation with the amount of aid an area receives from the Federal Emergency Management Agency (FEMA). The analysis suggests that increasing environmental hazard damages and wealth inequality are dynamically linked (Howell and Elliott 2019). Moreover, these marginalized groups are statistically less likely to receive FEMA assistance, exacerbating their vulnerability (Waters et al. 2024).

Broadly speaking while there have been many studies on environmental justice by academics and civil society groups, few have taken specific legislative or other steps. While certain states such as California, Massachusetts, and Washington demand certain environmental justice actions to be taken (Gerrard and McTiernan 2023). New Jersey adopted a specific environmental justice law in 2020 (Levin 2023). In 2021, the state of New York added an environmental rights clause that reads "Each person shall have a right to clean air and water, and a healthful environment" to its Constitution (Bratspies 2024: 100). In 2023, it adopted an environmental justice law (Gerrard and McTiernan 2023). The interrelationship of environmental justice and climate change is not dealt with. However, in 2019, the state of New York enacted the Climate Leadership and Community Protection Act (CLCPA) that deals with various issues including the mandating of a reduction in greenhouse gas emissions, but also reducing pollution and promoting energy justice (Salter 2023). It critically invests in disadvantaged communities and develops issues of environmental equity, and so ties it to environmental justice (Jones 2024). However, more ought to be done to make the connection clearer and more direct. They are deeply linked and will become more so over time.

The first city to undertake an environmental justice assessment was New York. It released that report in April 2024 (New York City Mayor's Office 2024). The state has also taken up legislation to deal with various green issues (Bratspies 2024). Interestingly, New York City's (NYC's) environmental justice study was conducted by NYC's Mayor's Office of Environmental and Climate Justice, an office that other municipalities should model to ensure that the important links between environmental justice and climate change are recognized country wide. More cities, states, and other authorities need to create such offices and carry out such studies to understand the environmental justice situation in their locales. It is also important that a greater consensus emerges concerning the situation in neighborhoods across the country. This is because, as the report notes, there is no comprehensive inventory of possibly contaminated sites in New York City or around the nation, nor is there a concerted attempt to identify legacy industrial zones where contamination already exists. Because of this, determining the precise location of polluted places, and how the contamination affects people living in the areas most affected, is challenging.

Usefully the NYC report develops the notion of Disadvantaged Communities (DAC) in the study, identifying where these communities exist using a scoring system that considers diverse sociodemographic and environmental circumstances in different census tracts. According to the classification, 44 percent of New York's census tracts, where 49 percent of the city's population dwells, are deemed areas in need of environmental justice action. This exercise identifies areas where remedial action needs to take place.

Unsurprisingly, the report finds that communities of color and low-income areas of New York are the sites of an unfair share of the city's deteriorating infrastructure, while also experiencing cuts to investments in environmental services like solid waste collection and parks and natural resources. Redlining, or racially biased real estate practices, still has an influence today. There are many instances of racial bias continuing (Kruczkiewicz et al. 2023). House appraisals are lower in majority-Black neighborhoods (Rothwell and Perry 2022). The effects of redlining also see variations in communities such as the number of trees, which is going to have a greater effect on climate change in those areas (Jung et al. 2024). There is still a disparity in park density in areas where these communities live. The report noted that commercial garbage routes, delivery routes, arterial highways, and medium- and heavy-duty fleet parking lots are more often found in communities of color than in more wealthy districts. There is a discouragingly long list of environmentally damaging infrastructure, industries, and activities in these zones, including waste processing facilities, hazardous waste generators, and various power plants, as well as small, chemically intensive enterprises like auto shops that generate and store hazardous waste. The report also found that historically redlined neighborhoods have the highest rates of lead paint violations and poor house maintenance. Overall, the people living in these communities are disproportionately exposed to emissions which have notable negative health consequences (New York City Mayor's Office 2024).

Usefully, the report also examines the impact of climate change on these communities and the specific areas where they live. For instance, the report finds a correspondence between high-heat-vulnerability neighborhoods (Vecellio et al. 2022) and redlined residential areas. These same areas are disproportionately vulnerable to flooding resulting from the present decade's high rainfall rates, persistent tidal flooding, and coastal storm surges, and this disproportionate vulnerability will likely continue. The report concludes that while all New Yorkers will be impacted by climate change, those who live in the areas designated in the report as environmental justice areas are particularly vulnerable.

The report therefore recommends investment in communities where environmental injustice has occurred, and that climate response budgeting be incorporated into agency decisions. It also recommends that there be an increase in data transparency, improved communication, and greater coordination by regulatory and permitting authorities, and that new approaches be found to include these communities in the necessary response work.

While many regions in the United States are facing environmental justice issues, they are at the same time grappling with major climate change issues (Bailey et al. 2014). Indigenous and minority communities have enormous difficulties coping with climate change because of a lack of resources and the fact that most state governments do not engage with these communities as they do with others. In this regard, it has been noted that African Americans are the second most impoverished group, with nearly 21 percent living below the poverty line (Buiquy 2022). Indigenous Peoples also experience an inordinate level of poverty. While they

> comprised … 1.7% of the United States population in the 2010 Census, yet seven of the eight poorest counties in the United States were majority Indian. [Indigenous] The average unemployment rate on Indian reservations is over 50%, over twelve times the national average. Few jobs exist in Indian country because there are few small businesses on reservations. Though the Indian entrepreneurship rate is at parity with their percentage of the population, over 40% of Indian- owned businesses are in the twenty lowest sales industries. (Crepelle 2019a; Crepelle 2022, references omitted)

As a result of discrimination which has reduced their purchasing power (Chawla and Kumar 2022) and limited their rental opportunities, minorities have been forced to live in places where land and houses are cheaper. These places often have minimal or no municipal services. While this scenario began because of racial segregation, it continues still for many other reasons.

Significant studies in the United States have concluded that minorities suffer excessively from living and working in places where environmental hazards exist (Al-Kohlani and Campbell 2022). The Commission for Racial Justice issued a report, "Toxic Wastes and Race in the United States," which found that the racial makeup of an area was the singular variable that indicated why profit-making hazardous waste industries were placed in a specific area (Lee 2019; Mohai and Bryant 2019). In fact, it has been argued that the placement of many environmentally problematic industries has been purposeful and has had further devastating effects on the health of the people of color who live in these poverty stricken areas (Terrell et al. 2024). As studies, including this book, attest, this is not an uncommon occurrence (Fos et al. 2021) and not only in the United States (Walker 2012). The placement of these dangerous industries in these areas is a result of a combination of the availability of cheap land, a local supply of minimum wage or low-paid labor, and less resistance to the placement of the facility in the area because of a lack of awareness and organizational skills among the affected population.

As a result, minority populations have had to endure harmful environmental conditions in their areas of residence. For example, the research indicates that "The burden of air pollution is not evenly shared. Poorer people and some racial and ethnic groups are among those who often face higher exposure to pollutants and who may experience greater responses to such pollution" (American Lung Association 2023). Thus, in 2023, nearly 40 percent of Americans resided in places where the air quality is so bad that their health is at risk (American Lung Association 2024). This is not abnormal as far as minorities are concerned, as numerous research studies have shown the differences in air pollution exposure across various ethnic and socioeconomic groups (Jbaily et al. 2022; Gardner-Frolick et al. 2022) including Black, Asian, and Hispanic communities (Thind et al. 2019). Pollution-related mortality rates (Josey et al. 2023) in African American communities are higher than for residents of predominantly White regions (Spiller et al. 2021). These groups are more likely than the general population to be exposed to "lead and allergens in their homes as well as harmful emissions from nearby bus depots, factories, and dumps when they live in or close to low-income or distressed areas" (Bahls 2011).

These types of problems occur because the political, legal, and social systems have permitted, and continue to permit, this discriminatory situation to exist. This can again be seen in various studies. Part of the problem is that municipal, state, and federal governments often do little to prohibit such situations from developing or continuing. One study indicated that:

> Penalties under hazardous waste laws at sites having the greatest white population were about 500 percent higher than penalties at sites with the greatest minority population. Hazardous waste, meanwhile, is the type of pollution experts say, is most concentrated in minority communities. (Coyle et al. 1992: 52)

It has also been argued that less time is taken to address problems in areas where minority populations predominate than in areas with a White majority population (Coyle et al. 1992: 52).

Another reason why environmental injustices occur in areas inhabited by such groups is that there is inadequate enforcement of environmental protection and related laws by the various authorities. Enforcement, one author argues, occurs less often in poor areas, but the study found no evidence that there was a racial basis for this (Konisky 2009). However, a correlation between the existence of environmental problems and the percentage of minority residents in an area was found in a different study (Scholz and Wang 2006). Yet another study determined that the inequality in the distribution of the impact of environmental degradation results from the fact that some communities are unable to mobilize or make demands on the polluting agencies when they have problems (Anderson et al. 2020; Anderson et al. 2023). Again, race, class, and other identities are significant aspects of the environment degradation/environment justice issue (Mele 2016). As was noted by a group of independent United Nations experts in 2023:

> Systemic racism ..., including as it relates to structural and institutional racism, is understood to be the operation of a complex, interrelated system of laws, policies, practices and attitudes in State institutions, the private sector and societal structures that, combined, result in direct or indirect, intentional or unintentional, de jure or de facto discrimination, distinction, exclusion, restriction or preference on the basis of race, colour, descent or national or ethnic origin. Systemic racism often

> manifests itself in pervasive racial stereotypes, prejudice and bias and is frequently rooted in histories and legacies of enslavement, the transatlantic trade in enslaved Africans and colonialism. Systemic racism ... exists in the United States ... The systemic racism faced today ... is deeply affected also by intersectionality, that is, in combination with several other identities, including sex, gender, gender identity, sexual orientation, nationality, migration status, disability, religion, socioeconomic and other status. (United Nations Human Rights Council Special Procedures 2023)

This situation is applicable to a range of problems, including environmental justice issues in the country. For example, a 2023 study found that it takes longer for noncompliant water systems in counties with higher percentages of Black and Hispanic residents to become compliant with the law. Contrarily, it was found water systems that serve a greater proportion of White residents are out of compliance for a shorter amount of time, and as the number of White residents in a region rises, the time to achieve compliance falls. The study argues that the results show that minority groups do not receive equal attention when it comes to monitoring water compliance, which is caused by the legacy of racialized urban planning and long-term underinvestment in water infrastructure (Bae et al. 2023). Thus, while it can be argued that overt racism is not the cause of inequality, the effects of environmental degradation are clearly greater for minorities. This issue will be returned to in the discussion concerning how the courts have dealt with this issue. What is clear is that an intersectionality approach would help ensure that the link between class and race is recognized and addressed.

The issue of the lack of access to water and having degraded water supplies are common for Indigenous and minority groups around the world (Henry 2023). Even in the developed world, this occurs regularly. In some places, the problems are very severe (Chambers 2018).

This is often the same for other issues, such as air pollution (Sprainer 2022). In the United States it has been found that "people of color are 61 percent more likely than their White counterparts to live in counties above the [accepted standards] for one pollutant and 300 percent more likely to live in counties exceeding [those standards] for three pollutants." That study showed the widespread disproportionate air pollution (Cormany 2024: 167) across areas in the United States inhabited by people of color (Revesz 2022: 187). This is supported by other studies as well (Tessum et al. 2021).

ENVIRONMENTAL INJUSTICES ON MINORITIES AND INDIGENOUS PEOPLES IN THE UNITED STATES

Massive violations of the rights of Indigenous Peoples and minorities have occurred for centuries all over the world (Glauner 2001; Piccard 2013; Ochs 2024), not least in the United States (Stark and Mills 2022). For a long time,

American Indigenous Peoples and minority groups have suffered from the effects of environmental harm to a significantly greater degree than what other people in the country have had to endure (Motairek et al. 2022). This includes the practice of redlining (Cohen et al. 2023), where people in a specific area are discriminated against by being systematically denied access to financial services, including insurance, mortgages, and other loans, on the basis of their race or ethnicity (Bullard 2021). While this is no longer legal, the legacies of it, and systematic racism generally, continue to exist to a greater or lesser degree in many places (Mitchell and Franco 2018). As has been noted:

> low-income people – and especially those of color – do not have much choice as to where they might live. They cannot find inexpensive rental housing in the suburbs, and in the aftermath of shameful racial covenants, redlining, and exclusionary zoning, people of color find themselves virtually confined to inner-city neighborhoods. (Papke and Papke 2023: 1)

The legacy of the past on these communities is therefore enormous (Rothstein 2017). For example, they have to endure higher insurance rates compared to other groups (Bradt et al. 2021). This will see them further not able to deal with various climate change effects.

More broadly, however, the disproportional violence (Crepelle 2023a) and discrimination meted out to them continues (Rothstein 2017). As has been argued, there is still tremendous segregation in US cities which:

> stem from "three interrelated and mutually" reinforcing forces in America: high levels of institutionalized discrimination in the real estate and banking industries; high levels of prejudice among whites against blacks as potential neighbors; and discriminatory public policies implemented by whites at all levels of government. (Shirazi 2016: 74)

There are major differences in the values of similar houses (Perry et al. 2018) in neighborhoods with different racial makeups (Dobbs and Gaither 2023). Borrowing costs (Eldemire et al. 2022) and particularly property taxes (Avenancio-León and Troup 2022) in Black neighborhoods are often higher despite this (Avenancio-León and Troup 2022). These affect home ownership rates (Martinez and Aja 2021) and the returns that minorities can earn on their houses (Kermani and Wong 2021). This has affected the wealth gap in the United States (Derenoncourt et al. 2024). It affects many other issues including school education in the different neighborhoods (Monarrez and Chien 2021), which has long-term negative effects on minorities overall. It will have dramatic effects as climate change effects are felt (Contat et al. 2024). Very problematic, from an environmental justice point of view, in comparison to otherwise comparable neighborhoods, historically redlined areas, and those with high concentrations of minorities, have almost twice the number of oil and gas wells (Gonzalez et al. 2023). This has had massive negative effects on the health of the people who live in these places (Lee et al. 2022). One study indicates that Indigenous Peoples today deal with demoralization, distress, and other impairments to sometimes overpowering degrees. Over the past 30 years, the idea of Indigenous historical trauma has emerged as a counternarrative in mental health studies that challenges the dominant understandings of mental diseases, addiction, trauma, and psychological impairment (Gone 2023).

Clearly, the mental trauma that these groups suffer (Stein et al. 2021) is in addition to the adverse effects on their physical health caused by the hazardous facilities where they live, which continue today (Lane et al. 2022). These dangers are enhanced because of the effects of climate change, such as an increased risk of flooding, which, as argued elsewhere, impact Indigenous and minority communities to a greater extent than the rest of the population (Napieralski et al. 2024). Oftentimes the people who live in these communities are unable to relocate for financial, cultural, legal, and practical reasons (Bravo et al. 2016).

This situation is exacerbated by the fact that in the United States, Indigenous Americans and other minority groups are exceptionally underprivileged compared to White people. So they live in less desirable areas where problems such as degraded land, insufficient potable water, and air pollution exist, and which often have fewer basic services than other communities. The evidence suggests that this disparity is often the result of systemic racism. Services provided to African Americans or other minority communities are often both substandard and more expensive. The effect of this is that millions of people do not have access to such services. This phenomenon is now commonly called environmental racism, because of the historical discrimination practices and the ongoing system of oppression at least as far as these matters are concerned (World Economic Forum 2021).

This racism against Indigenous Peoples and minorities is about the unequal distribution of power and resources (Pulido 2017). This plays out in various ways, including a common practice around the world, for example, which sees governments dump toxic waste on Indigenous lands. Often this is done in secret, without warning, or the people's consent (Dombey 2019). This is often done in their areas because these communities are less able to resist such practices. These types of events have numerous consequences including to further depress land and housing values in such communities which are already low. It also further locks them into these degraded areas as selling their properties becomes even more difficult. As I have noted elsewhere:

> Whilst racism and elitism undercut social welfare projects, the white settler population has continued to grow generally richer, whilst for Native-Americans, whose

> socio-economic circumstances have devastated that population, opportunities have stagnated, despite technically having legal equality. (Sarkin 2019)

As others have noted, "Native American tribes face serious and challenging environmental issues that affect their health, way of life, and economic well-being" (Dornan and Hearn 2022: 1). Thus, for example, the effect of the COVID pandemic on these communities was higher than that for other communities, and proportionately three times more Indigenous Americans died than people in the White community (Crepelle 2021a).

One consequence of the historic racist treatment of Indigenous and other minority groups is an intensification of resource conflicts (Menton et al. 2021) around the world (Deloria 2003; Temper and Martinez-Alier 2013), including in the United States, where they have become common (Adair et al. 2009). For example, the Navajo community in New Mexico has suffered various environmental injustices, including the effects of uranium mining, for decades (Atencio et al. 2022). This is not uncommon, as there is a long history of Indigenous lands being subject to resource extraction operations by outside operators (Johansen 2023). It has also been found that "much of the world's nuclear industry has been sited on or near Native lands" including reservations, treaty-guaranteed and sacred lands (LaDuke 1999: 97). The Navajo saw their water sources being depleted, and what water remained was significantly contaminated (Gilles 2020). The Navajo have experienced massive environmental pollution leading to enormous negative health consequences because of the limited restrictions placed on the uranium mining industry (Nwankwo 2022). Despite this, for many years the companies that caused these problems were able to escape responsibility and liability. This has caused much conflict. This came to a head in 2024 over the transport of uranium over the land of the Navajo.

These issues are not unique to the Navajo. They have had a long vintage and continue in many places today. In this regard, it has been stated that:

> Americans' fundamental belief in their divine Providence, coupled with the construction of Indigenous peoples' racial inferiority, justified their settlement of stolen lands and continues to legitimize the United States' history of genocidal violence in the name of progress … The ideological tenets of manifest destiny that were popularized two centuries ago have not disappeared, but have merely changed and adapted to modern times, continuing to affect the lived reality of Indigenous nations. (Freiin von Bibra et al. 2022: 23)

The treatment they receive today is a continuation of what was meted out to them in the past (Buiquy 2022). The "crude mistreatment of the tribes across America has continued to undermine Indigenous wealth and empowerment, leaving many Native people without proper housing, education, running water, healthcare, and telecommunications services" (Buiquy 2022: 3).

It has been recognized that the "poor conditions on tribal lands have made economic development more difficult than in any other parts of the United States" (Buiquy 2022: 3). These conditions include a lack of education, housing, municipal infrastructure, and access to resources, including water (Lively 2021). In this regard, it was reported in 2017 that about 12 percent of Native American houses lack safe water or basic sanitation services, which is 20 times the average of other households across the nation (Buiquy 2022). One research analysis indicates that one Indigenous group, the Navajo Nation, had a COVID-19 infection level that was 3.5 times higher than that of White Americans, and Indigenous Peoples were dying of COVID-19 at twice the rate of the latter (Buiquy 2022). The situations of many Indigenous groups have become more dire during the last few years, in part because of the COVID-19 pandemic, which hit these groups hard (Buiquy 2022). It added to the fact that "Indigenous communities continue to endure disadvantaged living conditions, leaving their people without adequate resources" (Oladipo 2021). They are the most indigent group in the country, with about a quarter of Indigenous People living in poverty (Crepelle 2023b).

While processes to reverse such problems exist, such as programs that support minority owned businesses, these are under attack because of conservative efforts to reverse affirmative action initiatives. As noted, there has been a weaponization of "wokeness," which has seen a push to undermine attempts to deal with historical injustices that minorities and Indigenous Peoples face (Hernández-Truyol 2023).

These issues have also been seen in the courts. For example, in March 2024 a federal district court judge in Texas ruled that the Minority Business Development Agency, which had existed for more than half a century, had eligibility parameters that violated the Fifth Amendment's equal protection guarantees (*Jeffrey Nuziard v. Minority Business Development Agency* 2024). The judge, Mark Pittman, appointed by Republican President Donald Trump, found that the rules assume that minorities are inherently disadvantaged and therefore discriminate against others. He agreed with White business proprietors that the eligibility restrictions victimized them noting that the:

> racial presumption fails strict scrutiny and thus violates the Fifth Amendment's equal protection guarantees. While the Agency's work may help alleviate opportunity gaps faced by MBEs, two wrongs do not make a right. And the MBDA's racial presumption is a wrong. "Legislation should never be designed to punish anyone." While the Agency may intend to serve listed groups, not punish unlisted groups, the very design of its presumption punishes those who are not presumptively entitled to MBDA benefits. (*Jeffrey Nuziard v. Minority Business Development Agency* 2024)

Judge Pittman enjoined the Agency from using race or ethnicity as a basis for assessing whether support could be provided by it. The judge had no time for precedent or the fact that this process had been running for more than half a century. He wrote: "What became the modern MBDA was set in motion this very day fifty-five years ago. To the extent the MBDA offers services pursuant to an unconstitutional presumption, that's fifty-five years too many. Today the clock runs out. 'Yesterday is not ours to recover, but tomorrow is ours to win or lose'" (*Jeffrey Nuziard v. Minority Business Development Agency* 2024).

This case continues the trend of conservative reversals of progressive programs, which began during the tenure of President Nixon. In contrast to Judge Pittman's actions, the Biden Administration increased funding for, and expanded and made permanent, the Minority Business Development Agency through the 2021 Infrastructure Investment and Jobs Act.

What is significant is that this case is a portent of what is to come, as other similar processes will be attacked to reverse them. For example, in 2023 the US Supreme Court ruled that affirmative action programs were unconstitutional in universities and other such institutions (*Students for Fair Admissions Inc* 2023). Another case disallowed an attempt by the government to provide minority farmer loan forgiveness as part of an initiative meant to remedy centuries of racial inequality in the agricultural sector (*Wynn v. Vilsack* 2023).

As noted throughout this study, the inequalities between groups in the United States is partly a legacy of the past, which saw extensive, targeted racist-based deprivation for some. Preventing affirmative action programs allows these inequalities to remain and continue to undermine the health and well-being of the affected groups (Bray 2021).

POLITICS AND CLIMATE CHANGE AND ENVIRONMENTAL JUSTICE MATTERS

There is an intimate connection between the political party in power in a particular government, and its leader's outlook, and how issues related to climate change, climate justice, environmental justice, and the protection of minorities, are addressed. This is particularly so in the United States. However, the extent of the effect on each of these matters in a country is specific to that nation. It depends on the political trends over time in each country and the extent to which such issues are at the forefront of a nations' attention. This can be seen, for example, in Brazil, where data produced by the Brazilian space agency showed that in the year that Lula da Silva became president again (2023), the rate of deforestation in that country decreased by half compared to the year before, and was the lowest rate in the previous five years, when President Bolsonaro was in power (Santos 2024). It was during President Bolsonaro's regime that Indigenous Peoples suffered significant violence (Gonzalez 2023) and the dire consequences of environmental and climate injustice (de Belmont 2024; Rapozo 2021). This is indicative of the different levels of importance each leader attributed to environmental protection and Indigenous rights.

In the United States too, there is an intimate connection between climate change issues and politics. Environmental issues in general, and environmental injustice specifically, have been on the US political agenda for some time. Progress on these issues, or the lack thereof, has depended a great deal on which political party was in power over the years (Kashwan 2023). To some extent, the position of the parties reflects voter sentiment on the issues. However, a survey conducted in 2023 saw the number of Americans concerned by climate change issues rise. Now those "Alarmed" or "Concerned" make up 56 percent of those polled. Only about a quarter of Americans are "Doubtful" or "Dismissive" about climate change (Leiserowitz et al. 2023). As far as climate justice is concerned, knowledge about it is limited. One survey found that two-thirds of Americans had not heard of it at all (Carman et al. 2022). Another survey in 2024 found that during the election year for president, the issue of climate change only ranked 19th out of 28 issues (Leiserowitz et al. 2024).

Again, in this section, the review focuses largely on the national level and not what has occurred in the various states. A lot has happened as far as both climate issues and environmental justice, but some states have done a lot more than others, such as California, which enacted most recently The Safe Drinking Water, Wildfire Prevention, Drought Preparedness, and Clean Air Bond Act of 2024. It is not the intention to deal with these laws and processes at state level, because that would be too much to deal with due to the many issues involved in each state.

Overall, after reviewing how environmental justice issues have been dealt with at the federal level by three successive administrations ending with the Trump Administration, one author wrote that the United States has:

> gone from an administration that paid little attention to Title VI administrative claims, ignored the disproportionate impacts of climate change on low-income communities and communities of color, rejected the principle of differentiated responsibility for climate change, and signaled no real commitment to EJ; to an administration that recognized the failings of the Title VI program and sought to remedy them, explicitly identified climate change as an EJ issue, embraced the principle of differentiated responsibility, and generally signaled a clear commitment to advancing EJ; and then back to an administration that has failed to carry forward the momentum built around the Title VI process, prefers to think of climate change as impacting "all people" rather than any particular group(s), explicitly rejects the principle of differentiated responsibility, and signals a flagging commitment to EJ overall. (Soloway 2021: 10053)

These issues have been on the table to some extent for decades. As early as 1994, President Bill Clinton signed an Executive Order to deal with environmental justice for vulnerable groups (United States Environmental Protection Agency 1994). During the Obama Administration, much was done to make progress on dealing with climate change (Craig 2022). In spite of this, many problems and conflicts concerning environmental justice continue to exist in the United States.

While the Democratic Party has sought to address climate change issues and ensure environmental justice for everyone, these issues have had less support during Republican party presidencies. This can be seen in the fact, for example, that the United States has not ratified the 1992 UN Convention on Biological Diversity because of Republican Party opposition to it.

It was, however, during the Trump presidential era that much was done to reverse prior gains related to environmental justice matters (Outka and Warner 2019). This is because Trump believes that climate change is a hoax. His view on this is seen in a 2024 interview where he stated:

> You know, when I hear these poor fools talking about global warming. They don't call it that any more, they call it climate change because you know, some parts of the planet are cooling and warming, and it didn't work. So they finally got it right, they just call it climate change. They used to call it global warming. You know, years ago they used to call it global cooling. In the 1920s they thought the planet was going to freeze. Now they think the planet's going to burn up. And we're still waiting for the 12 years. You know we're down almost to the end of the 12-year period, you understand that, where these lunatics that know nothing, they weren't even good students at school, they didn't even study it, they predict, they said we have 12 years to live … These poor fools talk about global warming all the time, you know the planet's going to global warm to a point where the oceans will rise an eighth of an inch in 355 years, you know, they have no idea what's going to happen. It's weather. (McKibben et al. 2024)

This has been a long term view of Trump. Thus, very shortly after he became president, Trump declared that he would take the United States out of the Paris Agreement. He appointed Scott Pruitt, "a prominent skeptic of climate science and an ardent foe of government action on climate change" (Malakoff 2016) to lead the Environmental Protection Agency (EPA) (Malin and Ryder 2018), an individual who had himself sued the EPA 14 times to limit the regulations the Agency had issued previously (Malin and Ryder 2018). The rollbacks that occurred then transpired across many environmental areas. For example, the Keystone pipeline was pushed through despite the attempts by affected groups to halt its plowing through sensitive areas (Klass 2021; Rulli 2021; Deloria et al. 2018). This decision was later reversed by the Biden Administration to promote a more sustainable approach to the environment (Baicker-McKee 2022; Cha et al. 2022; Spears 2022).

During the Trump era, much was done to block research on climate issues across the US government, and a host of steps were taken to reverse progress achieved in the past (Friedman 2020). In fact, former President Trump proved to be a major supporter of increased fossil fuel production. He sought to protect the fossil fuel industry including coal, but to no avail in that coal usage for energy purposes has kept on being reduced. In a range of areas, he sought to reduce or eliminate environmental protection policies and regulations, leaving the environment vulnerable to damaging industrial activities such as mineral extraction. It has been reported that he reversed more than 100 environmental regulations (Popovich et al. 2021). This caused much consternation and caused difficulties for state and local administrations in their processes to adapt to climate change (Craig 2022). Thus, much was done during the Trump presidency to roll back environmental protections adopted before him. To deal with the possibility of a return by Trump in 2025, various states set in motion actions to try and reduce his potential negative effects on steps previously taken to deal with climate change. California in particular, probably the state that has taken the most actions to deal with climate issues (Franklin and MacDonald 2024), did a lot in 2024 to attempt to guard against a potential Trump re-election and the effect that that would have on moves to reverse steps previously taken on climate change in the state.

THE BIDEN ADMINISTRATION AND ENVIRONMENTAL JUSTICE AND CLIMATE CHANGE

This section provides a snapshot of some of the issues that occurred during the presidency of Joe Biden as far as environmental justice and climate change issues. This examination of what was done in the United States during this period does not deal with all of that was done and does not deal with those developments in detail. The idea is to give an overview and see what was done so the lessons can be learnt for other places. It does need to be noted that he and his party were frustrated in their attempts to achieve more by the razor thin majority that they had in the Senate and because one senator, Joe Manchin, often did not support their agenda. The second half of Biden's tenure saw the Republicans take back the lower house, which saw their legislative agenda shelved.

Nonetheless, during his four years, Joe Biden's administration did act on both environmental justice matters and climate change. While the two issues are interlinked, often publicly the focus is declared as for one or the other.

However, dealing with both these issues was of such importance for him, that at the end of January 2021, just a week after being sworn in, he issued an executive order (The White House 2021a) dealing with climate change, which also touched on environmental justice (The White House 2021a). He specifically noted that his administration would deliver on environmental justice, saying: "We must deliver environmental justice in communities all across America" (The White House 2023a). He saw the interrelationship between climate change and environmental justice when it was noted that:

> Every person must have clean air to breathe; clean water to drink; safe and healthy foods to eat; and an environment that is healthy, sustainable, climate-resilient, and free from harmful pollution and chemical exposure. Restoring and protecting a healthy environment – wherever people live, play, work, learn, grow, and worship – is a matter of justice and a fundamental duty. (The White House 2023a)

As noted previously, the lead agency to deal with environmental justice issues in the United States is the EPA. Under the Biden Administration, it has already done much to promote environmental justice. Much was done symbolically, but also putting laws in place to take action on a variety of issues. For example, in 2021 Biden ordered Juneteenth to be a federal holiday. This day honors 19 June 1865, when it was announced that the Emancipation Proclamation would be enforced, freeing all enslaved African Americans.

While the issues will be taken up below, one piece of the process to develop strategies to deal with environmental justice was Biden's Executive Order 14096 of 21 April 2023, titled "Revitalizing Our Nation's Commitment to Environmental Justice for All" (White House 2023a). However, there are enormous tensions over this and other agencies, with the Supreme Court over time hollowing out its powers to regulate. This is having massive effects on the EPA's ability to deal with climate change but also environmental justice issues, as is discussed elsewhere.

As far as climate change is concerned, very shortly after he became president, Biden re-joined the Paris Agreement. The toing and froing on environmental matters in different administrations can be seen as far as that agreement is concerned. Thus, under President Obama, the United States joined it, under Trump the United States withdrew and then Biden re-joined it.

To further advance efforts to address climate change and its effects on vulnerable groups, the Biden Administration introduced a Climate and Economic Justice Screening Tool (CEJST) to determine which disadvantaged communities would be prioritized for certain federal investment programs to compensate for climate and environmental damage and improve the community's socioeconomic condition (Wang et al. 2023). These actions have, however, been criticized as insufficient because CEJST does not use race-ethnicity as an indicator in determining which communities are eligible for this assistance (Wang et al. 2023).

As far as environmental justice developments specifically are concerned, Biden did take the significant step of appointing the first ever Native American cabinet secretary with his appointment of Deb Haaland as Interior Secretary. This was part of an overall objective to achieve diversity in his cabinet. Thus, more than half of his cabinet were people of color and nearly half were women. There were many firsts in his executive, in that an openly gay person was appointed, as was an immigrant, and the first women to lead the treasury and national intelligence, besides others.

The appointment of Haaland was especially significant for the country, as the Interior Department oversees around 500 million acres of federal lands. Many of those acres are leased by its agencies for wind and solar farms, as well as for oil and gas exploration. In addition to providing scientific data regarding the effects of climate change, it also manages the nation's 574 federally recognized tribes, safeguards vulnerable and endangered species, and maintains the nation's national parks and wildlife refuges (Stolberg et al. 2024). It is important to bear in mind that US Geological Survey estimates that it is the burning fossil fuels, that are taken from federal lands, which accounts for roughly 22 percent of total carbon emissions in the United States. Thus, the role of these lands, and who manages them, are critical from the perspective of climate change, but also for a range of environmental matters, as well as for Indigenous Peoples (Stolberg et al. 2024). Thus, what happens on these lands are very impactful from a climate change perspective but also for various peoples who are dependent or have certain connections to them.

At the time of appointment, Haaland stated that:

> A voice like mine has never been a Cabinet secretary or at the head of the Department of Interior … I'll be fierce for all of us, our planet, and all of our protected land. (McIntyre-Mills 2022)

As can be seen later, the role she would play, as she announced at the time of her appointment, turned out to be true. However, her appointment also created a sense of greater diversity in such positions in government, and promoted further the issues of equality and inclusion. Clearly, inclusion was an important issue for Biden, and he wanted to set a standard in this regard.

To deal with the past discrimination and exclusion the Department of the Interior had various initiatives. Thus, for example, in June 2021, Secretary of the Interior, Deb Haaland, set out the Federal Indian Boarding School Initiative to acknowledge the problematic legacy of federal Indian boarding school policies. As a result, an apology was offered to Indigenous People by

President Biden for these policies in 2024 (The White House 2024a). This was offered because:

> Between 1819 through the 1970s, the United States implemented policies establishing and supporting Indian boarding schools across the nation. The purpose of federal Indian boarding schools was to culturally assimilate American Indian, Alaska Native and Native Hawaiian children by forcibly removing them from their families, communities, languages, religions and cultural beliefs. While children attended federal boarding schools, many endured physical and emotional abuse and, in some cases, died. (United States Department of the Interior 2022)

Another program specifically for previously disadvantaged groups was the EPA's Environmental Justice Thriving Communities Grantmaking Program. This saw $600 million being given to projects tackling climate and environmental justice issues in those communities.

As far as climate change issues are specifically concerned, there was a mixed approach to deal with Indigenous Peoples generally. In this regard, it has been stated that:

> Some of the agencies' climate action plans … addressed engagement and consultation with tribes and Indigenous Peoples to understand climate impacts and collaborate on solutions, support opportunities for Indigenous knowledge to guide adaptation, evaluate implementation to avoid maladaptation or disproportional impacts, and to make assistance and research opportunities accessible. Other agencies largely did not outline how they would engage with tribes … Others referenced tribes but did not emphasize their unique roles as autonomous governments or holders and generators of Indigenous knowledge … Tribes and Indigenous Peoples are key partners for adaptation solutions in pathways that should uphold their sovereignty. (Von Hedemann et al. 2023)

Thus, inclusion, in the general sense, is still necessary to achieve. It is also, however, about the inclusion of Indigenous knowledge. As has been noted: "climate adaptation cannot be effective if it is not ethically co-produced with Indigenous Peoples and IK-holders (IK means Indigenous Knowledge)" (Caretta and Morgan 2021).

INFLATION REDUCTION ACT OF 2022 AND OTHER STEPS TAKEN

The Inflation Reduction Act (IRA) was a landmark piece of legislation, which other countries can learn from and can replicate some of its provisions. It is designed to allow the United States to mitigate and adapt to the impacts of climate change (Daly et al. 2024). However, it also integrates dealing with climate change and dealing with environmental justice issues simultaneously.

Whilst this is clearly a goal, it is implicit in the provisions of the law rather than being explicit.

The role of politics can be seen in that not a single Republican supported its passage into law, and many of them have tried to overturn its provisions. It is important to remember that Kamala Harris, as vice-president, cast the tie breaking vote to pass the law, indicating her important role in getting the law passed. It also needs to be noted that much has been done to undermine the law politically, with Republican states taking various steps to reverse the steps taken, including suing to stop the law. However, it seems that it is particularly those states that have benefited from the funds allocated to the climate change features of the law.

Critically, few Americans understood what the IRA law did in general, and what it does to mitigate and adapt to climate change (Yale Program on Climate Change 2024). This is surprising, given that 2024 was the year of a presidential election, when the issue of climate change ought to have been firmly on the ballot. It did not, however, largely figure in the election campaign, and the advances made by the administration to deal with those issues hardly made a mention. This is also a reflection of a lack of messaging generally to the general public, about what the IRA did as far as climate change, although many people were advantaged and received lots of financial assistance, as far as dealing with climate change is concerned.

Importantly, the IRA builds on the Infrastructure Investment and Jobs Act, which also dealt in part with climate change issues. Specifically, on climate change, the IRA promotes climate resilience through attempts to reduce dependence on fossil fuels, repurposing power infrastructure, and promoting electric vehicle charging projects. It allocates substantial funds to increase domestic manufacturing and provides for consumer refunds by way of tax credits for home electrification and electric vehicles and the expansion of sustainable energy use. It has provisions dealing with a range of other issues which have not been dealt with much anywhere else, including on carbon capture and storage, direct air capture and clean hydrogen. These are important technologies that assist with climate mitigation strategies.

So extensive was the funding given to various marginalized groups, that by May 2024 over 56,000 specific projects and awards in more than 4,500 communities in all the states had been funded in terms of this. Thus, approximately 165,000 miles of roads, 9,400 bridge repair projects, 450 port and waterway projects to fortify supply chains, over 300 airport terminal projects to modernize and expand terminals, over 4,100 projects to assist communities to become more resilient to threats like the effects of climate change and cyberattacks, over 1,400 drinking water and wastewater projects, and the replacement of up to 1.7 million toxic lead pipes were all made possible by the funding allocation (United States Environmental Protection Agency 2024e). In 2023, the administration began the process of amending an EPA rule to ensure that all lead pipes in the country would be changed within ten years, (CBS News 2023) and in 2024 finalized that new rule (United States Environmental Protection Agency 2024c), and provided $2.4 billion to begin lead piping replacement (Smolinski 2024). Thus, implicitly the interconnection between climate change and environmental justice was recognized. Understanding the connection between dealing with climate change and the ills of the past caused by environmental injustices, the funding also saw the installation of high-speed internet in approximately 65,000 Indigenous People's homes, and about 8,000 abandoned oil and gas wells were sealed off to combat legacy pollution and restore contaminated areas and therefore promote environmental justice. Over 500 projects for water recycling, storage, conservation, desalination, and other uses occurred as well as funding to over 200 states and Indigenous People to enhance the robustness and dependability of the US electric grid. The money was also used to fund 12,000 miles of middle-mile high-speed internet infrastructure (The White House 2024b). There have, however, been criticisms, although before much of the plans were created, that argued that:

> [p]olitically sensitive issues such as housing and tax policy, food security, land inequality, rural and indigenous resource extraction, or long-term national and regional spatial planning have not been part of the national adaptation conversation. (Shi and Moser 2021: 2)

Some of these issues have not been sufficiently addressed. As has been stated, various "institutional barriers exist in the United States that severely limit their adaptive capacities" (Craig 2022: 9). The effect of not doing this was noted in an official 2023 report to Congress. That report, the Fifth National Climate Assessment (NCA5), mandated in law to be undertaken every four years, noted that:

> Despite an increase in adaptation actions across the country, current adaptation efforts and investments are insufficient to reduce today's climate-related risks and keep pace with future changes in the climate. Accelerating current efforts and implementing new ones that involve more fundamental shifts in systems and practices can help address current risks and prepare for future impacts. (US Global Change Research Program, 2023)

Usefully, however, in 2023 it was found that double the number of cities had adaptation plans than had ten years before (Zoll et al. 2023). However, at the state and local levels many problems exist. These include a lack of uniform planning and prioritization. It has been found that plans are unequally dispersed amongst US cities, with race an important factor in this regard (Liu and Fan 2023). Often adaptation strategies have not included equity standards sufficiently. Part of the reason is that:

> [p]eople of color, indigenous and immigrant communities, and other minority groups often face critical barriers to involvement in community planning due to historical disenfranchisement and have fewer resources for adaptions. (Lioubimtseva 2022: 2)

For example, it has been argued that "urban policymakers should engage with local grassroots groups and advocacy networks to better identify and broaden contextually appropriate equity and justice goals, facilitate internal institutional arrangements that embody these normative visions and criteria, and design planning and policy benchmarking tools to ensure that equitable plans are grounded in local needs and implemented consistently and inclusively" (Cannon et al. 2023: 30).

What can be said is that the law and the way its funds are being used recognizes the interconnection between environmental justice, climate, and climate justice. The Act directs community-based investments to improve resilience and sustainability in minority communities. However, whether this law will be sufficient to make a meaningful dent in the problems that climate change is causing remains to be seen, particularly in light of the backlash in the United States to dealing with these issues and the economic imperatives that counter some of these efforts, which are discussed below.

Other Steps Taken by the Administration

Besides, the IRA, Biden introduced several other measures to help address environmental justice issues. He launched several initiatives to alleviate the uneven burden that Indigenous and minority groups bear from environmental threats (Robert 2022). One of his executive orders required "Every federal agency … [to] take into account environmental and health impacts on communities and work to prevent those negative impacts" (The White House 2021b). In this regard, he made sure that federal agencies improved their efforts (Bachtel et al. 2021).

A further initiative he took to advance the rights of LGBTQ+ people, was an executive order that he issued in 2021 to attempt to set in motion an educational environment free from discrimination based on sex, including sexual orientation or gender identity (The White House 2021c).

Another set of new regulations in April 2024 sought to regulate power stations. These sought to reduce the air, water, and waste pollution emanating from fossil-fuel-fired power plants. This was done because fossil fuel power stations accounted for 90 percent of the top 50 pollutants. 45 power plants collectively contributed 27 percent of the nation's greenhouse gas emissions from the production of energy. Nearly 126 million metric tons of greenhouse gases were released in 2022 by just nine power stations. Therefore, new rules were enacted to set new greenhouse gas standards for such power stations, updating mercury and air toxics standards for coal power stations, limiting effluent in the wastewater of those coal power stations, and better regulating legacy coal ash sites at power stations that had already closed. Another regulation in May 2024 dealt with reporting on emissions from such facilities (United States Environmental Protection Agency 2024g). These new regulations should have had important positive climate and environmental justice effects, but also be critically important for the people who live near these facilities. Crucially important is the regulation on closed plants, which together with other disused facilities such as mines continue to cause many problems for the residents nearby. It does depend on whether they survive, as is generally the case with all these initiatives, into the second Trump Administration. Even before the end of Biden's term, there were legal challenges to many of the regulations. Regardless, they are significant from a comparative point of view for other countries to see what is possible to deal with such matters.

What can be seen is that the steps taken, and the developments by Biden on dealing with climate change, build on Biden's Justice40 Initiative. That provides vulnerable communities with 40 percent of the benefits from federal investments in clean air, water, flood avoidance, soil clean-up, and other environmental works. This is useful as "holistic climate policies can explicitly lay the groundwork for connecting climate-change mitigation and adaptation with community building, economic opportunity, and political power in ways that advance racial and economic justice and reduce health inequities" (Morello-Frosch and Obasogie 2023: 943).

President Biden also created a 25-member White House Environmental Justice Advisory Council (Cha et al. 2022) and established a White House Office of Environmental Justice (The White House 2021d; Beard et al. 2023).

A further initiative with an environmental justice goal was the creation of regulations designed to end the use of asbestos. This is because it continues to be used in various industries in the United States despite being a major health hazard causing cancer and other problems (United States Environmental Protection Agency 2024a). The environmental injustice issues related to asbestos use have been known about for decades (Qian 2023). Despite this, 40,000 Americans die every year because of the effects of asbestos (American Public Health Association 2019; Mir et al. 2023); the final rule that was issued phased out its use for certain types of asbestos (others are allowed to be imported with permission) over 12 years rather than the two years the draft rule proposed. This is problematic considering the dramatic negative health effects that asbestos has. These effects have an even greater effect on African Americans (Juon et al. 2021) and therefore phasing it out over a shorter time would have had a more significant environmental justice consequence on rates of lung cancer, for example (Markowitz 2015), which afflicts more African Americans (Juon et al. 2021; Wortzel et al. 2021) proportionally (Richards et al. 2017). This is exacerbated by low health insurance rates amongst this group, but also other minority communities (Lee et al. 2021; Bolibol et al. 2023). Thus, asbestos and other dangerous substances continue to cause health problems that have disproportionate effects in neighborhoods where minorities reside (Hull 2022). However, the ability of industry to lobby (Meng and Rode 2019) and push back on proposed regulations (Lantushenko and Schellhorn 2023) can be seen in a variety of issues (Schnell 2024). However, the climate lobbyists are being successful to some degree under the Biden Administration and elsewhere (Böhler et al. 2022).

Another environmental justice initiative, with also positive implications for dealing with climate change, occurred in April 2024 when the EPA issued a final rule (United States Environmental Protection Agency 2024f) to drastically cut harmful air pollution such as ethylene oxide and chloroprene that emanates from chemical facilities, such as those in cancer alley, mentioned above. To ensure the capture of the best possible information about the health risks associated with exposure to air toxics (United States Environmental Protection Agency 2024c), and how the proposal would work in practice, the EPA carried out a first-of-its-kind community risk assessment. The assessment examined the risks to communities from air toxics emitted by large facilities located within approximately six miles of the communities, including facilities that would not be covered by the rule, to obtain a more complete picture of the air toxics-related risk in these communities. The study concluded that the proposed rule would reduce the number of people in those communities who have elevated air toxics-related cancer risk by 96 percent (United States Environmental Protection Agency 2024c).

The new rule should reduce toxic air pollution emissions by about 6,200 tons annually. By doing this, the EPA hopes to address environmental justice matters, and significantly lower the number of individuals who are at an increased risk of cancer because of exposure to toxic air pollution. The new rules should cut emissions of harmful chemicals by roughly 80 percent. The affected facilities must also monitor for harmful substances and disseminate the data to the public for the purpose of better informing and protecting populations that live close by (United States Environmental Protection Agency 2024g). These are useful first steps to curb air toxics in neighborhoods affected by this blight over decades (Fos, Honore, and Honore 2021).

Usefully, a further $4.6 billion is to be given in grants to states, Indigenous Peoples, cities, and others to set in motion climate action plans under the Climate Pollution Reduction Grant program (Craig 2022). While most states seem to be cooperating, the politicization of climate change issues has seen some states (Florida, Iowa, Kentucky, South Dakota, and Wyoming) push back on these grants (Henry 2024).

In September 2023, Biden announced the creation of an American Climate Corps which would employ 20,000 young people to work on climate issues including to "focus on equity and environmental justice – prioritizing communities traditionally left behind" (The White House 2023b). These are programs and processes that other countries should look to and replicate to deal with environmental justice issues in their countries.

Another initiative Biden introduced in 2024 was designed to reduce methane emissions by charging a fee to industry for methane emissions. He also halted new Liquid Natural Gas (LNG) export terminals because, while when burned LNG only produces half the carbon dioxide of other fossil fuels (Al-Breiki and Bicer 2023), it has a high methane content (Davidson 2024). These steps were sorely needed given that large oil and gas fields in the United States may be emitting three times as much planet-warming methane gas as had previously been officially estimated (Sherwin et al. 2024). It is useful therefore that 200 countries agreed at COP28 to "substantially" reduce methane emissions by 2030.

Linked to the reduction of methane gases, but much wider in scope, was a strategy to reduce food wastage taken in June 2024 (United States Environmental Protection Agency 2024d). This was the administration's "National Strategy for Reducing Food Loss and Waste and Recycling Organics." It was announced to be a "part of President Biden's whole-of-government approach to tackle climate change, feed people, address environmental justice, and promote a circular economy" (United States Environmental Protection Agency 2024d). Thus, the integration of tackling climate change, as well as dealing with environmental justice issues, can be seen. The Strategy provides incentives and support for reducing food loss, preventing waste, and recycling. It does so in several ways to change the behavior of business and consumers, and by promoting research. A way to redirect food not consumed to others should be a goal, however.

To overcome the legacy of past discrimination and to deal with climate change, in March 2024 the Environmental Protection Agency announced a plan to allocate $20 billion to provide loans to incentivize the use of clean energy, predominantly to disadvantaged communities. This is a significant strategy for mitigating climate change, as residential energy use accounts for approximately 20 percent of greenhouse gas emissions in the United States (Wang et al. 2024). The clean energy initiative is particularly noteworthy because it targets historically marginalized groups, including minorities and Indigenous Peoples. This is critical, as these communities often spend a disproportionately large portion of their incomes on energy bills due to inefficient housing (Kwon et al. 2023). Many are unable to afford these costs (Memmott et al. 2021), and their energy burdens are typically higher than those of the general population, largely because of the systemic housing inefficiencies that leave them especially vulnerable (Hernández 2015). Additionally, these communities are more likely to rent rather than own their homes, which often excludes them from programs designed to improve housing resilience to climate change (Crago et al. 2023). As a result, greater focus is needed on increasing energy – efficient housing, retrofitting homes with solar panels, improving insulation, and installing efficient heating and cooling systems, such as heat pumps (Edwards et al. 2024). The IRA addresses these concerns by offering tax credits for upgrades, including more efficient doors, windows, insulation, and heating and cooling systems. It also provides funding for home energy assessments to improve efficiency. Similarly, the Low Income Home Energy Assistance Program (LIHEAP) helps families by providing financial assistance for energy bills, weatherization, and home repairs. While these government initiatives represent important steps toward addressing environmental and climate-related injustices, significant challenges remain (Graff et al. 2023). One of the primary obstacles is effectively reaching the most affected communities, ensuring that they can access and benefit from these programs.

ONGOING CHALLENGES IN DEALING WITH CLIMATE CHANGE AND ENVIRONMENTAL JUSTICE IN THE UNITED STATES

There are, however, a range of challenges when it comes to dealing with climate and environmental justice issues. One of those is the US reliance on fossil fuels. Despite President Biden's efforts to reduce the US reliance on these fuels, production in fact rose after it had plummeted during the COVID-19 pandemic. By 2023, production achieved record levels, reaching 13.2 million barrels a day, even higher than that of Russia and Saudi Arabia. At least 20 oil fields received development approval in 2023 as well, allowing for the extraction of the equivalent of 8 billion barrels. Companies hope to obtain approval for almost four times as many additional fields by the end of the decade. While this is a global issue, 40 percent of new oil production will be in the United States (Zimmerman 2024).

This increased production is at least in part an effort to tackle high inflation, which had seen the price for consumers of oil rise dramatically. It is in fact expected to rise even further (Krauss 2023). There was, therefore, a duality in the Biden Administration's position on climate change and fossil fuel. While, as noted above, there have been administration attempts to bolster production, at the same time steps have been taken to make it more expensive to extract such fuels and at times limit its extraction. This can be seen in a flurry of activity to deal with fossil fuel production and protect various sites across the United States. This was seemingly being done before the elections for the American presidency in November 2024 to ensure that some steps were taken to deal with those matters if Trump was elected.

These steps were taken because of the fear that if Donald Trump was re-elected, he would begin "one more hairpin turn in a zigzag approach to environmental regulation in the United States, a pattern that has grown more extreme as the political landscape has become more polarized" (Davenport 2024). The concern was that he would do much to reverse the gains made in recent times on climate and environmental justice issues. For that reason, some steps were taken across the country, particularly in a number of more Democratic orientated states, but also at the federal level, to "Trump-proof" some of the climate actions that had been taken. Thus, for example, to protect the integrity of the science processes, and to avoid future meddling, particularly as far as the research dealing with climate issues, the Biden Administration set out various processes to do so (The White House 2021e). Much was done by the administration during its tenure to try and exclude future interference in climate research, and similar matters (Tollefson 2023).

In this regard, these steps understood that if Trump was re-elected, the writing was on the wall as far as what he would do as far as climate change processes were concerned. This can be seen partly in the fact that the 16-page 2024 Republican Party platform has no mention of climate change. Rather it pronounced "Drill, Baby, Drill" and emphasized drilling for fossil fuels, "terminating the Socialist Green New Deal" and ending various regulations (Republican Party 2024). Those were meant to signal issues around climate and other similar matters.

The Democratic Party, on the other hand, in a press release noted that the goal of Democrats, as far as dealing with climate change in the future was concerned, would be about scaling:

> up solar, wind, and geothermal projects made possible by the Inflation Reduction Act and invest in clean energy R&D to build America's legacy as a nation of new frontiers and possibilities. President Biden will expand the clean energy workforce, with a goal of tripling the size of the American Climate Corps by the end of this decade and making sure the clean energy and manufacturing jobs of tomorrow are open and accessible to communities that have been too often left out in the past. We'll also eliminate tens of billions of dollars in unfair oil and gas subsidies and hold oil and gas executives accountable for potential collusion or price gouging. We will continue to implement the president's Executive Order on Environmental Justice for all. (Democratic Party 2024)

Regardless, Biden in 2024, before the elections, raised the royalty rates and bond rates that fossil fuel companies were to pay to operate on public lands.

These costs were raised for the first time in more than 100 years (Magill 2024). Another step taken to reduce the use of fossil fuels was the introduction of two rules to reduce pollution from cars and trucks by limiting their emissions. Steps were also taken to reduce emissions from fossil fuel powered plants (United States Environmental Protection Agency 2024b). However, this rule is quite limited, as it targets mercury and coal ash rather than all types of pollution including that stemming from methane. There was also criticism that part of the solution that was being resorted to was using carbon capture storage technology, which some argued is problematic (Deep South Center for Environmental Justice 2024). Carbon capture is deeply problematic when it is sought to be used to allow fossil fuel usage to continue at high levels.

Indicating the duality of the Biden approach, it must be noted that on the negative side, in 2023 new fossil fuel drilling was permitted in the 23-million-acre National Petroleum Reserve (NPR) in Alaska (United States Department of the Interior 2023a). Later, however, under pressure from environmentalists, the administration did announce new rules to protect parts of the area. Executive action removed from the NPR-Alaska over 2.8 million acres of the Beaufort Sea in the Arctic Ocean nearshore, designating them as permanently off-limits to future oil and gas leasing. This was done to protect it for polar bears, whales, seals, and other wildlife and consolidated what President Obama had done in 2016 to withdraw the Chukchi Sea Planning Area and most of the Beaufort Sea from access to companies that wanted to extract fossil fuels (United States Department of the Interior 2023b). This was done to further mitigate climate change.

To further protect the environment, the administration, in April 2024, instituted new rules to provide further protection to ten percent of US lands by allowing "mitigation leases" and "restoration leases" to be put up for auction to organizations that intend to preserve or repair public lands. This rule gives the Bureau of Land Management (BLM) the tools it needs to support responsible development, preserve significant wildlife habitat and landscapes, support public lands' resilience to climate change, and more. It also helps to identify the distinctive cultural and natural resources found on this land. The regulation also applies the current principles of land health to all the Bureau's programs, and sets out the procedures for designating and safeguarding Areas of Critical Environmental Concern (ACECs). The regulation will support the BLM's efforts to maintain land health protection while overseeing the development of clean energy projects and outdoor recreation on public lands (United States Bureau of Land Management 2024).

Regulations also provided more detail on the NPR in Alaska (United States Department of the Interior 2023b), that would apply to new oil production in the reserve (United States Bureau of Land Management 2022). This rule restricts future drilling and production on parts of the Alaska's 23-million-acre

NPR (United States Bureau of Land Management 2024). The rule provides that the BLM will issue no more oil and gas leases on about half the reserve. The rule further provides that where there are leases a more thorough evaluation will occur in the future before allowing more drilling. This was done also to protect the 40 Indigenous groups who still depend on the resources found in the Reserve for their daily needs, including caribou, birds, fish, and fauna. The Secretary of the Interior, as noted above herself an Indigenous Person, stated that:

> There is no question, using the best available science and incorporating Indigenous Knowledge practiced over millennia, that these decisions will help biological, cultural, historic and subsistence resources, safeguarding the way of life for the Indigenous people who have called this special place home since time immemorial. (United States Department of the Interior 2024)

Thus, having a person from the groups affected is important to put their issues more on the agenda. Thus, the needs of climate change and environmental justice were dealt with in parallel, indicating that both issues ought to be dealt with together to ensure that greater sustainability and biodiversity protection is attained.

The issue of promoting diversity has also been a goal. The Biden Administration also adopted several rules to protect various wildlife species. This was done partly by widening the definition of habitat. This allowed a number of other species to be protected, including wolves, the North American wolverine, the California spotted owl, and others.

On the issue of monuments, Biden designated two new national monuments in 2023. These were Avi Kwa Ame (also known as Spirit Mountain) in Nevada and Castner Range in Texas, which cover land that is considered sacred by Native Americans groups. Biden at the time noted that:

> Breathtaking deserts, valleys, mountain ranges, rich in biodiversity, sacred lands that are central to the creation story of so many tribes who have been here since time immemorial … it's a place of reverence, it's a place of spirituality, it's a place of healing and now it will be recognized for its significance as a whole and will be preserved forever. (The White House 2023c)

These two monuments will protect 514,000 acres. The Avi Kwa Ame is deemed a sacred burial ground by the Mojave, Chemehuevi, and Southern Paiute people but is also significant to other groups (Collins 2023).

Another step forward occurred in August 2023, when President Biden designated the Grand Canyon National Park in Arizona, covering about 4,000 square kilometers, a national monument. This objective had been sought for decades by American Indigenous peoples as well as by environmentalists. The designation protects a wide array of flora and fauna and forbids the mining of the uranium deposits there. President Barak Obama had wanted to do this but was unable to do so because of political opposition he faced at the time. He did later extend the protections accorded to the San Gabriel Mountains National Monument and the Berryessa Snow Mountain National Monument (Merenlender and Buhler 2021) in his endeavor to meet his pledge to provide protection to a third of the land and waters by the year 2030 (Simmons et al. 2021). So committed was Biden to extending these protections that even at the very end of his administration, in December 2024, he announced a new national monument in honor of Frances Perkins, who was the first ever woman appointed as a US cabinet secretary. Secretary Haaland also at the same time announced the creation of five national Historic Landmarks recognizing the historical role of women in the country (The White House 2024c). Even in January 2025, with just a few days left in his administration, Biden renamed the Pacific Remote Islands Marine National Monument as the Pacific Islands Heritage Marine National Monument. This was done to acknowledge the cultural heritage of Indigenous Peoples of the Pacific Islands (The White House 2025). He also created the Chuckwalla National Monument and the Sáttítla Highlands National Monument, protecting 848,000 acres in southern California's Eastern Coachella Valley. Thus, Biden continued to cement his legacy of providing for Indigenous Peoples nationwide and protecting the environment.

Another issue taken up was the return of Indigenous land taken during the colonization period. While cases have been brought to try and return land to its original Indigenous owners, they were unsuccessful. For example, in the 1956 case of *Federal Power Commission v. the Tuscarora Indian Nation,* the Supreme Court rejected land claims brought by the Tuscarora People. It has, however, been noted that:

> While *FPC v. Tuscarora* wasn't a watershed case for Indigenous land claims in American history, its existence helped to bring to life the later lawsuits in the 1970s and 1980s of the Oneida Indian Nation (414 U.S.661 and 470 U.S. 226). Those later cases were some of the few to grant rulings in favor of Indigenous nations' land claims in federal law proceedings. (Buettner 2022)

These issues will be taken up again later when the role of the Supreme Court is reviewed.

In the last two decades, there have, however, been processes that have helped build trust between the government and Indigenous Peoples and advanced land claims and other matters. In 2010, President Obama signed into law the Claims Resolution Act, allowing Indigenous groups to control their own tribal land leases, and the Land Buy-Back program (West 2016), both of which advanced efforts to provide restitution for land taken during colonialism, while helping to bring about both environment and climate justice (Racehorse 2023).

There have been some recent land returns as a result. The Onondaga had some of the land in New York taken from them during the colonization process (Oberg 2007) returned to them (Williams 2023) by the Biden Administration. This will have tremendous benefits for the group, including to ensure climate justice. The evidence suggests that the ills of climate change have been arrested to some degree on land controlled by Indigenous groups, but there is a long way to go to meet the challenges that climate change will bring over time (Liddell et al. 2021). It is clear, however, that dealing with climate change and providing some measure of climate justice can be attained by land being returned to these groups (Racehorse and Hohag 2023).

Biden also took steps in his final days in office to insulate large swathes of the ocean and coastal regions against drilling for fossil fuels. This occurred days before Donald Trump was inaugurated. This sought to prohibit new offshore drilling in 625 million acres along the entire eastern US Atlantic coast, the eastern Gulf of Mexico, and the Pacific Coast along California, Oregon, and Washington. He also sought to protect parts of the Northern Bering Sea in Alaska.

Overall, what is clear is that Indigenous groups, communities, and the environment would benefit significantly if other similar places obtained greater protection (Roberts 2023). More use should be made of legislation that permits the protection of land, such as the Antiquities Act of 1906, the Wilderness Act of 1964, and the National Historic Preservation Act (NHPA) of 1966. For example, the NHPA permits areas to be designated as a Traditional Cultural Property or Place· (Barcalow and Spoon 2018). Further studies, including consultations with Indigenous and other affected communities, should take place, and inventories of places in need of protection should be compiled. Other countries can learn from these processes as well, to find appropriate ways to achieve the best way forward so as to achieve many objectives, including cultural protection and environmental and climate justice, simultaneously.

The responses from environmentalists to these changes were positive, with an attorney from Earthjustice stating:

> It's no secret that the Reserve – a vast region of tundra and wetlands teeming with wildlife and globally recognized for its ecological value – has frequently landed in the crosshairs of the insatiable fossil fuel industry ... Today the administration has taken an important step to defend a cherished landscape from further fossil fuel development that would threaten these irreplaceable lands and waters and our climate. We applaud this move and call for even bolder action to keep the fossil fuel industry out of the Arctic, for the sake of the climate and future generations. (Earthjustice 2024)

However, taking this type of action does not always occur. Often, while there is a focus on climate change issues, and sometimes environmental justice issues, at the same time broader economic issues come into play. These issues include the effect on the economy, inflation, job creation or loss, when decisions are made concerning how to balance a range of economic imperatives with environmental issues concerning minorities and Indigenous Peoples. This can be seen in the fact that a court case dealing with parts of the Reserve not protected continues. This case was brought earlier to halt drilling in parts of the Reserve. In November 2023 a court in Alaska sided with ConocoPhillips and the Biden government refused to halt the drilling in the Willow oil drilling project in the western Arctic area of Alaska. The case was appealed by Earthjustice on behalf of Defenders of Wildlife, Friends of the Earth, and Greenpeace USA, with the Center for Biological Diversity and Natural Resources Defense Council on the basis that drilling in the region would have a catastrophic effect on the climate. A second case by the Trustees for Alaska, challenging Willow on behalf of the Sovereign Iñupiat for a Living Arctic and others, was also dismissed. The case was appealed to the Ninth Circuit (Earthjustice 2023).

Another issue, that affects these issues, is the rising consumption of electricity in the United States (Wilson and Zimmerman 2023). While this has critical implications for climate change, broader economic issues also come into play, such as job creation or loss, requiring a balancing of economic imperatives with environmental concerns. This issue is exacerbated by the rising consumption of electricity (Wilson and Zimmerman 2023). The massive needs for power production has meant that old coal and gas power stations lifespans are being extended, new power stations are being considered and, in some cases, already coming online. This will dramatically affect the goals of mitigating climate change (Ekins 2023). The goal of achieving clean energy and decarbonization (Depledge 2023) by 2035 (Bang 2021) seems unobtainable in that light. What is the case, however, is that renewables have increased threefold in ten years. However, they are not keeping pace with the increasing demand for electricity. To keep up with the increasing demand, power companies constructed many wind turbines, solar panels, and batteries in 2024, however, the usage of natural gas because of increased electricity usage reached all-time highs, while the use of coal only slightly decreased (Plumer 2025). The combined usage of solar and wind generation finally overtook coal as the main supplier of energy (Rhodium Group 2025). To deal with the increasing demand for electricity, what is being considered as well, is nuclear power stations, including reopening such facilities, such as Three Mile Island, in the state of Pennsylvania, which suffered an accident some years ago. However, the dependency on fossil fuels and the massive emissions they cause continues, and in some instances have increased. While emissions levels peaked in 2004, and have come down since then overall, with a substantial decline in 2023, in 2024, the US emissions overall only decreased 0.2 percent, far off the goals set (Rhodium Group 2025). Thus, more work is needed to tackle the issue of fossil fuel usage, and accelerate the move to renewables. That is unlikely because the Trump Administration will be in power in the United States and will not be interested in doing this and will in fact do the opposite. Tackling the issue will also depend on the 20 or more lawsuits that are being brought all over the United States suing fossil fuel companies (Sokol 2023). Various states have brought these cases arguing that the oil and gas companies have known for decades that fossil fuels have a large role in climate change and ought to be held accountable in this regard (Henricksen 2024). There are also private cases that are being brought (Ganguly 2024). It has been suggested that in certain cases there could be prosecutions (Arkush and Braman 2024). However, this depends on the appetite of the courts during a Trump presidency to rule favorably in such cases.

As far as mitigating climate change, the good news is that the growth in the use of renewables over the last decade has been phenomenal. While in 2014 solar sourced energy supplied less than half of one percent of the country's electricity, it now provides nearly 4 percent. During that time, wind sourced power increased from 4 to about 10 percent. When hydropower, geothermal, and biomass are included, approximately a quarter of the country's electricity was generated by renewable sources in 2023 (Climate Central 2024). However, poorer people, including minorities and Indigenous Peoples, are less likely able to afford this, and the take-up without government support has been limited (Edwards et al. 2024).

A challenge, however, was that efforts to address environment degradation are of necessity being undertaken in a colorblind way; the intent is to assist communities of color without designating them as such because using race to identify and aid such communities could lead to legal challenges that would thwart assistance efforts (Friedman 2022). As noted elsewhere, this is exactly the pushback that is generally occurring on matters of race and other identity issues. However, is it not possible to reverse the effects of centuries of discrimination without acknowledging the identities of the abused groups and admitting the ingrained racism that has resulted, and continues to result, in their being subjected to the worst effects of environmental degradation and climate change? As Robert Bullard noted: "When you look at the most powerful predictor of where the most industrial pollution is, race is the most potent predictor. Not income, not property values, but race. If you're leaving race out, how are you going to fix this?" (Friedman 2022).

However, a challenge for Biden is that there has been enormous pushback from conservatives in the country over these initiatives. Congress has not made the advances it could have because, as has been noted by one prominent environmentalist, author, and journalist: "political power has come to rest in the reddest, most corrupt parts of America. The senators representing a relative handful of people in sparsely populated western states are able to tie up our political life, and those senators are almost all on the payroll of big oil" (McKibben 2022). As suggested above, these issues are very politically contested, and the divide between those who favor taking more action on climate and environmental justice issues and those against is roughly the same as those who support the Democratic Party and those who support the Republicans. These divisions are stark and growing, and encompass issues such as immigration, rights of women and sexual minorities, and eliminating inequality through the use of diversity and affirmative action processes.

One result of this glaring divide is that these issues, including climate and environmental issues, are increasingly becoming part of the community discourse and significant and at times violently contested issues in state and local elections. Conservative states and local authorities are increasingly challenging a whole host of actions taken to mitigate or adapt to climate change and enhance environmental justice. The number of cases being litigated in the United States continues to grow. As has been noted:

> while "[m]any Indigenous peoples have been proactively identifying and addressing climate impacts," "institutional barriers exist in the United States that severely limit their adaptive capacities," including access to the funding necessary for successful adaptation. Increasingly, Tribes and other indigenous groups in the United States, as in other nations, are turning to litigation as a source of adaptation funding and capacity. (Craig 2022: 9, references omitted)

One court case in 2024, resulted in the Court siding with two oil and gas companies that opposed Securities Exchange Commission (SEC) regulations that require public companies to report the commercial risks they face from climate change (*Liberty Energy v. SEC*) These rules were actually significantly less demanding than the rules initially proposed by the SEC (Moss 2024). The SEC removed and weakened many of the proposed rules because of opposition from Republicans and various industries. These efforts at compromise were fruitless, as the new rules were challenged anyway and suspended by the Court.

As can be seen, efforts to address environmental and climate issues face a variety of problems that are difficult to navigate. As has been noted, there are "several key challenges that render planning in adaptation policymaking ineffective: lack of vision with a comprehensive approach, no agreement on solutions, not embracing all climate risks, and data deficiencies" (Eghdami et al. 2023). Probably the biggest problem is the political environment and the role played by Republicans to undermine any progress and in fact reverse any actions taken. However, much research is needed to determine what has occurred and what can be done to discover the truth about many industries who have hidden their roles in climate change (Gonzalez and Knox 2023). More needs to be done with such information to shame these industries and make them responsible for what they have done (Yadin 2023).

Trying to insulate the gains made during his term in office, at the end of 2024, the Biden Administration set new climate mitigation goals for the country, outlining steps the United States was expected to take to reduce its greenhouse gas emissions. This announcement was made despite knowing that Trump would likely attempt to reverse these climate objectives as soon as he assumed office (Plumer and Friedman 2024).

All these issues, and the ways in which they were addressed, represent significant positive steps that other countries can adopt to promote environmental justice, as well as climate change mitigation and adaptation. The fact that many of these initiatives may not endure and could be reversed by Trump does not diminish the major advances achieved or the scope of those projects.

THE ROLE OF THE US SUPREME COURT

Notwithstanding recent efforts by the Biden Administration to promote climate and environmental justice, the courts, especially the US Supreme Court also plays a critically important role in this area by establishing precedents for how such cases should be handled by American courts.

This section can, however, only touch on a small number of issues affecting the role of the courts on these matters. The role of the Supreme Court as far as climate change issues will only be touched on briefly because it is the broader role of the Court that is sought to be reviewed. The Court has played a major role in more recent years especially on a range of matters affecting environmental justice and various groups. It has played a critical role in deciding who decides what issues and how. Thus, for example, it has played a significant role recently by reducing the powers of various bodies and agencies (including both tribal authorities and government agencies such as the EPA) and giving itself and lower courts more authority to decide what powers other branches of government have. This has impacted the progress of environment-related matters, which is discussed below.

However, the Court's activities must be seen in a wider context as it has been going through a period of turbulence in recent times. This is in large measure the result of some of the controversial appointments to the Court that have been made recently, resulting in a Court that is much more conservative than it has been for decades if not for longer. The Court's legitimacy has been undermined because of the process by which these appointments were made, and because of the consequent rollback of certain well-established precedent cases such as the abortion decision of *Roe v. Wade* adopted by the Court in 1973. Thus, a case which had stood for five decades was overturned in 2022.

These issues are seen by some commentators, and by a segment of the wider public, as being reflective of political interference in what is supposed to be the neutral role that judges ought to play. In that context, some of the

Court's recent decisions are thought by many to reflect the political beliefs of the judges and are not based on US law. Thus, as was discussed above, who the judges are is crucial. Diversity and objectivity in appointments results in courts which have greater legitimacy and outcomes that consider and reflect the full range of viewpoints.

It does need to be noted that the current Supreme Court only has two people of color, and four women on it. This is surprisingly diverse in the American context. The problem, however, is that Judge Clarence Thomas (Black male) and Trump's three new appointees appear to have been chosen because of their extreme political leanings and their willingness to decide cases based more on their originalist reading (which allows them to adopt a conservative view) of the Constitution rather than on legal precedent. In fact, it can be argued that the originalism approach to constitutional interpretation masks political intent (Hausman 2024).

The issue of judicial appointments to the Court has been a major agenda item for both political parties for some years now. While the Democratic Party has sought greater diversity, that has not been an issue of importance for Republicans. Thus, President Biden, increased the number of appointments of women judges and judges of color (Means et al. 2019). In his campaign for the presidency in 2020, he vowed to appoint the first woman of color to the Supreme Court if he was presented with the opportunity (Johnson 2021); a promise he fulfilled by nominating Judge Ketanji Brown Jackson (Conklin 2022). Biden has also been nominating people of color and women in record numbers to the Federal Courts (Moyer et al. 2022). By the end of his term, Biden had appointed more people from a diverse background than ever before. Women made up almost two-thirds of appointees, while persons of color made up the majority. Many were non-traditional picks, as they were lawyers who had been public defenders, or worked in human rights or labor law.

The legitimacy of Courts generally depends on how they are viewed and whether they are deemed trustworthy by the public. The issue of diversity of the courts, while an important issue, has been taken over by a range of issues affecting the credibility of them. What is also significant is that a number of scandals have affected the courts, especially the Supreme Court, including judges receiving various types of gifts. However, the legitimacy of the courts has also suffered from a variety of attacks, that have sought to undermine US institutions in general. While much needs to be to done to reverse that trend, this is unlikely to occur in the short term.

What is clear is that, despite President Biden's efforts, the change in the composition of the Supreme Court over the last few years has moved it towards greater conservatism. This will lead to a decline in positive outcomes as far as environmental justice, and other issues of importance to Indigenous Peoples and minorities, are concerned.

What can be seen, in general, is that Indigenous Peoples have not had much success before the US Supreme Court. In this regard, in 2023, one of the newcomers to the Supreme Court, Judge Neil Gorsuch, wrote in *Haaland v. Brackeen* that as far as having their rights upheld in that Court: "Often, Native American tribes have come to this court seeking justice only to leave with bowed heads and empty hands" (*Haaland, Secretary of the Interior v. Brackeen* 2023: 21).

In a case that was handed down shortly after *Haaland*, *Arizona et al. v. Navajo Nation et al.*, where the Navajo community sought greater protection of their rights, the majority in the Supreme Court again declined to provide them relief. Again, Judge Gorsuch was scathing in his dissent (joined by three other justices), noting that: "To date, their efforts to find out what water rights the United States holds for them have produced an experience familiar to any American who has spent time at the Department of Motor Vehicles. The Navajo have waited patiently for someone, anyone, to help them, only to be told (repeatedly) that they have been standing in the wrong line and must try another" (*Arizona et al. v. Navajo Nation et al.*).

An area of contention that exists in the United States, which reflects attitudes towards Indigenous People, is the status of Indigenous law and institutions, including Indigenous courts. Although there has been some enhancement to the status of Indigenous People since the Nixon era, there is a long way to go (Crepelle 2019a). Sadly, there have also been areas of regression. One significant retrenchment recently has been the limiting of the status and role of tribal courts in the United States (Crepelle 2020a). It has been argued that the jurisdiction of these bodies and their ability to deal with matters concerning Indigenous People ought to be expanded (Florey 2013). This did occur to some extent in the 2021 case *United States v. Cooley*, when the Supreme Court seemed to accept tribal powers over non-members in limited contexts.

Indigenous People use several different types of community institutions, including "family and community forums, traditional courts, quasi-modern courts, and modern tribal courts" (Melton 1995: 126). However, their roles are limited. The roles of these institutions should be examined to determine which should be awarded greater jurisdiction to deal with the issues that affect the communities on their land. Limiting the jurisdiction to only dealing with issues between group members is problematic. This is, however, a complex area of the law, and the courts have interpreted when and how Indigenous institutions can deal with Indigenous issues.

If these Indigenous institutions are not provided with expanded jurisdiction and powers to deal with the panoply of issues that take place on their lands, or the right to deal with matters affecting the tribe, then the role of the ordinary courts needs to be reformed. The rule that where conflicts in jurisdiction occur, the case remains in the ordinary courts, needs to be reviewed.

The rules governing jurisdiction generally ought to be codified to simplify the determination of which court has jurisdiction. Legislation should ensure that Indigenous institutions have jurisdiction over the majority of cases related to or affecting Indigenous People or which have a territorial link to Indigenous land. This will provide these groups with a measure of self-determination regarding the issues that affect them most (Cirkovi 2007). In addition, there ought to be more reliance on the communities themselves to make decisions based on research conducted by their own people (Gewin 2021).

However, the Supreme Court has already demonstrated its intent to restrict the powers of Indigenous People on their own reserves. Despite this, there have been some advances in recent years in federal law, and in some cases, that recognized Indigenous rights, such as the reauthorized Violence Against Women Act (VAWA), which vested tribal courts with Special Domestic Violence Criminal Jurisdiction (SDVCJ) if they opted in on this and met a number of conditions (Hannon 2021).

There was, however, some hope of progress on this front in 2020 when the Supreme Court handed down the *McGirt v. Oklahoma* (*McGirt v. Oklahoma* 2020) decision, when the Court ruled that criminal cases that involved members of an Indigenous People where the crime took place on the lands of such a groups ought not be tried in the court of the state where the group live but in their own Indigenous courts. However, that hope was short-lived (Dornan and Hearn 2022). This relapse occurred in the 2022 case of *Oklahoma v. Castro-Huerta* (*Oklahoma v. Castro-Huerta* 2022). That case has set:

> … the stage for future sovereign conflicts (and likely tribal losses) by further overlapping state and tribal jurisdiction thanks to its broad assertion that reservations are geographically and politically part of the surrounding state. (Rusco 2022)

In this case, the Court restricted the powers of the Indigenous communities over crimes committed on their territory. In a narrow majority of five to four, the Court ruled that Oklahoma, and in fact all states that have concurrent jurisdiction with the federal government, "over crimes committed by non-Indians against Indians in Indian country, wiping away centuries of tradition and practice [this is because] … the court reversed the presumption against state jurisdiction, holding that unless Congress acts to preempt state jurisdiction, states can prosecute non-Indians for all crimes committed in Indian country." (Fletcher 2022). As has been noted, this case:

> represents further proof that this axiom of United States federalism and the Law of Nations is the root cause of the slow erosion of tribal sovereignty over time, … Castro-Huerta strongly suggests that tribal sovereignty will continue to be vulnerable until a solution is reached that results in either comprehensive sovereign separation as a result of respected tribal borders, or a selective division of authority

> between tribal, federal, and state government, … Anything less will continue the long-term war of sovereign attrition historically experienced by tribes. (Rusco 2022)

This is added to by another commentator who notes that this case has:

> … weakened Indigenous nations' self-government and refuted historic precedent that shielded them from "the arbitrary authority of states" by ruling that "state governments have the right to prosecute non-Natives for crimes committed against tribal members on reservation lands. (Freiin von Bibra 2022: 22)

The issue of jurisdictional conflicts between the Indigenous courts and state and federal courts remains highly contentious (Slaff 2023). It is in need of reform (Riley and Thompson 2022).

What is, however, clear is that the *Castro-Huerta* ruling illustrates the lack of respect that the modern Supreme Court affords Indigenous Nations and their judicial systems, emphasizing their ongoing subjugated status in the eyes of the US government (Freiin von Bibra 2022). This issue will be returned to later. However, this decision is not an aberration, but continues a trend of disrespect for these people, especially as far as their institutions are concerned. As was noted by Hannon even before the *Oklahoma v. Castro-Huerta* decision:

> The rationale of the decisionmakers at each of the stages of creation of the tribal court jurisdictional maze has been rooted in bias against indigenous people and the First Nations. Evidence of this bias and disdain for indigenous culture and history has been peppered throughout decisions rendered in the highest echelons of our judicial system, and in statements of the most advanced legislative body in this country. But this bias is pervasive. It is still present in the modern-day interactions between tribes and the federal government, in each of the judiciary and the legislature. (Hannon 2021: 257)

In fact, as Carlson noted in 2015, Native Americans (Indigenous People) lose in the US Supreme Court 75 percent of the time (Carlson 2015). The negative trend towards Indigenous rights may have continued in a case decided in 2023: *Brackeen v. Haaland*. In that case, the Supreme Court was called on to resolve questions surrounding the Indian Child Welfare Act. The Court had to decide whether on issues of adoption and fostering of Native children the Indian Child Welfare Act exceeds what is permitted by the Constitution (Benson 2022). It also had to decide if the Act unlawfully discriminates on the issues of race and puts the needs of the group above that of children (Sinha 2021). If the case had been decided against the group, it would have been a continuation of a trend (Fletcher and Singel 2021) that follows the decision by the Court in *Adoptive Couple v. Baby Girl* (*Adoptive Couple v. Baby Girl* 2013). However, the Supreme Court by 7 to 2 ruled that the Indian Child Welfare Act was constitutional. If this case had gone the other way, it would have had possible negative consequences for Indigenous Peoples and their rights (Roemer 2023). Their distinctive position in American law may be under threat in terms of what the Court may decide. Thus, as far as the United States is concerned, it has been argued that:

> Often perceived as guardians of minority rights, courts have not lived up to this reputation when it comes to Indian nations. As a result, Indian nations have optimistically turned to the political process, assuming that Congress can and will enact effective policies favorable to them. (Carlson 2015)

Thus, using the Courts in the United States may not advantage such groups over the next few years. This is the result of the Supreme Court's, and other courts where more conservative judges have been appointed under the Trump Administration, move towards conservativism and to the right, and therefore such groups may see their rights eroded even further.

The positions taken by the US Supreme Court on Indigenous People's rights has been mirrored in other areas where the Court has been reducing the power of federal agencies such as the EPA and the power of the US Congress to delegate more authority to the Agency. This was seen in the Court's decision of *West Virginia v. Environmental Protection Agency* (*West Virginia v. Environmental Protection Agency* 2022) handed down in 2022 (Smead 2022). In this case, the Court imposed significant limits on the powers of federal administrative agencies (Barsa and Dana 2024). Specifically, the Court held that the EPA did not have the requisite authority under the Clean Air Act to control greenhouse gas emissions that emanated from coal power plants. In a far-reaching finding, the Court held that when a "major question" exists concerning an issue that has economic and political significance, any action taken by an administrative agency is invalid without there being a clear direction on this issue by Congress. This decision has significant, wide-ranging ramifications (Gerrard et al. 2022). However, as Justice Kagan noted in her dissent:

> The subject matter of the regulation here makes the Court's intervention all the more troubling. Whatever else this Court may know about, it does not have a clue about how to address climate change. And let's say the obvious: The stakes here are high. Yet the Court today prevents congressionally authorized agency action to curb power plants' carbon dioxide emissions. The Court appoints itself – instead of Congress or the expert agency – the decision-maker on climate policy. I cannot think of many things more frightening. (*West Virginia v. Environmental Protection Agency* 2022 Kagan dissent: 32–33)

In response, at least in part, to the judgment, Congress enacted the IRA (Tollefson 2022). However, much can be learnt from other jurisdictions about what ought to be done, as the legislation has been criticized for not going far enough (Merner et al. 2024). The new Act tackles the major questions doctrine[1] by openly giving the EPA authority to regulate greenhouse gas emissions and related matters (Levy 2022). Whether that would be sufficient for the Supreme Court remained at that point to be seen. The Court did indicate where it is going in some respects, in the decision *Sackett v. Environmental Protection Agency* handed down in May 2023 (*Sackett v. Environmental Protection Agency* 2023). The ruling continued the trend that emerged in *West Virginia v. Environmental Protection Agency*, discussed earlier, which severely limited the ability of the EPA to protect the environment under the Clean Air Act specifically, and all other federal environment protection laws. The *Sackett* case significantly narrowed the definition of waters protected by the 1972 Clean Water Act and further curtailed the powers of the EPA to protect the environment. Commentators at that time noted that the EPA ruling was just one of a number that have seen the Court restricting the powers of various branches of government, as is noted below.

The Supreme Court did rule in 2024 further on these issues. In the momentous decision of *Loper Bright Enterprises v. Raimondo* (*Loper Bright Enterprises v. Raimondo* 2024), the Court ruled in June 2024 that courts when examining agency rules must no longer give deference to those guidelines as required by the case of *Chevron U.S.A., Inc. v. Natural Resources Defense Council, Inc.* (*Chevron U.S.A., Inc. v. Natural Resources Defense Council, Inc* 1984). The effect of this consequential decision, which has been the basis of much administrative law in the United States for 40 years, is to overturn the long-standing respect courts accorded to an agency's understanding of its authority in the law. As Cass Sunstein has argued this case "has a strong claim to being the most important case in all of administrative law" (Sunstein 2018: 1613). Thus, the effect of striking it down will be to continue the trend of further undermining the role such agencies can play in regulating a whole host of issues including environmental ones. As has been noted by Manish Bapna, President of the Natural Resources Defense Council:

> The Court's actions have put at grave risk the foundational safeguards we depend on, and even take for granted, to protect the mainstays of a modern society. Clean air and water, healthy wildlife and lands; safe food, medicine, workplaces, airplanes, bridges and cars; an economy unhindered by predatory lending, stock fraud and credit card scams. (Gelles and Andreoni 2024)

---

[1] The major questions doctrine is an American rule of statutory interpretation whereby it is assumed that executive agencies have not been delegated the power to make decisions with significant political or economic significance.

The effect of this decision was made retrospective by the Court's verdict in *Corner Post, Inc. v. Board of Governors of the Federal Reserve System,* handed down a few days later in another 6 to 3 judicial outcome (*Corner Post, Inc. v. Board of Governors of the Federal Reserve System* 2024).

In another case, also handed down in late June 2024, the Court again weakened the role of the EPA specifically as far as pollution is concerned. It ruled in the case of *Ohio v. Environmental Protection Agency* (*Ohio v. Environmental Protection Agency* 2024) that the EPA's "Good Neighbor" rule that aimed to reduce ozone emissions in other states, be blocked while a number of states contest the legality of the rule. This together with the case *Corner Post v. Board of Governors of the Federal Reserve System,* also handed down in June 2024, which ruled that businesses harmed by federal rules could contest their validity in court, regardless of how long the rule had been in place (*Corner Post v. Board of Governors of the Federal Reserve System* 2024). These cases will have massive effects on the ability of the EPA and other agencies to regulate.

Overall, these cases, together with the claims that the EPA has been generally ineffective (Dana and Tuerkheimer 2016), mean that the environmental protection is even more likely to be limited in the future in regulating environmental issues. Certainly, the EPA has recently played a lesser role in ensuring that environmental justice problems are dealt with overall, especially in areas where Indigenous Peoples and minorities live. The courts have played a major role in reducing the role that the EPA and other agencies can play.

More broadly as far as climate change issues are concerned the most important case that has come before the courts is *Juliana v. United States.* (*Juliana v. United States* 2020). This is among the most significant lawsuits pertaining to the consequences of climate change that wound its way through the courts. The case began in 2015, when more than 20 young people filed a lawsuit against the US government citing, among other things, the Constitution's Due Process Clause. It accused the United States of promoting, supporting, and permitting the production, use, and transportation of fossil fuels – all of which cause climate change. They asked for a court order compelling the government to develop a plan to phase out fossil fuels and get atmospheric carbon dioxide levels back to what they were before the industrial revolution. The claim was ultimately rejected after several appeals, including to the Ninth Circuit. Crucially, the courts decided that, in accordance with the Political Question Doctrine, the case could not be heard. It did, however, motivate the litigants to seek relief through the political system. Some are highly critical of this decision and contend that the court applied the law incorrectly. What is clear is that the case is significant because it raised awareness of the climate change issue. In 2025, the Supreme Court decided not to hear the case. Certainly, the political climate was not conducive the hear the case.

CONCLUSION

While outside the scope of this book, what would happen in the United States in 2025, and beyond, depended on the 2024 presidential election outcome. If Harris was elected the same direction on issues such as climate change and environmental justice would have continued, largely following the approach of Biden. This is because Harris stated many times that climate change is an "existential threat" that needs to be dealt with. She vowed to continue to invest in various types of renewables. She also understood the need to deal with environmental justice and avowed her support for these processes.

However, the fact that Trump was successful in the election meant that dealing with the climate change and environmental justice issues will be very affected in the United States, but also around the world. This is because, as has been noted, the elections were important as:

> This is perhaps the first truly existential election for the climate. The U.S. is the world's biggest historic carbon emitter, and its second-biggest right now after China. What it does at home will impact the entire world's ability to avert catastrophic climate change. – Somini Sengupta, international climate reporter. (Frost 2024)

Already within days of Trump being elected, it was reported that his plans already included putting people into various departments to reverse the directions of the agencies tasked with dealing with climate and a variety of environmental issues. It also seemed already then that much would be done, for example, to abandon various climate change initiatives, including the incentives that were introduced by Biden to promote clean energy. However, the environment will suffer in a variety of ways, as a goal of the new administration will be to reduce the size of national monuments in the West to permit greater mining and drilling on public lands. Another goal will be to remove all offices in every agency tasked with reducing pollution, especially that which disproportionately impacts impoverished communities. Another critical objective will be to withdraw the United States form various treaties and institutions, especially climate orientated ones, such as the Paris Climate Agreement. Further, there will be a process to "deregulate," which otherwise means the reversal of regulations that do much to control the air, water, and other types of pollution. He also was set to remove the Biden goal of attempting to ensure that 40 percent of clean energy developments went to underserved communities (Davenport and Friedman 2024). Thus, not only will the new administration reverse course on dealing with climate change, but it will also do so on environmental justice. Thus, again the interlinkages between the two can be clearly seen, from a negative point of view.

Regardless, of the change in administration, and the new direction on the issues under review, there is a lot to learn from the environmental and climate justice situation in the United States in general, and from the role of the US courts. This is despite the different situation that exists in the United States compared to other countries. What is the case, is that similarly to other countries, but maybe not to the same extent, these issues have become deeply politicized due to their link to issues of race and systemic discrimination, issues that have become even more controversial over the past decade.

However, overall, and what the US experience demonstrates, is that much can be done to deal with a range of environmental and climate issues by reducing or eliminating race-based discrimination and ensuring that marginalized groups are treated better. These issues need to be taken up in ways that ensure that there is a greater understanding of environmental justice, climate change, and climate justice, and their interrelatedness. The public needs to be educated about these issues and why protecting and providing greater support to marginalized groups is useful for the society as a whole (Liddell et al. 2021).

The required reforms of society in general, and the legal process in particular, can be tackled in a variety of ways. Governments at all levels can undertake legislative and policy initiatives that would lead to a better-informed population and a fairer legislative process. Roundtable discussions involving Indigenous and minority group leaders, government officials, and business leaders could be initiated, with an eye to the drafting and implementation of joint plans to tackle climate change, mitigate environment-related damage, and enhance environmental justice overall. All states should create agencies, equipped with the necessary tools and resources, to study and respond to these issues. The executive branches should also assign personnel and resources to monitor, police, prosecute, and take other steps to ensure that vulnerable groups are not trodden on or over. Legislation should be passed ensuring that non-state actors do not take advantage of these groups, and steps taken to monitor what is happening with resource utilization and extraction. A lot can be learnt from comparative practice (Nigam and Parikh 2023), as discussed elsewhere in this book.

What is clear is that many areas of the law, in the US and elsewhere, are in need of extensive reform (Ahmad 2023). The issue of structural racism (Alvarez 2023) which is built into the law (Crepelle 2019b), must be addressed. The law and various processes that deal with a range of issues require transformation (Lohmann 2022). This is because:

> Most of the laws impeding tribal self-government are predicated on outright racist ideals, so they are of questionable constitutionality. Additionally, the plenary power of Congress purports to assert over Indian tribes because of their status as wards is highly suspect. The federal government is one of limited, enumerated powers. Yet

> the federal government micromanages the daily lives of the original Americans. (Crepelle 2021b: 189)

To achieve environmental and climate justice for these groups requires that they be given more control over their own lives, that the trust-based restrictions concerning their use of their own lands must be eliminated, and the various regulations that overtly and subtly limit their ability to function in society (Crepelle 2020b: 563) must be removed. It also requires that control over "tribal affairs" by outside governments is relinquished so that they are more in control of their own lives. It is also about their role over various resources such as groundwater (Matsumoto 2024).

4. Climate change actions and environmental justice in South Africa

INTRODUCTION

South Africa is a very useful country to study with respect to climate change and environmental justice for a variety of reasons (Abraham-Dukuma et al. 2022). Part of the reason is that it has extensive problems with both.

What is the case, is that South Africa's environmental situation reflects its geo-political context as well as its history (Sarkin 2017a). Its location at the tip of Africa, and the fact that it was colonized in the seventeenth century, first by the Dutch and then later by the British, are important considerations in understanding the country, even today (Sarkin 1999a). From the time of its colonization, the Indigenous People of South Africa, as well as all its Black people, were colonized, dominated, and subjugated until political independence was achieved in 1994 (Sarkin 1998). During the Apartheid years, the White minority exercised their political power solely for their own economic benefit (Sarkin 1996). This resulted in significant economic, environmental, and other injustices for the Black, Indian and Coloured populations in the country, using South African nomenclatures (Sarkin 2017b). As a result of the discovery of minerals, such as gold, diamonds, coal, and others, mining became a key industry in the country. Those mines were exploited in ways that disregarded the rights of the people living in the affected areas. This has had major deleterious negative environmental and other effects which persist today (Sarkin 2015).

As with the rest of the world, South Africa too has problems caused by climate change. It is already suffering from more floods, drought, wildfires, heat waves, water problems, etc. Droughts have been a perennial problem but water levels in the country and the region at the end of 2024 were at dangerously low levels. Some large dams had levels that made them unusable. It is, however, often the places where the disadvantaged live that suffer the most from a lack of water and low-quality water. When there are rains these areas flood, which causes major difficulties for the inhabitants. One of the problems in this regard is changing rainfall patterns (Singh et al. 2022). These are already having a variety of impacts again especially in the areas where vulnerable groups live. So too is the increasing temperature, which is going to have many challenges that have not been sufficiently considered as to what it will cause for all communities and people (Vecellio et al. 2022), but especially the marginalized, going forward.

It is also important to consider South Africa because it is a developing country with a unique story in the Global South. That history saw the Apartheid system have dramatic negative effects on various environmental matters for the vast majority of South Africans. Regardless, much needs to be done to protect the rights and needs of the diverse communities in South Africa regarding climate change and environmental justice. In this regard, understanding and documenting issues of exclusion and marginalization are important.

This chapter deals with both climate change and environmental justice in the South African context. As far as environmental injustices are concerned, the country has without doubt had some of the worst consequences caused by a variety of environmental harms, including those caused by various industries and the mining sector. This was endured by the majority of the population because of the pigmentary political system which determined all aspects of their lives and was permissive as far as allowing capitalism projects to outweigh the effects on Black people. The Apartheid system determined where people could live, where they could study, what jobs they could have. This had deep effects on their economic situation. While politically that system ended in 1994, its legacy is deeply entrenched. The economic status of Black people today, has a lot to do with that past system.

This is deeply problematic, as climate change is going to deepen these problems and the levels of inequality in South Africa and around the world even further. One study in 2019 found that the effects of climate change globally are likely to worsen inequality by 25 percent (Diffenbaugh and Burke 2019). This is going to have major effects on countries where already inequality is enormous. South Africa is a case in point (Dasgupta et al. 2023). It could lead to increased violence, instability, insecurity (Mudzingiri et al. 2024) and cause conflict. The result of climate change will be devastating for the poor, who will see a range of effects including, for example, rising food costs as well as shortages (Kubik et al. 2023).

The context of these problems cannot be forgotten. It is reflective of the country's historical legacies, and the economic inequalities that exist (Webster et al. 2020; Jimoh et al. 2021) are highly correlated with race (Mosoetsa and Francis 2019). As Cathy Albertyn has noted:

> It is no surprise that the achievement of equality, human dignity and freedom under South Africa's Constitution is closely tied to the eradication of poverty and inequality. These goals are an essential part of South Africa's transformative constitutional

> project, part of the wider constitutional commitment to "improve the quality of life and free the potential of all persons." (Albertyn 2011: 149)

As a result of the inequities of the Apartheid system, the problems that are being brought on by climate change are not ones that most people have insurance for. Even those in wealthier neighborhoods are seeing insurance rates escalate as insurance companies assess what climate change will bring in the way of claims. In this regard, one South African insurance company noted in 2023 that:

> Climate change, infrastructure concerns and socio-economic challenges have created a tough environment for local insurers and global reinsurers. Short-term insurance companies not only have a responsibility to ensure that their businesses remain strong and are able to sustainably withstand the cost of the risks that are currently dominating the landscape. (Santam Insurance Barometer 2022/23)

Thus, insurance is not really going to help, and that added to a lack of resources means that while there are many adaption and mitigation plans, realizing them in the absence of funding is going to be difficult. As has been stated, "despite the existence of adaptation strategies, there is still a lack of urgency in implementation due to a lack of funding and capacity" (Anekwe et al. 2024: 479).

What is, however, clear, and is not being considered sufficiently, is the interconnection between socioeconomic circumstance and the ability to deal with climate change (Andrijevic et al. 2021). This is very true in the South African context, where people's livelihoods will make it even more difficulty for them to deal with new challenges and thus ought to be part of processes to institute mitigation and adaptation strategies.

As far as dealing with climate change is concerned, South Africa is making a lot of progress relatively speaking compared to other developing countries. In this regard, it has been noted that the country has "put in place one of the most elaborate and consultative climate governance systems observable among developing and emerging economies" (Averchenkova et al. 2019: 1). These will be examined below. The country does face many problems in this regard, including "overstretched human and technical capacity … structural issues, such as historical tensions between the main players, lack of clarity in the assignment of responsibilities, lack of ownership over implementation agendas, multiple ministries dealing with issues concurrently without sufficient coordination, and cumbersome and ineffective communication practices" (Averchenkova et al. 2019: 1; see also Sibiya et al. 2023 145). South Africa does face a huge task of tackling all the problems of climate change simultaneously including a lack of finance to tackle its daily challenges to resources such as water (Kwame et al. 2022) and electricity (Msimango et al. 2023), which are in short supply generally because of corruption and other issues (Adom and Simatele 2021). In that regard, the Presidential Climate Commission argued in 2023 that climate finance in South Africa needed to surge three to fivefold from the level it was at of R131 billion ($7 billion) per year (Meattle et al. 2023).

They estimated R334 billion ($18 billion) was required each year to reach net zero by 2050, and R535 billion ($29 billion) annually to reach its Nationally Determined Contribution (NDC) objective by 2030 (Meattle et al. 2023). However, South African documents on climate financing as with many others makes no mention of environmental justice. It is not covered in the funding that is needed. Climate funding also does not deal with other problems such as electricity, water in general including flooding, waste, and pollution, in general, although some are indirectly caught up in the climate financing proposals. Usefully, South Africa has secured financing from various donors, and some for adaption from African Development Bank and the Green Climate Fund (Chukwudum 2024). However, much needs to be done because there "has been little progress in effectively budgeting adaptation strategies and identifying appropriate financing solutions, a significant impediment to implementation" (Anekwe et al. 2024: 486).

As far as its climate change plans in general are concerned, the government of South Africa did begin to work on climate change issues some time ago, which means that it has made much progress in setting plans into operation. Laws have also been put into place (Ruppel 2022b). The courts have also become involved (Papacostantis 2021). However, much is needed to put plans in operation to secure its mitigation and adaption strategies.

However, it is both climate change and environmental justice that is being explored together here in this chapter. This is because it is argued that they are interrelated, and that climate change mitigation and adaption will not succeed if environmental justice issues are not dealt with as the same time.

What is the case is that in government processes to deal with climate change mitigation and adaption, environmental justice as a concept is hardly used. It is mentioned once or so in some documents, but many policies or other climate change documents do not mention it (South Africa Department of Forestry, Fisheries and the Environment 2018). Thus, for example, in its National Action Plan (NAP) document on a "just transition" it mentions environmental justice once and then says that the just transition processes are an investment in democracy and environmental justice (South Africa Secretariat of the Presidential Climate Commission 2021). However, there is no unpacking of the issues or the way that environmental justice could impact it.

The focus on a "just transition" is that of the Presidential Climate Commission (South Africa Secretariat of the Presidential Climate Commission 2021), a body whose purpose is to advise the President and the cabinet on the best practices for the country's climate change efforts while keeping in mind the interests of all South African citizens (South Africa Presidential Climate Commission 2024). The just transition framework as set out by the Presidential Climate Commission does, however, emphasize the importance of procedural, distributive, and restorative justice (South Africa Presidential Climate Commission 2024). In 2022, the Commission set out a 30-year plan to transition to net-zero carbon emissions by the year 2050, through a just transition framework (South Africa Department of Forestry, Fisheries and the Environment 2021). That was very necessary, as between 2010 and 2015 the country's emissions increased more than 20 percent (Merven et al. 2021).

Certainly, the conversion to a more environmentally friendly society by way of a "just transition" would be important. This is because:

> Climate change mitigation goals and solutions cannot exclusively be considered in the context of low-carbon and energy efficiency. Rather, they are tied to broader socio-economic implications. Therefore, they need to be embedded into a just transition framework for South Africa. (Hanto et al. 2022: 170)

This type of transition to using renewable energy sources, as stated in the NAP document, takes the interests of the poor and vulnerable into consideration and does not leave them worse off than they were before the transition (Annecke and Wolpe 2022).

Certainly, therefore, there are many benefits for using the notion of a just transition. In a country like South Africa, the effects of a just transition can be twofold (Lowitt 2021). First, a just transition is a way to address how climate change impacts the poor and marginalized communities in the country (Lowitt 2021). In addition, a just transition provides an opportunity to address the larger, systemic issue of inequality and can create meaningful social change in regard to the broader issue (Lowitt 2021). However, a just transition is more easily executed in theory than in practice. This is because achieving a truly just transition is not just about switching from fossil fuels to renewable energy sources, but also ensuring that vulnerable people, workers and others, including the people affected by the economies in mining towns, are not completely devastated by the switch (Nel et al. 2023). However, the focus must be much more than that. It is far from only about a just transition and switching to renewables. It is about the whole effect of climate change, particularly on the total lives of the most vulnerable and marginalized. It must also be all the other environmental injustices that these people face that have critical ongoing effects on their lives. Thus, there is a need to attain justice in a variety of conceptualizations in a just transition, including bringing to bear major reforms in the way climate change is dealt with and the incorporation of environmental justice issues. Thus, environmental justice ought to be a critical part of a just transition and ought to have been incorporated into it in South Africa (Swarnakar and Shukla 2024).

What is important more broadly, is that in dealing with climate change both issues (environmental justice and climate change/ justice) need to be dealt with collectively (Ziervogel et al. 2022), not separately, as the massive problems they are both causing will only grow more severe over time.

SOUTH AFRICA'S DEMOGRAPHY

To understand who is most affected by the issues under discussion it is important to understand a country's makeup so that there can be an understanding of who is most affected by these issues and why. Critically, South Africa is somewhat unique as it was a minority who oppressed the majority (Sarkin 2017a). While political independence arrived in 1994, economic freedom for the majority remains elusive (Sarkin 2017b). It is also the effect of Apartheid, including the legacy of environmental injustices which still have massive effects on the majority of people living in the country because of what Apartheid did to them. That legacy is a critical aspect of matters concerning dealing with climate change. This is taken up more below.

As far as South Africa's demographics are concerned, its racial/ethnic makeup in 2024 was 81.4 percent Black African, 8.2 percent Coloured (people of mixed-race identity), 7.3 percent White, 2.7 percent Indian/Asian, and 0.4 percent were other (Central Intelligence Agency 2024b). There are two Indigenous groups: the Khoekhoe and the San (Sapignoli and Hitchcock 2013). However, despite South Africa recognizing these groups' existence, South Africa calls them "vulnerable marginalized communities," as opposed to Indigenous Peoples because of the difficulties in determining who is Indigenous in the African context (Sapignoli and Hitchcock 2013: 355). The recognized languages spoken in South Africa as of 2018 are as follows: 25.3 percent Zulu, 14.8 percent Xhosa, 12.2 percent Afrikaans, 10.1 percent Pedi, 9.1 percent Tswana, 8.1 percent English, 7.9 percent Sotho, 3.6 percent Tsonga, 2.8 percent Swati, 2.5 percent Venda, 1.6 percent Ndebele, and 2 percent which includes South African sign language and Khoi/Khoisan languages (Central Intelligence Agency 2024b). The religious makeup of its citizens as of 2022 are as follows: 85.3 percent Christian, 7.8 percent African traditional, 1.6 percent Muslim, 1.1 percent Hindu, 1.3 percent other, and 2.9 percent who declared none (Central Intelligence Agency 2024b).

APARTHEID'S LEGACY FOR ENVIRONMENTAL INJUSTICES AND THE IMPACT OF CLIMATE CHANGE

The Apartheid system was the major reason for the many environmental justice problems that exist in the country. It produced inequality under the law and allowed people to be treated inequitably (Goodrick and Redelinghuys 2022). However, other causes of many injustices of all types were colonialism, capitalism (Brisman and South 2017) and patriarchy (Humby 2013).

Critically, poverty is a major legacy of Apartheid and in 2023 was set at 63 percent by the World Bank (The World Bank 2023). This is particularly important as, in the post-Apartheid era, South Africa remains one of the most unequal countries in the world (Goga and Mondliwa 2021). The GINI coefficient, or the rate of inequality in the country, has worsened over the last quarter century (Finn 2015) and the World Bank rates South Africa as the worst country in the world as far as inequality is concerned (World Bank 2022). Half the country lives in poverty (Francis and Webster 2019). This is due to a legacy of colonialism and Apartheid (Khanyile and Culwick Fatti 2022), as noted above, coupled with the high levels of grand government corruption of the past decade and a half (Francis et al. 2021). The World Bank highlights that skewed asset bases for the poor perpetuate the inequality divide (Sulla and Zikhali 2022). As was noted in a 2024 by a group of University of Cape Town researchers "[f] or decades, South Africa has grappled with deep-rooted social and economic challenges. Widening inequalities, crumbling infrastructure and lack of access to basic services are prevalent across the country and show little sign of relief" (Johnston et al. 2024: 6).

Climate change is going to have many negative effects (Adom et al. 2022), including but not limited to critical challenges for food security and the livelihoods of many South Africans, many of whom are already vulnerable (Mthethwa and Wale 2023). In 2024 a report found that:

> Through rising temperatures and more extreme weather, the impact of climate change on South Africa has direct consequences for millions of citizens by threatening food and water security, raising the cost of living, contributing to poor health and reducing incomes. These risks are not distributed equally; they stand to widen existing inequalities. (Johnston et al. 2024: 13)

In this regard, a range of strategies are important, including using Indigenous knowledge to deal with the harmful effects of climate change (Kom et al. 2023).

CLIMATE CHANGE

This section focuses on climate change and its effects on the country. However, it is not possible to separate climate change and environmental justice into strictly separate sections. Keeping with the argument throughout this book, it must be recognized that the issues are intrinsically intertwined. Thus, the effect of one on the other and how issues affect each other will be noted.

As far as climate change in the country is concerned, there is no longer much scientific debate about whether southern Africa's climate is warming. The concentration of greenhouse gases in the atmosphere is increasing, and the sea level surrounding the wider continent is consistently on the rise (Climate Transparency 2022). In particular, since 1990, South Africa has seen an average temperature rise at a rate more than twice that of worldwide temperature escalations (South Africa Department of Forestry, Fisheries, and the Environment 2021). It has been estimated that 53 percent of South Africans are currently affected by climate change (Abel et al. 2023).

Despite its status as a developing country, South Africa is a comparatively significant contributor to climate change. For the past decade, South Africa has positioned itself amongst the world's top 15 polluters in per capita greenhouse gas emissions (Ntombela et al. 2019). Thus, while South Africa was formerly not a major contributor to climate change-related activities, it is now contributing to climate change at a rate greater than ever before.

The pollution-related problems facing South Africa are enormous, and they are accentuated by environment injustice. There are many dimensions to that. In this regard, Greenpeace published a study in 2024 that found that nine out of ten of Africa's top nitrogen dioxide sources, and six of the world's largest emitters of nitrogen dioxide, are in South Africa. South Africa is also where some of the largest sulphur dioxide emitters are found (Abbas et al. 2024). While these are not always connected to mining, the health effects of sulphur dioxide emissions for South Africans are massive. As the 2022 State of Air Quality and Health Impacts in Africa Report notes, air pollution is the second greatest cause of mortality across the region (Addaney et al. 2023). It is estimated that in 2019, 1.1 million deaths could be attributed to air pollution (Health Effects Institute 2022).

A major contributor to global warming is the use of fossil fuels. South Africa's dependency on coal, for example, which it uses in large amounts (Mirzania et al. 2023) continues to be a major problem (Mirzania et al. 2023). It is, however, only responsible for one percent of world-wide carbon emissions as compared to other countries, such as the US and China, which produce 40 percent of global emissions (Abraham-Dukuma et al. 2022). South Africa is one of the highest coal use-related emitters in Africa, but there are plans in place to reduce this over time to stem the climate change effect (Abraham-Dukuma et al. 2022). South Africa is experiencing difficulty weaning itself off coal for a variety of reasons, including ongoing problems with the provision of electricity to the nation (Tladi et al. 2024).

The coal mining sector poses major challenges related to environmental justice. It is highly environmentally destructive, leading to many serious effects on the people living near the mines. This continues today, as South Africa's energy sector is heavily reliant on coal for power production. Despite the problems caused by coal, the country is having enormous difficulties weaning itself off it and transitioning to a greater use of renewable energy sources. This is one example of the difficulties in dealing with climate change and adopting mitigating strategies.

These problems do, however, need to be contextualized by the fact that the coal mining industry today is a major employer in the country. In 2021, South Africa's coal sector employed roughly 93,000 workers (Matola and Connock 2023). Thus, a rapid shift from coal as the country's primary energy source to more renewable energy sources would cause widespread unemployment and supposedly infringe on the "just" aspect of a "just energy transition" (Matola and Connock 2023). This needs to be weighed against the climate-related costs for the country, and the effect greenhouse gas emissions from coal power stations have on tens of millions of people. Another consideration is the effects that coal mining and coal usage have on the workers in the mines and those who live near the mines, or close to coal power stations. This issue is not dealt with often and then only by those who work on the issue of coal and climate change (Mirzania et al. 2023). Thus, moving away from coal usage also needs to be examined in the context of environmental justice and the problems coal usage causes beyond the effect on greenhouse gas emissions.

The use of coal also needs to be understood as a human security issue (Jayaram and Ajayan 2022) in the country (Motswaledi et al. 2023). This is because of the effect of coal mining and coal usage on people's lives. It has been found that:

> The continuous use of coal and fossil fuels has led to negative effects on health due to direct or indirect exposure, such as cancer, respiratory, and cardiovascular problems, and to negative environmental and socio-economic effects such as polluted air, soil and water. Coal mining further increases the risk of water scarcity, a problem expected to worsen in the years to come. (Hanto et al. 2022: 170, references omitted)

This is why a climate change approach needs to be addressed using a climate justice and environmental justice approach. This is dealt with below.

While the country has ramped up its use of renewals, funding the move away from coal and other fossil fuels remains a major problem (Msimango et al. 2023). Thus, dealing with these issues and the legacy of environmental injustices that affected (and continue to affect) Black people more than others, needs to be resolved. Determining how to do this is not easy for a variety of reasons, beginning with an insufficiency of resources (Lekunze 2024). Efforts still do not incorporate an environmental justice approach which, together with climate justice action could, at least in part, resolve past environmental injustices which have left a damning legacy of inequality affecting people in many ways. Advancing on these issues is vital as climate change will continue to have many devasting and wide-ranging effects including to:

> exacerbate the deep inequality in South Africa. It has far-reaching impacts for all rights, including the rights to: equality, human dignity, life, access to sufficient food and water, and the rights of children. (Loser 2024: 75)

SOUTH AFRICA'S STRATEGIES TO MITIGATE AND ADAPT TO CLIMATE CHANGE

In light of global and national studies projecting, amongst other things, further temperature increases, rising ocean levels, and an increasing frequency of severe weather events, South Africa has pledged to tackle climate change through mitigation and adaptation efforts (United Nations Framework Convention on Climate Change 2022). It has been working on these issues for some time. In 2011, its Climate Change Response Policy recognized:

> that South Africa is extremely vulnerable and exposed to the impacts of climate change due to our socio-economic and environmental context. It also recognizes that climate variability, including the increased frequency and intensity of extreme weather events, will disproportionately affect the poor. (Loser 2024: 69)

However, properly addressing climate change raises major challenges with respect to human rights, justice, and equity. South Africa's commitment in this regard is crucial (Fløttum and Øyvind 2013). Grounding equity at the forefront of its just transition is even more imperative, considering that one of the many legacies left by the Apartheid era is the profound skepticism that a significant portion of the Black community has about environmental issues (Patel 2014). The skepticism is the result of the legacy of environmental damage caused by government and industry and the inability to correct that damage over the last 30 years. Nevertheless, as noted by Garth Myers, the approach adopted is a result of the government's adherence to:

> [n]eoliberal ecological modernisation, wherein marketisation, new technology and focus group stakeholder meetings are the panacea that will rescue Africa's cities

> from themselves and from environmental calamity, magically reducing poverty and instituting democracy. (Myers 2008: 696)

While market-oriented attitudes on the part of environmental policymakers have been subject to criticism, where South Africa's post-1994 environmental policies do merit praise is related to their insistence on a governance approach towards climate change mitigation and adaptation (Keary 2016; Meadowcroft 2016). However, South Africa's large population, many of whom are unemployed and dependent on state grants, means that the country does not have the resources it needs to do more to assist its people in dealing with climate change. This problem is enhanced by the fact that the majority of the people have faced enormous environmental injustices which means that they are already in problematic situations as far as the environment is concerned. They continue to suffer various negative environmental effects that are a legacy of the past (Sarkin 2000). They already have lingering health problems that will be exacerbated by the effects of climate change.

Recognizing that enduring structural inequalities render specific communities in the country acutely susceptible to the effects of climate change, the integrated and collaborative nature of a governance approach ensures that the various vulnerabilities of the people most affected are recognized and protected (Patel 2014). Moreover, a governance approach enables the outcomes of environmental policies to be attributed to a broader array of actors – public, private, and voluntary – rather than just the state, a crucial breadth of attribution given the multifaceted nature of climate change (Patel 2014). However, despite the consistency with which governance approaches are set forth in post-1994 environmental policies, South Africa's de facto response to climate change remains ambivalent. Illustrating this point, independent scientific analysis by the Climate Action Tracker project notes that:

> South Africa's energy system remains at a crossroads: as its energy crisis continues, it is moving forward with the Just Energy Transition Investment Plan (JET IP), and is in the process of adopting a renewable energy plan. However, it is continuing to pursue controversial fossil gas capacity and potentially faces delays in its shift away from its heavy coal reliance as it props up its failing state-owned company ESKOM. (Climate Action Tracker 2024)

However, while South Africa is taking major steps to deal with climate change issues, at least as far as its policies and documents are concerned, it does not have a process to really deal with the legacy of the environmental injustice that has been caused in the country.

Thus, as noted in the introduction to this book, Naomi Klein in her 2016 Edward Said lecture "Let them Drown: The Violence of Othering in a Warming World," operationalizes her "othering" and applies it to the climate crisis (Klein 2016). Elaborating on how "othering" insidiously operates in environmental policy discourse and development, Klein lays bare the justifying ideologies behind the human and ecological cost of growth, noting how environmental "othering" sustains dispossession, violence, and culturicide (Klein 2016). While Klein does not focus on South Africa, her argument might well be said to extend to pre-1994 South Africa, where the dominant conservation-based approach to environmental policy led to the dispossession and displacement of countless Black South Africans (Klein 2016; Hamann and O'Riordan 2000). It also othered many South Africans who were dispossessed of their lands and suffered disproportionately from the many environmental harms resulting from government policies. Thus, the successor government ought to distance itself from its predecessor.

The post-1994 South African state sought to shift the environmental discourse by providing for an environmental approach to ensure justice and equity, in consideration of Section 24 of the Constitution (Sarkin 1999b) which provides that:

> Everyone has the right to an environment that is not harmful to their health or well-being; and to have the environment protected, for the benefit of the present and the future generations through reasonable legislative or other measures. (South Africa Constitution of the Republic 1996)

Since that provision was incorporated into the 1996 Constitution, South Africa has been seen as a leading player in the environmental space. It has ratified a litany of global environmental agreements (Patel 2014). Thus, on the international level, in 1997 the country acceded to the United Nations Framework Convention on Climate Change (UNFCCC) and ratified the Kyoto Protocol in 2002. It later joined the Paris Climate Agreement and other related international instruments, including the Stockholm Convention on Persistent Organic Pollutants, the Rotterdam Convention on the Prior Informed Consent Procedure for Certain Hazardous Chemicals and Pesticides in International Trade, the Basel Convention on the Control of Trans-boundary Movements of Hazardous Wastes and Their Disposal, the Vienna Convention for the Protection of the Ozone Layer, the Minamata Convention on Mercury, and the Montreal Protocol on Substances that Deplete the Ozone Layer.

However, the extent to which these have been translated into progress on the ground is questionable. Its role on environmental matters can also be seen by the fact that South Africa hosted the 2002 World Summit on Sustainable Development and the 2003 World Parks Congress. It has played a significant role in forming the international agenda on sustainable practices and conservation strategies (Patel 2014). Furthermore, the ambitious greenhouse gas emissions targets defined by South Africa at COP15 in 2009 were said to have set a target for the developing world, and COP17, hosted in Durban in 2011, provided significant impetus in taking action to deal with climate change (Patel 2014).

Thus, South Africa is seen by the media to have played a successful role in global diplomatic efforts to address climate change. That role significantly influences its national policy goals, as demonstrated by South Africa's 2021 NDC. These are national-level commitments made under the UNFCCC, under the Paris Agreement (Zahar 2024). This instrument aims to keep global warming below an increase of two degrees celsius from mean temperatures in the pre-industrial era (United Nations Framework Convention on Climate Change 2022: 1).

A core component of the NDC involves a commitment to a "peak, plateau and decline GHG emissions trajectory range" (United Nations Framework Convention on Climate Change 2022: 14). This target period sets levels for the country to assess how effective its mitigation strategy is (Madondo and Nkwana 2021). This benchmark projects greenhouse gas emissions to peak (2020–2025), plateau (2025–2035) and decline (2036 and beyond) (Madondo and Nkwana 2021: 195). To reach these quantifiable greenhouse gas emission reduction objectives, the NDC outlines the importance of decoupling economic growth from greenhouse gas emissions by investing in renewable energy and shifting economic growth to "less energy-intensive sectors" (United Nations Framework Convention on Climate Change 2022: 4). However, as has been noted, "South Africa's climate institutional form is therefore one that can deliver high level, internationally impressive targets and trajectories, but, upon scrutiny, is revealed to be powerless over the country's main emitting sector" (Tyler and Hochstetler 2021: 5199).

Thus, while most view such documents as the NDC as aspirational, "they are also narrative documents that can be read as persuasive discourses that reveal deeper tensions, ideas, and values about international climate policy, national identities, and aspirations" (Mills-Novoa and Liverman 2019: 10). This is extremely relevant in the South African case as a great deal is revealed both by what is included and what is not included. Thus, while NDC documents in general discuss various issues (Höhne et al. 2024), such as "responsibility and fairness or emphasize vulnerabilities and adaptation; they may focus on land use and biodiversity" (Mills-Novoa and Liverman 2019: 10), the South African NDC does not deal with many adaptation issues, in particular how the government plans to assist the many vulnerable people and groups in the country. This is seemingly the case for many African countries (Zhu and Shang 2024). Notably, the South African NDC does not include issues of justice, equity, human rights, and other similar concerns. It mentions inequality, unemployment, and poverty, but only as contextual matters, not as consequential issues to be dealt with and considered in addressing the impact of climate change. The NDC should include other matters like inequality, for example, as inequality will be magnified by the effects of climate change (Schapper 2018). Further, climate change will have dire consequences on education as poor schools, which are where most Black children are schooled, will not have the same ability to regulate the heat in classrooms (Nkrumah 2021). The result will be that these children will receive less schooling, with the knock-on effect that the reduced education will result in a lower likelihood of meaningful employment.

Along these lines, the NDC ought to have addressed the issue of human rights, for example, as some of the mitigation and adaptation processes adopted generally around the world have had negative effects for the most marginalized and vulnerable groups (Calzadilla 2018). Meeting these challenges is fundamental to addressing climate change effectively. The NDC should have included these, as the Paris Climate Agreement, which South Africa is a party to, states that "climate change is a common concern of humankind, Parties should, when taking action to address climate change, respect, promote and consider their respective obligations on human rights, the right to health, the rights of Indigenous Peoples, local communities, migrants, children, persons with disabilities and people in vulnerable situations and the right to development, as well as gender equality, empowerment of women and intergenerational equity" (United Nations Framework Convention on Climate Change, Paris Agreement 2015).

These issues are not taken up in the South African NDC despite the interplay between race, poverty, and the environmental harms stemming from Apartheid era policies and actions and which still endure. The NDC does, however, outline some adaptation strategies, including the adaptation of water resources, health systems, and other infrastructure to ensure resilience to the negative effects of climate change (United Nations Framework Convention on Climate Change 2022: 1) These strategies are directed towards crucially need resources, which South Africa is short of. In general, however, the focus of governments and other key players around the world is on the mitigation of climate change-related environmental damage, rather than adequately addressing the impact on specifically vulnerable people through adaptation. In this regard, in the South African context, it has been noted that:

> The vulnerable populations and the apartheid legacies of social inequalities are repeatedly addressed with the inclusion of vulnerable people, women, children and disabled people; however, black people and other disadvantaged racial minority groups are not emphasized. Moreover, the effectiveness of adaptation strategies among women and vulnerable population groups in the country is not mentioned, and strategic actions to adjust adaptation accordingly are not addressed. (Khine and Langkulsen 2023: 3425)

As has been noted concerning addressing climate change generally, "the focus is on an equitable distribution of the costs of adaptation rather than the specific vulnerabilities and needs experienced by those at risk" (Lyster 2015: 131). This is the case even though the Paris Agreement provides that "adaptation action should follow a country-driven, gender-responsive, participatory and fully transparent approach, taking into consideration vulnerable groups, communities and ecosystems" (United Nations Framework Convention on Climate Change, Paris Agreement 2015).

A key concern regarding the NDC is its impact on those already marginalized in the country due to past discrimination, poverty, unemployment, and the like. In this regard, the question is: "What will be the overall effect of substituting fossil fuels with other energy sources on the most vulnerable?" (Gambhir et al. 2018). The problem for the country is how to balance its overall fiscal needs, and the needs of its people, against the resources required to implement the processes to mitigate and adapt to address climate change. Tackling the environmental injustices of the past and their endearing legacy seems to be something that the country considers unaffordable. However, the relationship between the two, (climate change and environmental justice), and its related concept of climate justice (Barnwell and Wood 2022) ought not to be discounted, as properly addressing climate change will not be possible if most people cannot afford to take the required mitigation and adaption steps. What must also be considered is the ongoing negative effects climate change has on people's livelihoods, including their health challenges.

More broadly, while the NDC represents just one component of what is one of the most "elaborate and consultative climate governance systems observable among developing and emerging economies," South Africa's overarching efforts to respond to climate change are afflicted (as others are) by several enduring domestic impediments that must be addressed (Averchenkova et al. 2019: 3). Among these impediments, three core issues present the most formidable challenges to addressing climate change, environmental justice, and climate justice, effectively; namely, a lack of policy alignment, capacity shortages, and distrust.

As regards a lack of policy alignment, Averchenkova et al. suggest that the incoherence between climate change objectives and other key developmental strategies represents a systematic issue impeding the implementation of the NDC (Averchenkova et al. 2019: 3). While South Africa has instituted a system to ensure vertical (national to local) and horizontal (interdepartmental) coordination and obligates its government departments to align strategies, the fragmented nature of climate policy responsibility has hindered coherence and, therefore, the effect of policy formulation and implementation (Averchenkova et al. 2019: 4). In this regard it has been noted by Sibiya et al. that:

> there is inadequate coordination across all government levels and that this affects the planning, implementation, and monitoring phases of adaptation. … there is a need for government officials working on climate change adaptation issues to advocate for coordination across government levels and sectors, as this will allow for resources mobilization to address the wicked problem we are faced with. (Sibiya et al. 2023: 145)

Second, limited financial and human resource capacity impedes the ability of the state to tackle climate change (Averchenkova et al. 2019: 4). As the climate situation worsens, and the requisite complexity of policy responses increases, these capacity constraints are further exacerbated (Averchenkova et al. 2019).

Lastly, given the aforementioned history of environmental policy in the country and the resultant distrust of environmental interventions, the implementation of policies is often delayed, and the requisite level of civil society engagement is often absent (Averchenkova et al. 2019: 5). In addition to these general issues afflicting, to varying degrees, all environmental policies, specific legal and policy responses to climate change face unique challenges.

An insufficient amount has been done to incorporate an environmental (and climate) justice approach which does justice to the rights contained in the Constitution and the needs of the people. Thus, the Constitution's requirement to protect all rights, including socioeconomic and environmental rights, in a way that "improves the quality of life … of all persons," needs to be given far more focus in the country's approach to dealing with climate change.

It is also vital to incorporate the rights and needs of people who are usually voiceless (Brown 2011). This has been a major problem in the country, especially as far as climate change is concerned (Sibiya et al. 2022). They can be better included by ensuring processes are more people-centered, incorporating and allowing traditional communities to be involved, and being gender-sensitive while respecting human rights (Jodoin et al. 2021). In short, the question is how can the "poorest, most vulnerable, and those furthest behind" be better involved in such processes (Segone and Tateossian 2017)? For meaningful change to occur in these sectors (Szablowski and Campbell 2019), opportunities must be provided to vulnerable and excluded communities and groups to engage in and make significant contributions to the processes (Huizenga 2019). Key issues include determining the strategies and means which need to be adopted to combat these problems. For environmental and climate justice challenges and problems to be dealt with in ways that achieve real equity and fairness (Schrock et al. 2015), intersectionality conceptually ought to be brought to bear on these environmental matters in the wide sense.

It does need to be noted that in South Africa, but also elsewhere around the world, if there is really to be no one left behind, then data is needed to determine the real impact of climate change in the broadest sense. It is therefore necessary to disaggregate population data, particularly with respect to identity and groups, to determine who is being affected by environmental injustices, but also crucially who is not. This is the case despite the reality that many states are averse to amassing such data. This data is essential, as identity is a key determinant of who should be included and who is not being included in the discussions regarding climate change issues. Critically, Target 10.2 of the Sustainable Development Goals (SDGs) seeks to empower and promote the social, economic, and political inclusion of all irrespective of age, sex, disability, race, ethnicity, origin, religion, or economic or other status. This can only be achieved if it is understood who is being included or excluded.

SOUTH AFRICA'S PROCESSES AND INSTRUMENTS TO DEAL WITH CLIMATE CHANGE

To shine a light on the efficacy of South Africa's legal and policy responses to climate change, various core elements of South Africa's integrated climate change response framework are examined. It is also important to examine these in the context of ameliorating the legacy of past environmental harms. Doing this will help millions of people in the country be better able to cope with the climate change problems they face.

Examining South Africa's response to climate change within the context of the vestiges of the Apartheid regime is critical, as the Climate Action Tracker monitor found that South Africa's response to climate change has been insufficient across the board (Climate Action Tracker 2024). This is despite a flurry of recent policy developments, some of which are the: Revised National Water Resources Strategy (1 September 2023); National Biodiversity Offset Guideline (23 June 2023); National Biodiversity Framework 2019 to 2024 (29 August 2022); Liquefied Petroleum Gas Rollout Strategy 2022 (13 May 2022); Mine Community Resettlement Guidelines 2022 (13 April 2022; Artisanal and Small-Scale Mining Policy 2022 (30 March 2022); National Water and Sanitation Master Plan (28 November 2019); Draft National Botanical Garden Expansion Strategy 2019–2030 (15 November 2019); Final National Climate Change Adaptation Strategy (13 November 2019); Integrated Resource Plan 2019 (Corrected Version) (18 October 2019); and the Draft National Climate Change Adaptation Strategy (6 May 2019).

The first evaluation of climate action taken by the country, as assessed by the Presidential Climate Commission, identified the primary obstacles encountered were incoherent strategies, weak governance arrangements, inadequate financing and "inconsistent actions" by government and others (South Africa Presidential Climate Commission 2024). The Presidential Climate Commission Report also noted that the effects of climate change have "disproportionately impacted vulnerable groups, including women and young people, older people, the unemployed, historically marginalised groups and those living in informal settlements" (South Africa Presidential Climate Commission 2024: 3). This is critical as not often are the historically marginalized groups and those in informal settlements mentioned in these reports. However, in the 104-page Report, marginalized groups and the effect of climate change on them are mentioned only three times, including the instance noted above. Another reference to them is the statement that "Poor and marginalised groups face disproportionately high levels of inadequate water and sanitation access" (South Africa Presidential Climate Commission 2024). However, these problems are not blamed on environmental injustice issues. All that is said is that significant water is lost because of old and poorly maintained water and sanitation infrastructure. It is noted that water is unfairly allocated. While in another part of the Report it is noted that South Africa has made progress in recent years to deal with water and sanitation deficiencies (South Africa Presidential Climate Commission 2024: 41), the Report would have been a perfect opportunity to highlight the need to deal with water and sanitation concerns more broadly, to rectify the problems of the past.

While the term "justice" is used several times in the Report, it is mostly to discuss how the principles of procedural justice, distributive justice, and restorative justice underpin the climate response. Very little is said about what this means in reality, nor is there any discussion about what practically can and will be done to overcome these challenges faced by the marginalized. There is no mention of environmental justice or even climate justice. Much more needs to be done to specifically and practically take up these concerns, which although mentioned at times, are not often explicitly articulated or discussed in ways that engender meaningful commitments to deal with them.

In this context, the three instruments examined below are the 2011 National Climate Change Response White Paper (NCCR-WP), the 2019 Carbon Tax Act, and the newly enacted 2024 Climate Change Act, which came partly into force in March 2025. Beyond their centrality in South Africa's climate change response framework, these legal and policy responses have been selected for analysis due to their collective cross-sectoral implications, that is, their combined embodiment of a governance approach towards climate change.

However, it should be noted that South Africa has done very well at the policy and legislative level. As far as instruments and documents are concerned, it is more advanced than other similarly situated countries. This is because it adopted its National Climate Change Response Strategy in 2004 and later its NCCR-WP in 2011. In 2012 climate change became an important part of the National Development Plan. In 2020 it adopted its National Climate Change Adaptation Strategy (NCCAS), and in 2024 its Climate Change Act.

However, while the intent to address climate change is real, several problems exist which limit the government's practical ability to follow through on the intent. For one, "the country has for too long lacked good governance, an essential ingredient in the development of transformative climate change adaptation policies and strategies" (Sibiya et al. 2022: 7111). As a result of the fragmented nature of climate policy responsibilities among government agencies, developing coherent policies and then executing them remains a serious challenge. The policies' efficacy has been hampered by their lack of coherence and clarity, particularly when it comes to implementation (Kwame et al. 2022). Another study, already in 2023, which was echoed in the Presidential Climate Commission report of 2024 discussed above, noted that a "lack of policy coherence, skill scarcity, a lack of political leadership, and corruption are some of the major barriers encountered in promoting effective climate change adaptation in South Africa" (Sibiya et al. 2023: 145). There is also a detrimentally fractured coordination between the levels of government regarding climate change adaptation measures, and this is affecting all phases of successful adaptation, namely planning, implementation, and monitoring of adaptation strategies (Sibiya et al. 2023).

In each different sector of the economy, there are an array of issues. Consequently, it has been argued that there is also "a weak set of formal climate institutions, and poor progress on institutionalizing decarbonization in the systemically important energy sector" (Tyler and Hochstetler 2021: 5201). Problematically, the laws are not often implemented and are not always complied with very well when they are put into effect. Even when they are operationalized, the laws are not well enforced when they are not complied with. In fact, often the full extent of the progress of the various instruments produced by the government is their adoption, with little work done beyond that on the ground to implement them. This means that the harmful effects are to be found almost everywhere in the country. Those most affected are the poor and the vulnerable, which are mostly one and the same.

THE NATIONAL CLIMATE CHANGE RESPONSE WHITE PAPER (NCCR-WP)

Published by South Africa in 2011, the NCCR-WP took seven years to complete. It began in 2004 with the publication of the National Climate Change Response Strategy for South Africa (Fløttum and Øyvind 2013) and was one of the policy actions taken to give effect to Section 24 of the South African Constitution of 1996 (Madondo and Nkwana 2021). Implemented by the Department of Environment, Forestry and Fisheries (DEFF), the NCCR-WP sets out its objectives as being climate change mitigation and climate change adaptation (Madondo and Nkwana 2021). Rather than target specific sectors, the objectives of the NCCR-WP are cross-sectoral. It is a country-wide policy, outlining a course of action that curbs greenhouse gas emissions and intervenes in damaging environmental practices (Smith 2013).

The NCCR-WP recognizes that the country's mitigation approach is underpinned by notions of contributing fairly to a world-wide push to limit greenhouse gas emissions while managing the country's domestic priorities, such as development and poverty eradication (South Africa National Climate Change Response White Paper 2024). However, as in other environment-related documents, the notion of environmental justice, or even climate justice, is not addressed, directly or even indirectly. The term "just transition" does appear once or twice in the document. Unfortunately, the phrase is not defined and there is no unpacking it to explain how this concept should be applied to ensure a green economy in ways that are fair and inclusive and leave no one behind. The document does, however, note that climate change "has the potential to undo or undermine many of the positive advances made in meeting South Africa's 'own development goals' and the Millennium Development Goals (MDGs)" (South Africa National Climate Change Response White Paper 2024). These goals are left undeclared in the NCCR-WP, but both South Africa's development goals generally and the MDGs specifically do in fact recognize poverty alleviation, achieving equality, environmental sustainability, health promotion, and other related targets as being essential to South Africa's development as a modern state. These objectives are directly and specifically included in the White Paper. The objectives should also have included environmental and/or climate justice. As a result of this omission, the White Paper is overly narrow in its construction. While it uses some buzz words, and the correct rhetoric at times, such as dealing with climate change to assist the poor and the vulnerable, there is insufficient focus on what could really make a difference in mitigating and adapting to climate change (Ismail 2023) in ways that would affect millions of South Africans who are deeply afflicted by environmental harms already, never mind what is still to come.

One of the major problems with the White Paper specifically is that informing South Africa's mitigation strategy as outlined in the NCCR-WP is the fact that coal provides 90 percent of South Africa's energy (Cock 2019). As a result, South Africa has recognized its need to strike a balance between limiting greenhouse gas emissions and domestic priorities, particularly its energy requirements (Chamberlain and Fourie 2024). To achieve this, the NCCR-WP outlines various mitigation strategies such as the National GHG Emissions Trajectory Range, the introduction of carbon budgets for heavy greenhouse gas emitting sectors and establishing a greenhouse gas inventory to assist with monitoring and evaluating greenhouse gas emissions. Sadly, the document neglects the equally important goal of achieving justice, including both environmental and climate justice, (Barnwell and Wood 2022) by recognizing and addressing the impact that the mining sector has on the environment and particularly on the communities located near the mines. Thus, one of the priorities ought to have been environmental justice (as well as climate justice) and the impact that the mining sector has on the environment and on the communities, particularly those living near the mines. However, those are not part of the strategy. That should have been the case for both mitigation and adaptation, but especially adaptation. This is because while mitigation is focused on the national level, adaptation is primarily concerned with the local level (Madondo and Nkwana 2021). In fact, from an environmental and climate justice point of view, it is at the local level that many of the environmental challenges exist. In addition, the NCCR-WP acknowledges that, relative to mitigation, the benefits of adaptation strategies may be much easier to achieve and are more effective (South Africa National Climate Change Response White Paper 2024). It is clear that the NCCR-WP ought to have an environmental and climate justice component to deal with those issues. The NCCR-WP establishes early- medium- and long-term requirements (South Africa National Climate Change Response White Paper 2024) to achieve its goals. This approach could easily have been used to address the country's extensive environmental damage.

Despite representing the "South African Government's vision for an effective climate change response and the long-term, just transition to a climate-resilient and lower-carbon economy and society," the effectiveness of the present approach to adaptation and mitigation in the NCCR-WP is hindered by various factors (South Africa 2024). Notably, the NCCR-WP contradicts other programs. For example, the National Energy Act (NEA) (34 of 2008) aims to "ensure [the] uninterrupted supply of energy to the Republic" (South Africa 2008). Given the reliance on coal, a reduction of greenhouse gas emissions, as called for by the NCCR-WP, seemingly violates the NEA.

Further, as noted above, the NCCR-WP does state that its guiding principles include equity, uplifting the poor and vulnerable, and informed participation to "promote participation and action at all levels" (South Africa National Climate Change Response White Paper 2024). However, there is little discussion about how to actually achieve these goals. Recognizing and addressing environmental justice issues and achieving climate and environmental justice would have been useful, and in fact critical, in meeting those goals. The guiding principle of "informed participation" is even more important given that in its absence a governance approach to mitigation and adaptation would be impossible. Despite this, Shelley Smith notes that the NCCR-WP makes no mention of how the government will incorporate climate change into training initiatives and education, nor does it note how public awareness ought to be raised (Smith 2013). As Smith states:

> [The NCCR-WP contains] no tangible, practical steps of how to involve the most important and influential party to bring *action* to the cause, namely the citizens of the country. (Smith 2013: 48)

Important climate change mitigation and adaptation strategies must include climate change education (CCE). Such education must also occur in schools (Nkrumah 2021). In South Africa, CCE has emerged as an initiative to acknowledge the climate change crisis that is occurring in the country (Sibanda and Manik 2023). However:

> [d]espite the myriad of South Africa's CCE responses, it is worrisome to note that the country continues to be a significant emitter of GHGs worldwide ignoring the voices of NGOs highlighting the trajectory of the country. Thus, the existing CCE for climate action needs to be improved to incorporate the findings of civil society organizations so that GHGs emissions in South Africa can be reduced. (Sibanda and Manik 2023: 1829)

As noted, there is an enormous need to educate all South Africans about climate change and what they can do to make themselves more resilient to it. This is because, to some extent at least, local communities will need to take many of the required remedial steps themselves. For example, "[t]he development of community-owned renewable energy (core) projects in South Africa plays a critical role in seeking environmental justice for rural communities" (Madhanagopal et al. 2022: 7). In this regard, there are glaring gaps in CCE efforts, and to date it has been ineffective in achieving widespread climate change mitigation or adaptation (Sibanda and Manik 2023). Given South Africa's past and the problems that communities face from past environmental injustice, CEE ought to be given much more attention.

Linked to this is the analysis by Kjersti Fløttum and Gjerstad Øyvind suggesting that the "legal rights of the poor" are all but absent in the NCCR-WP, and the poor themselves are cast as passive and lacking in agency while the government frames itself as the hero (Fløttum and Øyvind 2013). Their examination of the wording revealed numerous instances of the use of words such as "poor," "poverty," and "vulnerable," but no examination of the issues of environmental and climate justice (Fløttum and Øyvind 2013). There ought to be much greater focus on what can be done in the environmental space to alleviate the suffering of the poor and vulnerable. There should have been an unpacking of who the poor and vulnerable are, the challenges they endure environmentally, and what can be done through mitigation and adaptation actions to rectify some of the problems that will be exacerbated further by climate change. As far as poverty is concerned, it has been argued that an intersectionality approach ought to be adopted (Basson 2023). Critically, once again, while the rhetoric may be present, the practical actions required to achieve change and reform are not identified.

Lastly, Madondo and Nkwana note that the NCCR-WP lacks enforcement and compliance directives and powers on two levels (Madondo and Nkwana 2021). Firstly, South Africa's two largest greenhouse gas emitters, ESKOM and SASOL, have lodged successful applications for postponements and exemptions from the mitigation plans in the NCCR-WP (Madondo and Nkwana 2021). In April 2024 the Minister of Forestry, Fisheries and Environment allowed Sasol greater leniency, in an already weak legislative framework (the National Environmental Management: Air Quality Act No. 39 of 2004), in relation to its toxic air pollution emissions caused by sulphur dioxide and other dangerous chemicals (Centre for Environmental Rights 2024a). This permissive decision has been criticized for its lack of transparency, but also because it will result in more premature deaths and higher illnesses in the very heavily polluted places where SASOL activities are located. Secondly, the data required for establishing a comprehensive greenhouse gas inventory is primarily submitted on a voluntarily basis (Madondo and Nkwana 2021). Consequently, climate change reports are often inaccurate or delayed due to the nonsubmission of data, hamstringing the attainment of the NCCR-WP's environmental objectives (Madondo and Nkwana 2021). Whereas, ostensibly at least, the NCCR-WP is cross-sectoral, involving the public, private, and voluntary sectors in its mitigation and adaptation strategies, much is left out and ought to be rectified so as to ensure that environmental justice and/or climate justice is a full component of the strategies.

THE CARBON TAX ACT

Another distinct response to the climate change issues in South Africa is evidenced in the 2019 Carbon Tax Act. It is one of many market-based tools for tackling climate change. However, the Carbon Tax Act is applicable only to the private sector. It was signed into law to "... reduce the greenhouse gas (GHGs) emissions in a sustainable, cost-effective and affordable manner" (South Africa National Treasury 2019). By levying a tax on firms' carbon emissions, the Carbon Tax Act seeks to incentivize firms to switch to more suitable, less energy intensive, sustainable technologies (Masondo and Nwosimiri 2023). However, while the "implementation of carbon tax has featured prominently in climate change agenda ... its suitability as a response to climate change is controversial" (Nemavhidi and Jegede 2023: 27). One aspect of the controversy is the legislation's human rights implications, which some argue have not been sufficiently taken into account.

In theory, the Carbon Tax Act follows the "polluter-pays-principle." This is included in the NCCR-WP to ensure that businesses "take the negative adverse costs (externalities) into account in their future production, consumption, and investment decisions" (South Africa National Treasury 2019). However, various scholars have problematized the effectiveness of the polluter-pays-principle in general and the Act more specifically. For instance, Mmatlou Kalaba argues that the Carbon Tax Act generates co-benefits; that is, the taxes it generates help fund investments to meet the SDGs (Kalaba 2020). However, a significant weakness in this approach is that the polluter-pays-principle extends only to payment of the amount needed to reduce emissions, not payment to compensate those who the greenhouse gas emissions may have harmed (Masondo and Nwosimiri 2023).

Given that the history of South Africa has resulted in many of the most vulnerable communities being situated downwind of adverse externalities (Leocadia 2022), it is likely that this oversight in the Carbon Tax Act only serves to reinforce those vulnerabilities (Patel 2014). It does nothing to reverse and mitigate the past effects of environmentally damaging activities. Furthermore, in addition to not fully accounting for the entirety of the social cost of greenhouse gas emissions, the Carbon Tax Act will see cost-shifting practices by producers when prices of carbon intensive goods rise as compared to less carbon intensive products (Masondo and Nwosimiri 2023). The consumer will ultimately shoulder the tax burden rather than the producer. Again, this is problematic as it is those who have already suffered who need to continue shouldering the burden.

Despite various concerns about the equity of the Carbon Tax Act, from a purely environmental perspective, the Act could be said to have been effective. Indeed, between 2019 when the Act was signed into law, and 2022, carbon dioxide emissions in South Africa did decrease (Tiseo 2023). As this time period (2019–2022) covers the first of three phases of actions identified in the Carbon Tax Act, future carbon dioxide reductions as a result of the Act may occur (Masondo and Nwosimiri 2023). Without decrying the efficacy of the Act, however, there is cause for skepticism. Namely, as the government recognized in the 2021 NDC, the South African economy, largely because of COVID-19, contracted significantly (South Africa Department of Forestry, Fisheries and Environment 2021). As a result, the carbon dioxide reductions recorded between 2019–2022 may well have been a result of lower economic growth rather than the Carbon Tax Act per se. Thus, while the carbon tax is a step in the right direction towards achieving climate change mitigation goals, the tax itself could be more effective if, for example, the tax rate was set at a higher level (Pooe 2022). The tax should also be extended to other types of pollutants (Abel et al. 2023).

CLIMATE CHANGE ACT

An important law that completed South Africa's legislative process in 2024 to address climate change is the Climate Change Act. It was enacted on 23 July 2024, when President Cyril Ramaphosa assented to this much-anticipated Act (Centre for Environmental Rights 2024b). Signed into law significantly more than a decade after the National Climate Change Response White Paper, the final product of a bill first tabled in Parliament in 2018, marks a significant departure from South Africa's previous responses to climate change. It comes at a moment of acute political fragility and against a backdrop of escalating climate crises. However, it was only in March 2025 that the law came into force, but then only in part. When the rest of the various parts of the law will come into force were left unannounced.

Critically, the Act provides a legal framework designed to bring many of the issues concerning climate change under its umbrella to improve the nation's ability and capability to gradually reduce greenhouse gas emissions and improve climate resilience. It gives the government the authority to impose more stringent policies to address climate change. The state also acquires the power to carry out the more expansive policies contained in the NDC. Consequently, furthering the country's efforts to reduce its greenhouse gas emissions as outlined in its 2021 NDC, created in response to the Paris Agreement, the Climate Change Act develops and embeds (1) adaptation; (2) mitigation; and (3) institutional arrangement strategies into the edifice of South African law (South Africa Department of Forestry, Fisheries and the Environment 2021; South Africa Climate Change Act 2024). To give effect to the values and aims codified in the Act, Section 7 stipulates that all existing climate-related policies, measures, laws, and programs must be reviewed and, where necessary, amended to ensure a concerted response to the climate disaster (South Africa Climate Change Act 2024).

The Climate Change Act take precedence over all of South Africa's previous responses to climate change. In this regard, for a country that, save for the 2012 carbon tax, has exclusively sought to mitigate and adapt to climate change through regulations, policies, and top-down strategies rather than legislation, in *de jure* terms, the Climate Change Act amounts to a "significant stride towards a sustainable future" (South Africa Climate Change Act 2024). Whether or not this amounts to a de facto improvement in South Africa's response to climate change and how the response signaled by the Act attends to the historical and contemporaneous vulnerabilities of the South African population is another question, and is reviewed below.

Usefully, the new law defines, manages, monitors, and implements South Africa's actions to deal more broadly with climate change. It establishes processes to ensure that adaptation and mitigation occur, and identifies the authorities responsible for implementing the processes (Averchenkova et al. 2019). These include national, provincial, and local governments. In this regard, the law orders every organ of state to review, revise, amend, coordinate, and harmonize all their policies and measures, programs, and decisions to ensure that the risks associated with climate change, and the damage caused by it, are taken into consideration in all government actions. The law formally recognizes the Presidential Climate Commission, gives it legal status, and empowers it in a number of ways, including to commission science-based research reports, facilitate difficult negotiations between government, business, labor, and civil society aimed at decarbonizing the economy and society in a fair and just way, and, more broadly, determine how to implement a "just transition."

Importantly, the law incorporates climate change governance, assessments, and responses into the provincial and local government structures to ensure coherence and institutional competence across the board (Rumble 2019), recognizing that climate change issues should not fall solely within the competence of the national authority, but are also the responsibility of regional and local governments. While a lot of the focus is on mitigation, procedural adaptation provisions abound in the Act. It calls for the development of a National Adaptation Strategy, including the adoption of National Adaptation objectives and scenarios. The law mandates identified officials within the local, regional, and state level governments to establish viable targets and goals within one year of the law coming into effect.

As noted above, while the law has been finalized, it has not, at the time of this writing, come into effect completely, which has delayed the implementation of these important provisions. The review of the Act below firstly presents a detailed discussion of the Act's provisions regarding (1) adaptation; (2) mitigation; and (3) the institutional arrangements deemed necessary to alleviate the risk posed by climate change in South Africa. The Act's requirements regarding each of these three pillars are discussed in relation to the state's previous efforts relating to each pillar.

Subsequently, the attention, or lack thereof, afforded by the Act to issues of environmental justice are examined. Specifically, the deficiencies of the Act in accommodating and legislating for unique adaptive measures to protect the historically marginalized populations in South Africa are reviewed.

Specific Provisions of the Climate Change Act

In the Act, "adaptation" refers to any modification in human or natural systems in response to anticipated or actual climatic stimuli which is undertaken to avoid harm or to exploit advantageous opportunities (South Africa Climate Change Act 2024). In mandating adaptation in response to the effects of climate change, the Act centers around four intertwining adaptive measures: first, Adaptation Objectives; second, Adaptation Scenarios; third, a National Adaptation Strategy and Plan; and fourth, a Sector Adaptation Strategy and Plan. Crucially, the Act stipulates that each of the four measures be developed and published within specified timeframes: the first, second and fourth adaptive measures are mandated to be published within one year, while publication of the third measure is required within two years (South Africa Climate Change Act 2024).

The commencement of each timeframe, however, begins at the moment that the various parts of Climate Change Act become operational; that is, when the President submits a proclamation to this effect (Tucker and Jordaan 2024). That was done in part in March 2025. Prior to a review of the adaptive measures, therefore, it is instructive to recognize that rather than instituting adaptation strategies, the Act has simply called for them to be published within a timeframe ranging between one and two years that starts from the moment the Act becomes operational – that is, an unspecified point in the future. It probably would have been better for the strategies to have been included in the Act, but that might have taken even longer to complete the Act. This does mean that certainty does not exist with respect to the strategies, and that the power to make these regulations is in the hands of the executive, rather than the legislature, which would have been preferable.

The first adaptive measure mandated by the Act is the creation of Adaptation Objectives, which will inform the country's adaptation to climate-related impacts and promote resilience. They are due to be issued within one year of the Act becoming operational (South Africa Climate Change Act 2024). To supplement the Objectives, indicators must be established to measure progress towards meeting the Objectives, and a date determined upon which all Objectives must be incorporated into all national strategies, programs, and planning mechanisms (South Africa Climate Change Act 2024, Section 19).

The second adaptive measure included in the Act is the creation of Adaptation Scenarios, the publication of which is expected within a one-year timeframe (South Africa Climate Change Act 2024, Section 20). Simply, Adaptation Scenarios are intended to foresee the most likely effects of climate change over the short, medium, and long term (South Africa Climate Change Act 2024, Section 20). In turn, these scenarios, which are to be based on the best available science, are to be used to guide the development of adaptation measures and to determine the most effective way to reduce the Republic's vulnerabilities (South Africa Climate Change Act 2024, Section 20).

Subsequently, the third adaptative measure codified in the Act is the development and publication of a National Adaptation Strategy and Plan (NASP) within two years of the Act becoming operational (South Africa National Adaptation Strategy and Plan, Section 21). Envisioned to guide the management of climate change adaptation in the country, the core purpose of the NASP will be to reduce the country's vulnerability to climate change by enhancing the resilience of the environmental and socioeconomic systems (South Africa National Adaptation Strategy and Plan, Section 21). Lastly, the fourth adaptive measure outlined in the Act is the development of a Sector Adaptation Strategy and Plan within one year of the Act becoming operational (South Africa National Adaptation Strategy and Plan, Section 22).

The Climate Change Act includes two principal mitigation measures: first, Carbon Budgets, and second, Sectoral Emissions Targets (South Africa Climate Change Act 2024, Section 32). The Act defines Carbon Budgets as a set portion of GHG emissions allocated to specific entities, such as businesses or entire industries (South Africa Climate Change Act 2024, Section 36). Thereafter, the entity to which a carbon budget was assigned is obligated to submit a plan on the mitigation measures that will be taken by the entity to ensure that the relevant carbon budget is not exceeded (South Africa Climate Change Act 2024, Section 36). Second, the Act introduces Sectoral Emissions Targets which are to be imposed on all of those greenhouse gas emitting sectors identified by a list which is to be published within one year of the Act being made operational (South Africa Climate Change Act 2024, Section 32).

Lastly, in addition to measures aimed at adapting to and mitigating the effects of climate change, the Act outlines various institutional arrangements to align policies and government departments with its provisions (South Africa Climate Change Act 2024). Prior to the Act's enactment, despite efforts to institute vertical (between national and local) and horizontal (inter-departmental) climate policy coordination, efforts to realize the objects of past climate policies have been thwarted by a lack of strategic alignment between government departments (Averchenkova et al. 2019). As the effects of climate change grow ever graver and the responses become more complex, the Act's enactment of a clear institutional arrangement for managing adaptation and mitigation goals is an important step forward. However, while the Act amounts to a crucial piece of legislation designed to define, implement, and monitor South Africa's response to climate change, the Act and its drafting process raise important concerns in relation to environmental justice.

The Climate Change Act and Environmental Justice

While the provisions codified in the Act obviously find their basis in the Bill of Rights in the Constitution of South Africa, 1996 (South Africa Constitution of the Republic 1996), little is mentioned in the Act about the Constitution and about its human rights provisions. The few constitutional references are administrative in nature, simply defining various organs of state and addressing issues of cooperative governance. There is only one reference to the protections contained in Section 24 of the Bill of Rights, which ensures the right of everyone to "an environment that is not harmful to their health or well-being" and obligates the state to enact legislation to protect the environment for present and future generations (South Africa Constitution of the Republic 1996, Section 24). That reference is in the Preamble to the Act. The Preamble, however, has no force of law, but can be referred to when the Act is being interpreted. Problematically, no other rights are taken up in the Act, or even mentioned.

The law is also silent on the structural barriers which hinder sustainable development in the country. This is particularly problematic, as it is marginalized groups who are often without adequate access to a range of key services such as water, sanitation, and shelter. This is known as "infrastructure orientation" (Zhou 2022: 1). In many areas, this orientation is characterized by inadequate drainage systems, limited waste management and overall, and more generally, a lack of infrastructure. Where it does exist, it is limited in nature (Zhou 2022). Thus, the adaptive capacity of these communities is substantially lower than for other, more privileged communities (Zhou 2022). As a result, many people live in perpetual resource insecurity. These people are the most vulnerable to climate change impacts. Their vulnerability is disproportionate to the resilience of more affluent groups living in more developed areas. In South Africa, this vulnerability is blatantly racialized. The Act is silent on these issues, and offers no guidance as to how, practically speaking, vulnerable communities can bolster their adaptive capacity and climate resilience.

The Climate Change Act's silence in relation to this reality flouts South Africa's constitutional commitments to substantive equality, human dignity as a right and a value. It flies in the face of the Constitution's commitment to ensuring a healthy natural environment in which all people can flourish – in the truest meaning of the word. This silence amounts to a denial of the lived realities of vulnerable groups across the country. Failing to identify these key issues and provide guidance on how they ought to be addressed to make all South Africans more resilient and be better able to adapt to climate change is a significant lacuna in the Act. The gross levels of inequality that exist in the country, as noted above, mean that the legacy of colonialism-Apartheid is alive and well. In reality, South Africa's material condition is defined by spatial Apartheid – which is a guided policy of underdevelopment. It has facilitated exclusion and inequality well after the formal end of the Apartheid regime. Consequently, it is a structural driver for climate vulnerabilities related to issues such as food and water insecurity. It is also an obstruction to building the Act's purported climate-resilient vision for the country. While urban areas with concentrated wealth and service provision remain disproportionately better off than rural and underdeveloped areas, it is underdeveloped areas, even and particularly those in urban areas where millions of people live because of

Apartheid, that are left to suffer the brunt of enviro-economic insecurities and vulnerabilities.

It is therefore problematic that the legislation is so silent on the critical climate change vulnerabilities faced by marginalized groups. Underdeveloped communities continue to bear the brunt of climatic decline. In more arid regions which are disconnected from essential service infrastructure, Indigenous People are specifically negatively afflicted (Samuels et al. 2022). Climate change adversely affects both the health of these people directly, and in many cases their livelihoods (Samuels et al. 2022). The impact of climate change and insufficient adaptive abilities is devastating for such marginalized communities.

However, it must be noted that this is not simply an issue of urban-rural divide. One interesting study examined Durban in KwaZulu-Natal, the third most populous city in South Africa (Zhou 2022). This study showed that while it is generally the case that rural areas are disproportionately adversely affected by climate change, the presence of inequality, unemployment, and poverty significantly diminish any area's adaptive capacity (Zhou 2022). The study reports that "poverty has been listed as the greatest limitation in adapting to climate change" (Zhou 2022). Thus, the fact that the Act is silent on this reality is lamentable. These vulnerabilities are also problematic because the legislation offers no engagement with Indigenous knowledge systems used by Indigenous People to sustain their own climate resilience.

The various vulnerabilities that exist amongst South Africa's communities need to be assessed critically and through an intersectional lens, in order to determine and develop the appropriate means to address the multiple and interrelated climate issues. None of this is included in the Act. The Act also ought to have recognized how past environmental injustices severely impacted the majority of people across the country in their (in)ability to protect themselves from the impacts of climate change. This reality is ignored. These circumstances deeply affect issues such as resilience to climate change.

Problematically, while the Act is meant to ensure that mitigation and adaptation occurs, those who are designated to effect the necessary changes (particularly in the business sector) have few incentives in the law to do so, as the enforcement mechanisms contained in the Act are weak and insufficient (Natural Justice 2022). Another criticism is that it does not go very far in substantively addressing the issues to be solved, and that the emphasis is far too much on coordination, an issue noted in early versions of the draft law (Sweet 2018).

A major flaw in the Climate Change Act, as with most South African climate instruments and documents, is that it contains nothing about environmental justice. This means that various groups will be left behind, a key theme of this book. In fact, the law does not mention justice in any form at all. Like other climate related legal instruments, the Act does use the phrase "just transition," but only in the context of dealing with poverty. The law does intend to reduce the possibility of employment losses while fostering new job opportunities in the green economy, but it does not address the legacy of the past and the consequences of past environmental injustice. While the term "vulnerable" is used a handful of times, it is used in very broad contexts and does not make reference to specific groups. The only section which refers to identified vulnerable groups is in the principles section, where Section 3(f) states that the application and interpretation of the Act must be guided by:

> … the need for decision-making to consider the special needs and circumstances of localities and people that are particularly vulnerable to the adverse effects of climate change, including vulnerable workers and groups such as women, especially poor and rural women, children, especially infants and child-headed families, the aged, the poor, the sick and persons with disabilities. (South Africa Climate Change Act 2024)

The Act acknowledges the socio-economic impacts of climate change and mitigation measures, and stipulates that the interpretation and application of the Act must be guided by considering the special needs of localities and vulnerable people. The Act acknowledges the particular vulnerabilities present in South Africa's society, and states that an awareness of those unique vulnerabilities must guide the Act's application (South Africa Climate Change Act 2024, Section 14). However, according to the law, the vulnerability of such groups is only to be taken into account for the purposes of decision-making. It is not mandated that their specific needs be taken up and addressed on their own. Further, the language is narrow and could be interpreted as meaning that the decision-makers have no obligation to directly account for the particular vulnerabilities of affected groups. Hence, there is potential for this provision to have little effect when providing for the needs of the vulnerable. The issue of how to meaningfully respond to the vulnerabilities of particular groups ought to have been addressed in numerous sections of the Act. This ought to be a key issue.

Others too have noted that vulnerability is a factor that needs to be considered in South African climate change processes (Ziervogel and Taylor 2023). However, again, as with other laws, such provisions are absent in this law. While ameliorating the situation of the vulnerable could be read into the law, perhaps based on the broad constitutional commitments to substantive equality and human dignity, it will need decision-makers to actually do it and courts to actually accept and legitimatize it. The fact that such provisions are not specifically and explicitly included in the Act is an egregious oversight. In this regard, each section of the Climate Change Act ought to have been much more specific on the need to prioritize and assist vulnerable groups. More specific goals and objectives ought to have been delineated in the Act to ensure that vulnerable and marginalized groups are not further left behind. Already climate change-related information available to such groups is limited (Fombad 2024).

Regarding public participation, which is major issue in environmental justice, the lack of such involvement in the Act's drafting process seemingly metastasizes into the complete omission of provisions in the Act detailing how public consultation and participation must be ensured and carried out. In this regard, the NCCR-WP, published in 2011, which lays the foundation for future climate change legislation, clearly articulates the necessity of informed and active public participation in climate change legislation (South Africa National Climate Change Response White Paper 2011). However, despite the NCCR-WP underlining the importance of public participation, commentary on the drafting of the Act found that there had "… not been proper consultation and participation of communities in the process leading to the writing of [the] Bill" (Natural Justice 2022: 41).

Provision for public participation in the development of the government's response to climate change, and the specific mentioning of South Africa's people in that process, is contained in Section 7. However, the section refers to traditional leaders only:

> In order to give effect to the principles and objects set out in this Act, organised labour, civil society, business, traditional leaders and other relevant stakeholders *may advise* on the Republic's climate change response, the mitigation of climate change impacts and adaptation to the effects of climate change towards the attainment of the just transition to a climate resilient and low carbon economy and society. (South Africa National Climate Change Response White Paper 2011, my emphasis)

Thus, the section allows various groups to advise but does not set out participation as a right of the people or a duty of the state to facilitate. Further, while traditional leaders and civil society are included in the list, their inclusion is too limited to ensure sufficient views are heard. This is because traditional leaders do not represent all South Africans, and civil society is usually understood to be nongovernmental organizations (NGOs). However, in many parts of the country, there are very few NGOs and those that do exist have limited ability to make their voices heard. Clearly, mandatory processes which provide opportunities for individuals, particularly from impoverished or marginalized communities, to express their views and have them listened to ought to have been included in the Act.

As far as specific measures are concerned, some of those outlined in the Act appear in this context to be tentative. For instance, Adaptation Scenarios are required to be published within one year of the Act becoming operational, a date which remains unknown and within the arbitrary decision-making powers of the state government. The scenarios must somehow anticipate the likely impact of climate change while considering the (unidentified) vulnerabilities of the Republic. While the text on Adaptation Scenarios and the National Adaptation Plan and Strategy may note the existence of "particular vulnerabilities," nowhere in the operative text are these vulnerabilities identified nor is community participation and consultation mandated to aid in the later identification of vulnerabilities.

Broadly, in South African society, the cumulative effect of omitting any express consideration of public participation as well as simply deferring the identification of particular vulnerabilities to ambiguous timeframes, only serves to further subjugate the historically marginalized populations. One example of this is evident in the absence of a loss and damage funding mechanism in the Act. Given that the Apartheid-era legacy of spatial segregation endures and results in many peri-urban areas lacking basic infrastructure, the effects of severe weather events, induced by anthropogenic climate change, are particularly severe in these areas (Busayo et al. 2019).

Despite this, the Act makes no reference to climate-related loss and damage. Thus, when climate-related weather events, such as flooding, result in structural damage to schools, hospitals or other public infrastructure, the relevant departments have to apply for funding through the National Treasury (University of Cape Town and the African Climate & Development Initiative 2022). Had community participation and consultation been more central in the drafting of the Act, then a loss and damage funding mechanism may have been seen as a crucial component of South Africa's first sweeping climate change legislation and a necessary measure to ensure that the acute vulnerabilities facing the historically marginalized were accounted for.

However, more importantly, the Act is problematic in that it does not directly address the problems that vulnerable groups face today, and will continue to face in the future, especially bearing in mind the environmental justice problems they already endure. As has been noted:

> Climate change amplifies existing risks and creates new risks for natural and human systems. While such risks are unevenly distributed and generally greater for disadvantaged people and communities, South Africa – ever since the end of apartheid – has been struggling to find and implement a roadmap to address distributive injustices of the past. (Ruppel 2022b: 203)

This is the case despite some arguing that:

> South Africa largely focuses on its ramifications for the lives and livelihoods of people, particularly vulnerable groups. Although South Africa sees climate change as a "threat multiplier" country-wide, it is becoming increasingly apparent that it will have a disproportionally greater effect on the poor. South Africa's risk-based approach to addressing climate change is therefore fundamentally informed by the

> country's development imperatives, particularly the triple challenge of poverty, inequality, and unemployment. (Ouweneel and Simpson 2023: 248)

While this might be a sentiment expressed in various places by South Africa's government, its formalization is often done implicitly rather than being explicitly contained in the various climate change-related legal and policy instruments. It does not seemingly find its way sufficiently into the laws and processes that actually deal with climate change mitigation and adaptation. More needs to be done to assist the adaptive capacity of engaged groups (Vogel et al. 2022). Therefore, an environmental justice approach also ought to be adopted.

CONCLUSION

The enactment of South Africa's Climate Change Act marks a milestone in the country to address climate change (Climate Transparency 2022). By embedding adaptation, mitigation, and new institutional arrangements into the legal framework, the Act signifies a marked break from previous climate polices, which were largely strategic or regulatory rather than legislative. The Act's emphasis on structured adaptation measures and the introduction of carbon budgets and sectoral emissions targets reflects a coordinated and legally binding approach to managing the nation's climate response.

Nonetheless, while the Act is an important step forward, it is not without its shortcomings, particularly when it comes to environmental justice. The absence of provisions to assist vulnerable people and address the effects of past environmental injustice, the lack of comprehensive public participation and community consultation during the drafting process, and the omission of specific provisions to ensure community consultation and the identification of vulnerabilities, raises concerns about the inclusivity and equity of the Act's implementation as well as its role in advancing equality for all South Africans. The Act's lack of an environmental justice focus, its failure to specifically address vulnerability-related issues, and the lack of a loss and damage funding mechanism further underscores the need for an enlarged holistic approach that directly addresses the vulnerabilities of historically marginalized communities, especially in light of South Africa's enduring legacy of spatial and socio-economic inequalities. Thus, while the Climate Change Act is a commendable advancement in South Africa's legislative response to climate change, its effectiveness will ultimately depend on how well those measures that it calls for incorporate the voices and needs of all segments of society, particularly the historically marginalized.

This review cannot do justice to all the laws that were or are being implemented to address climate change and environmental justice. New laws are being developed in what is an ongoing process to address the broad impact of climate change. For example, the National Biodiversity Economy Strategy was issued for public comment in 2024 and is supposed to align with the White Paper on Conservation and Sustainable Use. Another important process was the creation of a high-level panel on the management of rhinos, lions, leopards, and elephants. This panel recommended important steps to be taken on a range of issues including breeding of lions and rhinos (South Africa Ministry of Environment, Forestry and Fisheries 2020). Many were taken up in April 2024 (South Africa Department of Forestry, Fisheries, and the Environment 2024).

Overall, much needs to be done to find an approach that, while moving forward to deal with climate change, also reaches backwards to provide justice for the past environmental injustices that have occurred. This is not just for the sake of providing justice for the past, but also to ensure that justice is done for the present and the future when dealing with the continuing challenges faced by the people of South Africa in their efforts to mitigate and adapt to climate change that will worsen over time. Millions of people are already experiencing these challenges due in no small part to past and ongoing environmental injustices. To deal with climate change effectively and fairly this reality needs to become part of the climate change processes in the country.

5. Climate change actions and environmental justice in Canada

INTRODUCTION

As with the rest of the world, Canada too has and will continue to experience increasing challenges resulting from climate change (Bush and Lemmen 2019). This will include changing weather patterns, increased flooding (Mohanty and Simonovic 2021), droughts, wildfires, heat waves (Meadows 2024), destruction of land, loss of water supplies (Bhatti et al. 2021), and rising sea levels (Viseh and Bristow 2023). These alterations of the normal climate and environmental patterns will impact people's health in a variety of ways (Harper et al. 2023) including, for example, through an increase in types and severity of disease (Otten et al. 2020). The numerous effects of climate change (Odimegwu 2022) also include reduced food production (Mapfumo et al. 2023; Adekanmbi et al. 2023) and consequential food insecurity (Reza and Sabau 2022) and the loss of biodiversity (Cadieux et al. 2020), including a decline (St-Laurent et al. 2022) or loss of various species (Woo-Durand et al. 2020). While some industries will be affected more than others as climate change occurs, all businesses will be affected in some way (Hurlbert et al. 2024). One significant area of concern is the mining industry, both because of its grievous impact on the overall environment (air, land, and water) (Bresson et al. 2022), but also because of the industry's impact on the lands and culture of Indigenous Peoples (Scholten et al. 2023).

A broad and deep approach must be taken by all levels of the public and private sectors to reduce climate change and mitigate the damage already caused. One example of the complexity of these issues concerns the release of greenhouse gas emissions. Consideration must be given not only to the activity of heavy industry, but also to the impact of methane emission reduction efforts on the animal husbandry industry (Neethirajan 2023) and consequently food production and security. In addition, often overlooked are the secondary and tertiary consequences of climate change, such as the potential security implications related to Canada's geo-location (Greaves 2021). It is clear that the costs of reduction, mitigation, and adaptation strategies will be enormous (Williams et al. 2021) and a range of financing processes will be needed (Yirenkyi and Vodden 2024).

Climate change and environmental justice work in Canada is multidimensional (Kuh and Leach 2022). Like other countries it is deeply affected by politics (Greaves 2021). This chapter focuses only on those issues relative to the Indigenous Peoples of Canada. While other minority communities suffer as well (Acharibasam and Datta 2024), Canadian Indigenous Peoples constitute a significant minority of Canada's population. The bulk of the Indigenous population live in rural areas which have been, and continue to be, disproportionately affected by climate change (Guyadeen and Henstra 2023). In addition, their history, significant social isolation, cultural attributes and practices, and their reduced access to social benefits such as education and employment opportunities, makes their experience of climate change different from the general population (Ford et al. 2016).

The chapter focuses on both climate change and environmental justice as it is argued that both need to be tackled if adequate adaptation to climate change is to occur. These issues will be examined from the vantage point of Indigenous Peoples (Fayazi et al. 2020), to include the issue of the legacy of environmental and social injustices suffered by Canadian Indigenous Peoples that continues to plague Canada.

CANADA'S DEMOGRAPHICS

As of 2024, the ethnic makeup of Canadians, as denoted by Canadians in terms of their origins, was reported to be 15.6 percent Canadian, 14.7 English, 12.1 percent Scottish, 12.1 percent Irish, 11 percent French, 8.1 percent German, 4.7 percent Chinese, 4.3 percent Italian, 1.7 percent First Nations (Indigenous People), 3.7 percent Indian, 3.5 percent Ukrainian, and 1.5 percent people from the Indigenous People of the Metis (Central Intelligence Agency 2024a). The religious makeup of its citizens in 2024 was 53.5 percent Christian, 4.9 percent Muslim, 2.3 percent Hindu, 2.1 percent Sikh, 1 percent Buddhist, 0.9 percent Jewish, 0.2 percent Traditional (North American Indigenous), 0.6 percent other religions and traditional spirituality, and 34.6 percent no religion (Central Intelligence Agency 2024a).

The languages spoken among its citizens in 2024 were 87.1 percent English, 29.1 percent French, 4.2 percent Chinese, 3.2 percent Spanish, 2.6 percent Punjabi, 2.4 percent Arabic, 2.3 percent Tagalog, and 1.5 percent Italian (Central Intelligence Agency 2024a).

Canada adopted the Indigenous Languages Act in 2019, which officially acknowledges Indigenous languages and provides resources to enable and expend the teaching and use of the languages to ensure their continued vitality. This undertaking should protect the 70 or so Indigenous languages in Canada

(Meighan 2023), many of which are currently endangered (Metallic 2023). It is an example that ought to be replicated in other countries.

INJUSTICES COMMITTED AGAINST INDIGENOUS PEOPLES

During the colonial past, Indigenous Peoples around the world suffered grievously at the hands of the colonial powers, and were killed, injured, and harmed in many ways (Basso 2024). Indigenous Canadians, whilst not murdered on the same massive scale by the British as their African counterparts were, suffered enormously. They had to endure, for example, an assimilation program which forced children into reserve schools, where the authorities attempted to strip them of their cultural identity and they were abused and died in significant numbers (Corntassel and Holder 2008). That legacy of the past still exists to some extent in Canada and elsewhere (Harrington 2023). As was noted by a Special Rapporteur in 2024:

> One of the expressions of the injustices and human rights violations that the Indigenous Peoples of Canada have historically suffered is the constraint of First Nations in Reserves. This constraint has resulted in harsh living conditions that have generated and continue to generate the tragedy of suicide among children and young people. The development of certain extractive activities, mainly mining and hydrocarbons, has caused risks and impacts of toxic contamination of the water sources that supply the First Nations communities. (United Nations Special Rapporteur Water and Sanitation 2024)

How these issues play out is discussed below, but suicides are a major problem (Hajizadeh et al. 2021). Land claims processes (Nagle 2017) are still required to sort out the many and diverse historic land claims agreements entered into between Indigenous Nations and British and Canadian Governments. More research is required to inform ongoing and future negotiations over these treaties (Rodon 2018).

The Federal Government is slowly assuming greater responsibility for protecting the Indigenous communities. One attempt was the introduction of Bill C-230, an Act respecting the development of a national strategy to redress environmental racism to Federal Parliament that sought to enhance the ability of the government to address environmental racism/justice issues (Venkataraman et al. 2022: 567) The Bill, however, failed to pass, as a result of the dissolution of the Canadian parliament in 2021 (Venkataraman et al. 2022: 567). Subsequently, Bill C-226, retitled An Act respecting the development of a national strategy to assess, prevent and address environmental racism and to advance environmental justice (Canada Parliament 2023) reintroduced the aforementioned Bill C-230 to Parliament (Canadian Human

Rights Commission 2024). Bill C-226 had the same intention as Bill C-230: to empower the Canadian Government to come up with a strategy for addressing environmental justice issues (Canadian Human Rights Commission 2024). Additionally, on 13 June 2022, Bill S-5 received Royal Assent (Purohit 2023). The Act renews and modernizes the leading environmental legislation in the country, the Canadian Environmental Protection Act (Purohit 2023). Bill S-5 specifically establishes the Government's focus on environmental justice (Canada, Government of 2023a). These are important steps on the road to Canada fully and fairly addressing climate change issues, including past environmental injustices. Hopefully, "[a]lthough environmental justice was slow to emerge as a frame in Canada, it appears to be steadily gaining momentum in influencing government policies and priorities" (Giang et al. 2022: 444). It does, however, now need to be applied more specifically to climate change issues. It also should be applied taking into account past environmental injustices.

DEALING WITH CLIMATE CHANGE

Of importance is that Canada has recognized the potentially devastating effects of climate change (Health Canada 2022). It accepts that the frequency and intensity of health hazards (Durand-Moreau et al. 2022) associated with excessive heat, wildfires, floods, air pollution, decreasing water quality and availability, vector-borne illnesses, and chronic diseases are growing due to climate change (Berry and Schnitter 2022). It sees that the risks to food safety and the advent of new infectious diseases are being driven by climate change (Shafiee et al. 2022). There is recognition that food insecurity risks (Mattar et al. 2021) are rising because of disruptions to food systems (Berry and Schnitter 2022). It realizes that climate change is adding to the burden and raising expenses for individuals, institutions, and initiatives that collaborate to safeguard people's health (Canada, Government of 2021b). It also recognizes that climate change exacerbates access disparities, especially for Indigenous Peoples (Berry and Schnitter 2022) that will have further effects on the health inequalities they face (Kinay et al. 2023). What remains to be seen is how committed the Government is to initiating the significant social and economic changes that addressing this crisis wil entail.

To examine just one issue in some more detail, and area of contention in Canada is the forests which cover half of Canada. Canada has for a long time positioned itself as a global leader in safeguarding forests, work that is essential in the battle against climate change (Austen and Isai 2024). However, Canadian forests, like forests everywhere, have suffered a variety of challenges over time (Devisscher et al. 2021). Both forests (Aygün et al. 2022) and trees in general (Boulanger et al. 2022) have been adversely affected by climate change, including by climate change-enhanced wildfires (Chavardès et al. 2021). One study of the forested areas of Canada's two largest provinces (Quebec and Ontario) revealed that over the course of just under five decades, 35.4 million acres of their forests had been removed (Mackey et al. 2023). This left only about 56 million forested acres in these two provinces. The percentage of boreal forest removal is very high in proportion to the total size of the forests.

The effect of the decimation of these forests on climate change and biodiversity is severe. These forests provide a vital natural habitat for a wide variety of flora and fauna, some of which are threatened by the slow destruction of the forests. For instance, the study found that 19 out of the 21 caribou herds that the researchers examined in these boreal forests were in danger of not being able to sustain their numbers (Austen and Isai 2024).

Boreal forests such as those in Canada trap enormous amounts of carbon dioxide, one of the primary gases causing global warming, through a process called carbon sequestration. Severe logging of these forests has a significant damping effect on nature's ability to reduce carbon dioxide. Typically, less research is being conducted on these types of forests as compared to that being done on tropical forests. However, their role in mitigating the effects of climate change is critical. It is estimated that the soil and trees of the world's boreal forests store 703 gigatons of carbon. In contrast, the tropical forests of the world store roughly 375 gigatons of carbon (Austen and Isai 2024).

In addition to the threat posed to the environment directly from the continued reduction of these forested areas, First Nations people (Fa et al. 2020) who live and work in the forests are also at risk (Wyatt 2008), despite the important role they play in reducing deforestation (Sze et al. 2022). While cases such as *Tsilhqot'in Nation v. British Columbia* (2014), which acknowledged Aboriginal land rights (Morse et al. 2017), and collaborative governance agreements concerning the use of these forests (Nikolakis and Hotte 2020) have resulted in greater protections for the First Nations people, this has not seemingly reduced ongoing logging operations. While Indigenous Peoples play a critical role in protecting the forests in Canada and around the world (Fa et al. 2020) non-Indigenous-controlled land areas continue to suffer from irresponsible logging practices (Garnett et al. 2018).

And, while Indigenous Peoples do have interests in the logging industry, which seems to have grown over time (Beaudoin 2016), they seemingly derive very little benefit from it (Nikolakis and Nelson 2015). Despite this, tensions within Indigenous communities have arisen concerning balancing protection of the forests and benefiting from their economic development (Booth and Skelton 2011). Thus, conservation has come up against development (Smith 2015). There are indications that the First Nations are dissatisfied with the results of the many initiatives that they thought they would better benefit from.

One research project focusing on First Nations people living in the Canadian province of New Brunswick found that they feel that they have not obtained the benefits they desired or hoped for from governance agreements, and that authority over their natural assets stays firmly in the hands of the public and private sectors (Wyatt et al. 2015). The government claims that its policy is to focus on "conservation, recreation, habitat, water quality, economic development and the relationship Indigenous Peoples have with the land and forests" (Austen and Isai 2024). However, the problems that Indigenous Peoples have concerning the forests seem to continue to grow rather than diminish.

One striking example of the negative impact industry can have on Indigenous Peoples and the environment began in 1962 and lasted until 1975. During this period, at the minimum, 9,000 kilograms of mercury were released into the Wabigoon-English River, in an area known as Grassy Narrows (Wang and MacDonald 2022). This had a devasting impact on the Asubpeechoseewagong people living along the river below the release site (Schindler 2019). The community was purposely not informed about the possible health effects of the mercury exposure (Cecco 2023). They therefore continued to use the water and eat the fish from the river, resulting in serious, negative health and related social consequences, including the spread of Minamata disease among the people, that continue to impact the community today, three generations after the initial event (Philibert et al. 2022). This environmental catastrophe is a classic example of industry's willingness to sacrifice a marginalized people for profit, and a government taking a hands-off approach to industry. The ultimate result was the destruction of the native community, the severe physical suffering of hundreds of people with no way to counteract the effects of the Minamata disease they suffered from, and the mostly government-borne obligation to pay hundreds of millions of dollars in damages. And the suffering continues (Porter 2017). The problems caused to the people were exacerbated by limited industry and government interventions to deal with the effects of spillage. In this regard, belated attention to the ongoing catastrophe only began in 2020, when the Government approved the financing of a treatment center to assist the Indigenous People with the effects of the mercury poisoning (Giang et al. 2022). A case before the Inter-American Commission on Human Rights that is ongoing saw community members testify to the massive effects the industrial pollution has had on the people and their community. It has taken more than 60 years for these people to be heard and their needs to be at least partly met. There are, sadly, many other cases about other types of environmental harms that were perpetrated against these people.

Climate Change Mitigation

On paper, as with many other countries, Canada seems to have been making much progress in dealing with climate change. However, in 2018, one author argued that Canada was a "laggard amongst the international community in the fight against climate change" (Klaudt 2018: 189). As late as 2024, the Canadian Auditor-General reported that:

> *Repeated commitments, strategies, and action plans to reduce emissions in Canada have not yielded results.* According to Canada's 2021 National Inventory Report, Canada's emissions were 730 megatons of carbon dioxide equivalent in 2019, while its target for 2020 was 607 megatons. Canada's new target for 2030 equates to approximately 406 to 443 megatons. Despite progress in some areas, such as public electricity and heat generation, Canadian emissions have actually increased by more than 20% since 1990. (Canada, Office of the Auditor General of Canada 2024, my emphasis)

As the above comment suggests, Canada has a consistent record of not meeting its international obligations on climate change matters. Notable in its dismal climate record, Canada failed to achieve the critical goal of reverting to the 1990 levels of greenhouse gas discharges by the year 2000, as enshrined in the UN Framework Convention on Climate Change (UNFCCC). It failed to meet the legally binding objective of 6 percent under 1990 greenhouse gas levels by 2012, as stipulated in the Kyoto Protocol. It also fell short of the Copenhagen target of a 17 percent reduction in emissions from the 2005 levels by 2020 (Campbell-Duruflé 2023). Indeed, in lieu of meeting these international climate commitments, Canada's greenhouse gas emissions have actually increased by over 20 percent since 1990 (Canada, Office of the Auditor General of Canada 2024). Thus, it has been argued that "[o]verall, in spite of clear desire by some past and the current government to show leadership with respect to climate change mitigation in Canada, efforts to date have had limited results so far, and are still not in line with what is needed for Canada to make a fair contribution to the global effort" (Doelle and Dremliuga 2022: 273).

Mirroring Canada's long history of not meeting its internationally set climate commitments, Canada has seemingly missed *every* national greenhouse gas emissions target that it has set since 1990 (Wright 2023). Given the parallel failures of Canada to meet its international and national greenhouse gas reduction targets since 1990, it ought to come as no surprise that, since the 2015 Paris Agreement, "… Canada has performed worse on GHG reduction than any other G7 country" (Winters 2024).

Analyzing Canada's enduring inability to achieve its national and international climate change targets, scholars have underlined the "implementation gap" as the primary reason for the government's climate failures (Wright 2023). This "implementation gap" refers to the disjunct between stated climate aims and the actual environmental outcome achieved (Fransen et al. 2023); a result of a sluggish application of the various adaptation and mitigation practices in the country (Parry et al. 2022). The reasons for the "slow uptake of adaptation and mitigation practices include the challenges of integrating practices within a complex socioecological system, as well as uncertainties regarding the extent of their impact and potential risks associated with implementing specific adaptive responses" (Antwi et al. 2023: 43).

While Canada has recently adopted various strategies to deal with climate change, reducing fossil fuel usage remains a problem. Canada ranks fifth in the world for natural gas production and fourth for crude oil (Dike and Okparaibea 2023). In fact, there has been an increase in Canada's extraction of liquefied natural gas (LNG) (Sabti 2024), much of it going to exports (Chen 2023). One disconcerting twist is that the LNG terminals are being built with the cooperation of Indigenous Peoples on their land on the Pacific Coast. While some of the local people will benefit financially from this development, it has become a controversial and even divisive issue in the country, including for other Indigenous Peoples, because of the climate change consequences (Baird et al. 2024). While various groups protested, the projects went ahead regardless, with pipelines crossing some Indigenous land (Lajoie-O'Malley 2024). Critically, the goal of reducing fossil fuel emissions will suffer. Some effort has been made to find alternative methods to counter the negative consequences of expanding fissile fuel extraction. However, it has been noted that:

> While the Trudeau government's climate policy marks a notable break with three decades of federal inaction on the issue, the policy's central component (a small carbon price in exchange for new oil pipelines) is arguably worse than total inaction ... Worse still, in its promotion of ever-greater oil sands expansions, the policy exacerbates a shameful history of illegally exploiting Indigenous lands and resources without permission, threatening the very continuity of First Nations in the surrounding areas. (MacNeil 2021: 124)

The country has taken various steps to deal with fossil fuels and other climate change issues. It has two distinct carbon pricing schemes: a consumer carbon levy that is applied to the purchase of fossil fuels, and a program for large industry that charges firms based on a portion of their actual emissions. However, a controversial aspect of this, it seems, is that some large producers are freed up to some extent from the scheme's full effects. Individuals, small- and medium-sized enterprises, First Nations, and public sector organizations like hospitals, colleges, schools, and municipalities are all impacted by the consumer levy (Rabson 2024). The levy is applied to more than 20 products, and increases regularly, most recently in April 2024, and currently stands at, for example, 12.38 cents per liter for propane, 17.6 cents per liter for gasoline, 21.39 cents per liter for diesel, and 15.3 cents per cubic metre for natural gas (Campbell 2024). While this is a national tax, there are many rebate programs that result in consumers having some of the paid-out taxes returned to them. Despite extraordinary efforts on the part of the Federal Government to alleviate the burden of the tax on those least able to afford it, and explain its purpose and importance, there remains a 70 percent consumer resistance to the price increase (Canadian Taxpayers Federation 2024). These issues are also being exploited by some Canadian political parties for electoral purposes (Citizen's Climate Lobby 2024). As noted above, climate issues are deeply affected by the political environment. Despite the controversy, the levy has resulted in a reduction in emissions of eight percent since 2019 (Canada, Government of 2024a). However, alternative methods for obtaining a more significant reduction in emissions must be found (Plourde 2022).

Broadly speaking, much more work is needed to educate consumers on the purpose and benefits of the carbon tax, in addition to what still must be done to mitigate and adapt to climate change. As one study opines: "[f]inancial and logistical support and data, information, and knowledge sharing are key factors that enable the successful implementation of climate adaptation/mitigation practices" (Antwi et al. 2023: 56).

The critical role that Indigenous Peoples play in this work must be further acknowledged. As was noted by Patty Hajdu, Minister of Indigenous Services and Minister Responsible for the Federal Economic Development Agency of Northern Ontario, in 2024:

> Indigenous Peoples are on the front lines of climate change in Canada, and as the original stewards of our lands and waters, they are uniquely positioned to lead solutions to ensure a sustainable and prosperous future for generations to come. (Canada, Government of 2024b)

While the narratives of Indigenous People's role in addressing climate change are often clear, they are not often reflected in the practice on the ground. While there has been some development, such as the LNG projects discussed above, they still face many challenges.

A tough issue is what should take precedent: either the development of extractive and transport practices related to fossil fuels continues if Indigenous Peoples benefit financially from them, or should Indigenous Peoples forego these business opportunities in favor of reducing emissions? In some ways this is a representation of the growth versus sustainability argument that pervades all the debates around climate change. However, there has to be a move away from fossil fuels quickly to save the planet. Already, there are many obstacles to doing so. Indigenous Peoples need to be assisted to move to other developmental projects, and be given the resources to do so. Their roles in other areas of the economy need to be supported and assisted more often to allow them to thrive.

Climate Change Adaptation

While a 2023 paper reported that "in Canada and elsewhere, local strategic planning and policy remains focused on mitigation, with adaptation receiving significantly less attention" (Bonnett and Birchall 2023: 142), some see that more has been done recently as far as adaptation is concerned (Parry et al. 2022: 2).

At the federal level in Canada, many processes have been undertaken and policies and laws established to further climate adaptation action. These include the Climate Adaptation Platform (2012); the Pan-Canadian Framework on Clean Growth and Climate Change (2016); the creation of the Disaster Mitigation and Adaptation Fund (2018); and the National Adaptation Strategy and Adaptation Action Plan (2022) (VanderZwaag et al. 2023). The Adaption Action Plan implements the National Adaptation Strategy, setting out the federal plan across the 22 federal departments and agencies, and bringing all the programs and proposals together (VanderZwaag et al. 2023). Some other processes include the Greening Government Strategy (Adamson and Medeiros 2023), the Northern Transportation Adaptation Initiative (Tretheway et al. 2021), the Arctic and Northern Policy Framework (Canada, Government of 2022a), and the Federal Sustainable Development Strategy.

In addition to the many and various problems and challenges associated with all of these initiatives, there appears to be an overall implementation gap which hinders effective action being taken. It has been argued that "[s]uccessful adaptation must therefore be an ongoing, iterative process that addresses immediate threats and prepares for long-term changes" (Birchall et al. 2022: 2).

One foundational problem with these initiatives appears to be that certain societal sectors have been insufficiently included in their planning and execution; including Indigenous Peoples. There are, however, various programs that are assisting Indigenous Peoples in dealing with the effect of climate change on their lives. This is critical, because as was noted by the Canadian Minister of Energy and Natural Resources in 2024:

> Indigenous Peoples face significant and unique climate challenges that threaten livelihoods, cultures, knowledge systems and lives. In the face of these threats, Indigenous communities from coast to coast to coast are taking remarkable action to fight climate change and adapt to its impacts. Indigenous-led climate solutions offer better resiliency, better outcomes and more space for expert voices as we move

> together along the path to a clean and prosperous future. (Canada, Government of 2024b)

One program that existed until 2021 was the Climate Change Preparedness in the North Program, which focused on Indigenous groups in the northern part of the country only (Canada, Government of 2020). A corresponding effort, the First Nation Adapt Program focused on the groups living south of the 60th parallel (Canada, Government of 2022b).

Another program is the Northern REACHE Program to support renewable energy for these Indigenous Peoples. Usefully, renewable energy projects for Indigenous communities are being supported by governments at different levels, prioritizing projects that partner with Indigenous Peoples (Hoicka et al. 2021: 3). This has meant that in some cases Indigenous ownership can account for up to 50 percent of some Canadian renewable energy projects, which is useful also for promoting equity (Rioux-Gobeil and Thomassin 2024: 2).

Another project is the Indigenous Community-Based Climate Monitoring Program that provides $6 million per year to support Indigenous-led projects that monitor climate and the environmental effects of climate change within community boundaries and on traditional territories using Indigenous Knowledge Systems and Western science (Canada, Government of 2023c). The program is intended to run for ten years from 2018 (VanderZwaag et al. 2023). However, it ought to be a long-term project. Other similar programs that focus on Indigenous Peoples include the First Nation Adapt Program, the Building Regional Adaptation Capacity and Expertise Program, and the Aquatic Climate Change Adaptation Services Program.

While it seems that these programs have been supporting and funding research, what the effects and outcomes of these are is unknown. Hopefully, the effectiveness of these programs will be the subject of research as well as community attitudes towards them. Certainly, the number and breadth of these programs does seem to indicate a positive attitude by the authorities to deal more positively with the issues than was the case in the past. However, more broadly, it has been argued that "Indigenous peoples' concerns and aspirations have not always been central in the climate debate" (Johnson et al. 2022: 1542). Often, they have not been included in the processes (VanderZwaag et al. 2023). A 2021 report by Indigenous Climate Action, an Indigenous-led organization concerned with the environment, entitled "Decolonizing Climate Policy in Canada: Report from Phase One" found that "Indigenous Peoples in general, and our rights, our knowledges, and our approaches to climate change are systematically excluded from the creation and implementation of climate policies and plans" (The Decolonizing Climate Policy Project, 2021). Gricius and Martel in 2024 noted that: "[m]ore broadly, Indigenous women are extremely underrepresented in environmental decision-making, but many Indigenous women hold knowledge about the environment that is deeply rooted in their physical environment" (Gricius and Martel 2024: 1). The exclusion of Indigenous Peoples continues to be the case in some respects and in certain sectors. For example, the fisheries sector is especially important for both coastal and inland Indigenous Peoples. The industry has been, and will continue to be, significantly affected by environmental pollution and climate change. Despite these factors, it has received insufficient mitigation and adaptation-related attention that has resulted in worse than necessary negative economic and social impacts, particularly on Indigenous groups (Boyce et al. 2023). Other sectors also need greater focus.

Another problem area where the exclusion of Indigenous Peoples occurs is with climate change adaptation research. Generally, it has been argued that "climate adaptation research is being championed by external organizations and senior orders of government, there is a lack of effective and on-going collaboration between the municipality and the organizations providing climate-relevant expertise" (Boyce et al. 2023: 7). However, attempts are being made to bring together Indigenous Peoples and scientists on climate change initiatives. One such initiative is taking place at the Nunavik Research Centre in Northern Quebec, which operates under the Resource Development Department of Makivvik, an Indigenous political organization operating as the legal representatives of Quebec Inuit. The Centre monitors the effect of climate change on the Inuit People and recommends adaptation methods (Byskov and Hyams 2022).

As has been noted above, climate change research in Canada is lacking in certain respects. In particular, there has been an insufficient inclusion of Indigenous Peoples in the development and implementation of processes designed to reduce climate change and address the damage already done, and more specifically, a lack of attention to where Indigenous Peoples live, such as in the Arctic regions (Vogel and Bullock 2021). While efforts have been made, and are continuing, to enlarge the participation of Indigenous Peoples in climate protection and mitigation plans and policies, more needs to be done in this area. As has been noted:

> Canada has been slow to recognize the necessity of ensuring Indigenous peoples can effectively participate in climate action planning including in relation to adaptation, putting Indigenous rights at risk … some but not all Canadian federal and provincial government adaptation law and policy initiatives have sought to effectively include Indigenous peoples and for those have that, it is an evolving process. (VanderZwaag et al. 2023: 8)

Unfortunately, the engagement of Indigenous Peoples has not often been framed in terms of human rights issues; nor have the colonial practices and processes that marginalize them been acknowledged (Tormos-Aponte 2021:

126). As a result, many of the same problems persist. Properly addressing these issues is not simply a matter of enhanced inclusion. It is also about the effectiveness of the mitigation and recuperative activities. Indigenous Peoples have already been deeply affected by environmental injustices and climate change, and unless effective action is taken to address past wrongs, current actions to address climate change will be ineffective, to the continued detriment of the Indigenous Peoples and the country itself. As has been argued, the inclusion of Indigenous Peoples in climate change issues remains an evolving issue (VanderZwaag et al. 2023).

There are a range of other issues that need to be addressed to ensure that adaptation processes sufficiently benefit Indigenous Peoples. One of these is funding. While the Government has created a fund to assist nearly 100 "Indigenous Guardians" initiatives for "protecting and conserving ecosystems, developing and maintaining sustainable economies, and continuing the profound connections between natural landscapes and Indigenous cultures" (Canada, Government of 2024c), the fund only provides a total of $25 million stretched over a four-year period (Canada, Government of 2024c). While this constitutes important support for adaptation programs (Vogel et al. 2022a), much more is needed. In addition to a shortage of necessary funds, the Government funding processes themselves make long-term action planning and implementation difficult. An analysis of empirical data conducted in 2023 determined that:

> Northern First Nation and Inuit communities in Canada, for example, must navigate short-term funding cycles that can hinder adaptation progress and community participation. … The lack of available funding for climate change adaptation projects serves as a barrier for communities to act, even when communities have the agency and knowledge to move a project forward. (Taylor et al. 2023: 29 and 14)

Usefully, the Government of Canada is providing additional resources to assist Indigenous Peoples in their efforts to adapt their living conditions to reduce their impact on the climate. Indigenous Peoples predominately live in rural areas and use diesel for heating and electricity. Their dependency on fossil fuels is especially costly because of their remote locations, where often the only way to bring in supplies is by air. To address this issue, the Government has recently provided additional funds to alleviate Indigenous Peoples' overreliance on diesel by providing resources to allow them to transition to clean energy. This will have dramatic effects for these communities (Canada, Government of 2024b). Programs such as these have enabled Indigenous communities to move to the forefront of Canada's clean energy evolution (Laboucan 2022). Although not overtly addressed by the program, this progress towards clean energy reliance will also help mitigate past environmental injustices. The connections between climate change and environmental justice should be articulated more often. Additional, similar undertakings that assist Indigenous communities to adapt to climate are essential.

Historically, Indigenous women have been insufficiently involved in the development and implementation of climate-related processes. As has been noted:

> [by] centering Indigenous women's perspectives and experiences, Canada can foster more comprehensive and effective climate adaptation strategies that are ecologically sustainable and socially just, reflecting the rich tapestry of Indigenous wisdom and the imperative of gender-inclusive solutions in navigating the complexities of the climate crisis. (Datta et al. 2024b: 6)

There is a need, therefore, to focus on Indigenous women, particularly in the rural areas where many of them live. In fact generally, there does still need to be a greater focus on gender. For example, it is often the case that women from an affected community are marginalized (Horowitz 2017). The evidence to date is that ensuring that women take on an expanded role in variety of areas, including, for example, negotiations with mining enterprises has a positive impact on final results (Horowitz 2017).

Recognizing the many distinct Indigenous cultures and practices, as well as the remarkably different environments that Indigenous communities in Canada inhabit, those working to address climate change require a "flexibility to adapt approaches to local circumstances and [direct their efforts towards the] co-creation of climate response strategies based on regional or local context" (Antwi et al. 2023: 58). This flexibility is an essential component of an environmental justice approach, which would help rectify past and ongoing environmental injustices (Giang et al. 2022).

There does, however, seem to be a growing amount of research that is helping to conceptualize and "decolonize climate adaptation in line with Indigenous peoples' lived experiences" (Johnson et al. 2022: 1542). This is important to ensure their inclusion in the development and implementation of policies and practices that ensures their needs are met.

CLIMATE CHANGE POLICIES AND INSTRUMENTS

To understand some of the Canadian climate change-related laws and policies in more detail and to identify and discuss some of the issues that emerged from them, the section below analyzes four Government initiatives; three at the federal level and one from Canada's largest province, Quebec. Each of the initiatives was meant to address the implementation gap while also setting out distinct climate targets for a network of actors.

It does need to be noted that the provinces have taken action in this area. For example, in 2021 Nova Scotia enacted its Environmental Goals and Climate Change Reduction Act and in December 2022 updated its climate change plan. In Ontario, adaptation is progressively being integrated into standing plans and undertakings, "making the development of climate change adaptation plans a key approach to supporting municipal climate change adaptation" (Donoghue and Katz-Rosene 2023: 44). It must be noted that overall, across the country, especially at the local level, there are major differences in what is being done and how (Bonnett and Birchall 2023: 148). This is problematic as municipalities are at the coalface of dealing with climate change (Birchall et al. 2022: 2). Although all levels of government can make plans, local adaptation is best carried out at the municipal level (Birchall et al. 2022). Municipalities are "well positioned to increase resilience to climate change through adaptation planning, as they are the closest level of government to citizens and many climate change impacts are experienced locally" (Donoghue and Katz-Rosene 2023: 44). As a consequence of this localization of activities, results vary from place to place. For example, vulnerable communities are at times insufficiently catered for and therefore equity does not result. This has been found to be a problem, for example, in Toronto (Regier and Conway 2022). There have also been problems with the coordination of adaptation plans, especially in protected areas (Barr et al. 2021: 153). More is needed to ensure that standards are applied equally across the provinces to achieve effective results, even if each province diverges in what they do. There needs to be some standardization, but at the same time allowing some divergence, understanding that each municipality faces different problems and is capable of bringing different resources to the work.

The four instruments examined are the 2016 Pan-Canadian Framework on Clean Growth and Climate Change (PCF), the 2019 Impact Assessment Act (IAA), the National Adaptation Strategy, and the provincial 2020 Québec ZéN Roadmap. In addition to evaluating the environmental impact of each intervention, issues surrounding Indigenous Peoples' rights and broader environmental justice are discussed. The central argument is that while each of the four environmental interventions embody positive steps in Canada's response to climate change, they lack the urgency demanded by climate science and Indigenous Peoples' climate warnings. Moreover, each of the discussed interventions entrench power matrices that deny environmental justice and suppress the rights of Canada's Indigenous population.

The 2016 Pan-Canadian Framework on Clean Growth and Climate Change (PCF)

Presented in 2016 at a First Minister's Meeting, the PCF is a national government process, a top-down effort to reduce greenhouse gas emissions (Reed et al. 2021: 3). Somewhat ironically, just days prior to it being revealed, the Federal Government approved two oil pipeline expansions; namely, the Enbridge's Line 3 expansion and Kinder-Morgan's Trans Mountain Expansion (Lee 2024). Nonetheless, the PCF represents itself as a "historic step in the transition to a clean growth and resilient economy" (Canada, Government of 2016). It is intended to allow Canada to meet its emissions reduction target of 30 percent below the 2005 levels by 2030 (Reed et al. 2021).

Informed by the principles of the Vancouver Declaration, the PCF sought to ride the wave of momentum generated by the 2015 Paris Agreement and touted itself as a "collaborative approach between provincial, territorial, and federal governments to reduce GHG emissions and enable sustainable economic growth" (Canada, Government of 2016: 5) Accordingly, the Government requested four federal-provincial-territorial working groups made up of people from the federal, provincial, and territorial levels "to work with Indigenous Peoples; to consult with the public, businesses and civil society; and to present options to act on climate change and enable clean growth" (Canada, Government of 2016: 10). These four groups provided the foundations of the PCF in the form of four "pillars": "pricing carbon pollution; complementary measures to further reduce emissions across the economy; measures to adapt to the impacts of climate change and build resilience; and actions to accelerate innovation, support clean technology, and create jobs." (Canada, Government of 2016: 10). Collectively, these four pillars were envisaged to support an all-inclusive plan to reduce greenhouse gas emissions to a degree capable of "meeting or exceeding" the aforementioned 30 percent reduction target (Canada, Government of 2016: 4).

Despite talk of the PCF enabling Canada to meet or exceed its greenhouse gas commitment of a 30 percent reduction below 2005 levels by 2030, Robert MacNeil astutely points out:

> This commitment [i.e., 30% reduction below 2005 levels by 2030] places Canada well towards the back of the pack among its peer nations and is not remotely consistent with the international community's goal of limiting global warming to 1.5°C. (MacNeil 2019: 117)

Alarmingly, despite the inadequacy of the 30 percent reduction commitment itself, analysts who have studied the PCF have argued that the Framework is incapable of meeting even the inadequate 30 percent reduction target (Lee

2016). Thus, while climate conscious Canadians may have expected the Federal Government's PCF to compensate for the weak 30 percent reduction target and offer a more ambitious target, the PCF failed to live up to these expectations, ultimately drawing the conclusion from commentators that the PCF: "lacks the urgency called for by climate science and fails to confront the entrenched power of the fossil fuel industry" (Lee 2016). In fact, within the context of the oil and gas sector, the PCF does little to protect the environment (Williams 2022). Thus, it has been argued that the "PCF lacks clear sectoral pathways to reduce emissions outside of pricing carbon pollution" (Williams 2022: 26).

With regard to the fossil fuel industry, the PCF re-iterated previous commitments that Canada had made alongside the United States and Mexico in 2016, to (1) phase out subsidies for fossil fuels by 2025; and (2) reduce fossil fuel oil and gas sector methane emissions by "40 to 45 percent below 2012 levels by 2025" (Canada, Government of 2016). Despite the oil and gas industry contributing one-quarter of the country's greenhouse gas emissions, beyond re-stating the two previously made commitments, the industry was largely given a pass in the PCF, with "… the growth of the oil and gas sector taken for granted, and … no measures [offered] to keep carbon in the ground" (Lee 2016). In 2021 it was argued that the PCF did not have a sufficient plan to reduce greenhouse gases and actually permitted an unceasing increase and expansion of fossil fuels development, relying almost exclusively on carbon taxes and other offsets to achieve the goals of net zero by 2050 (Deranger 2021). More recently, some have begun to argue that the targets are in fact too ambitious, and various sectors, and particularly the agricultural sector, have criticized the government for the plan (Vinco et al. 2023).

Besides criticisms regarding the perceived inadequacy of the environmental measures outlined in the PCF, scholars have problematized the Frameworks attention to Indigenous Peoples' rights, involvement and the incorporation of Indigenous knowledge systems (Reed et al. 2021). It has been argued that the:

> critical importance of Indigenous-led adaptation is highlighted in the national frameworks (e.g. Pan Canadian Framework on Climate Change, 2016) … Yet, existing adaptation frameworks rarely meaningfully include Indigenous Peoples' worldviews; Indigenous peoples continue to be structurally excluded from the development of such frameworks; some of the proposed climate solutions ignore the structural inequalities and 'may lead to disproportionate negative impacts on Indigenous Peoples and violations of Indigenous rights. (Yumagulova et al. 2023: 3, references omitted)

While the PCF called for "robust engagement" with Indigenous Peoples, there were no Indigenous representatives in the various four working groups that were established to create the four pillars of the Plan (Reed et al. 2021).

Further, the Plan does not recognize Indigenous knowledge systems. Again as has been noted by authors, this disregards "the role of Indigenous Peoples in climate governance, but it also perpetuates a framing of Indigenous knowledge that can be integrated only when convenient, disregarding the power, rights, and governance imbalances raised in knowledge integration literature" (Reed et al. 2022: 525). More generally, it has been argued that Indigenous knowledge and traditional ways of adapting to change have not been sufficiently stressed in climate change initiatives and policies (Vogel and Bullock 2021: 2561). There are, however, some signs that this is changing with a greater cognisance of the value of Indigenous input into climate change responses emerging (Reed et al. 2022).

Interestingly the language of the PCF belies this lack of attention to Indigenous Peoples. For instance, while the term "fossil fuels" is found in the PCF a mere seven times, the term "Indigenous" is invoked in the Framework a total of 83 times (Lee 2016). However, while there are many references within the PCF to Indigenous Peoples, it has been argued that "there remain no tangible steps to actualize this empty recognition" (Deranger 2021: 1). Indeed, the Framework waxes on about the participation and inclusion of Indigenous Peoples, beginning with the identification of the significance of "free, prior, and informed consent" in the introduction to the PCF (Canada, Government of 2016: 4). Despite these repeated references to Indigenous Peoples with a clear recognition of Indigenous rights in the Framework, and calls for "... robust, meaningful engagement" (Canada, Government of 2016: 4) with Indigenous Nations and communities, as noted above, the PCF failed to include Indigenous Peoples in the four working groups set up to deal with the issues of Pricing Carbon Pollution; Mitigation; Adaptation and Climate Resilience; and Clean Technology, Innovations, and Jobs, to craft the Framework (Canada, Government of 2016).

In their analysis of the PCF, Reed et al. contend that the exclusion of Indigenous Peoples from the policymaking process constitutes a violation of the notion of free, prior, and informed consent and, by implication, a violation of Canada's obligations under UN Declaration on the Rights of Indigenous Peoples (UNDRIP) (Reed et al. 2021). It will be recalled that in 2021, Canada finally announced its support for UNDRIP, enacting the *U*nited Nations Declaration on the Rights of Indigenous Peoples Act, along with the requirement to obtain free and prior informed consent, for doing a range of activities on their land, from Indigenous Peoples in Canada (Bellamy 2023). The acceptance of UNDRIP was important as Canada's refusal to recognize it earlier was seen to be deeply problematic. Thus, UNDRIP's adoption and the National Truth and Reconciliation Commission, whose mandate was to inform all Canadians about what happened in residential schools, were important steps in moving the relationship with Canada's Indigenous Peoples forward. The National Truth and Reconciliation Commission was a process that included almost 100 "Calls for Action related to the reconciliation of relations with Indigenous peoples, provides an essential sociopolitical backdrop to the broader landscape for innovations in conservation partnerships occurring in Canada" (Vogel et al. 2022a: 8).

Beyond constituting a violation of their obligations under UNDRIP, the exclusion of Indigenous Peoples from the PCF policymaking and the Frameworks invocation of the term "Indigenous" arguably represents a more subtle, malign form of neocolonial power dynamics in settler states like Canada. In his book, *Red Skin, White Masks*, Glen Sean Coulthard challenges the concept of recognition as a means to achieve equality and justice for Indigenous Peoples (Coulthard 2014). Rather than achieve these ends, Coulthard argues that the politics of recognition fails to address the underlying power imbalances in settler states and instead perpetuates colonial relationships (Coulthard 2014). Accordingly, for Coulthard, the "Politics of Recognition" designates a relationship between the settler state and Indigenous communities where the former grants a tokenistic inclusion to the latter, while denying the latter the means with which to achieve complete self-determination and autonomy (Coulthard 2014). Extending Coulthard's analysis of the connection between the settler state and Indigenous Peoples in Canada to the PCF, the Framework's insistent invocation of the term "Indigenous," yet ultimate exclusion of Indigenous voices from the policymaking table, functions to solely acknowledge Indigenous Peoples within the same power matrices that have suppressed Indigenous rights, voices, and knowledge systems. Another Canadian environmental intervention that could be said to represent an additional case of Coulthard's Politics of Recognition is the 2019 Impact Assessment Act, which is dealt with below.

The 2019 Impact Assessment Act (IAA)

Another step by the Canadian Government to address environmental justice was the implementation of the IAA 2019. The IAA is a Federal Government framework for establishing firmer rules to safeguard Canada's environment and re-affirm public confidence in the decision-making process around resource development (Offenheim 2020). It was built upon the foundation of two previous Assessment Acts: the Canadian Environmental Assessment Act of 1992 and the Canadian Environmental Assessment Act of 2012 (Tsuji 2022). However, shortly after it was adopted the IAA was criticized because, while it "responds to some of the identified obstacles in ways that improve upon past federal EA Acts, … it falls short of guaranteeing a measurable improvement" (Eckert et al. 2020: 81).

Nevertheless, this is an important piece of legislation, as regulatory impact assessments inform decisions regarding proposed development projects that are within the government's jurisdiction (Offenheim 2020). In essence, impact assessments ensure that decision-makers are aware of issues, such as, for instance, the biophysical or social implications of a proposed project, when deliberating over a particular project. Simply put, as Morrison-Saunders et al. state, impact assessments encourage decision-makers to "… think before they act" (Offenheim 2020: 3; Morrison-Saunders et al. 2014: 2).

Thus, the new IAA was implemented to provide a streamlined way to assess the potential environmental impacts that proposed projects would have on all Canadians, especially the poor and Indigenous communities across the country (Paquin 2023). The IAA sought to prioritize the public interest in the government's environmental protection efforts through an assessment of five components: "significant adverse environmental effects, mitigation measures, impacts on Indigenous people and their rights, climate change, and sustainability" (Tsuji 2022: 3501).

The IAA identifies several primary objectives including: the fostering of sustainability; the promotion of cooperation between federal and provincial arms of the government; the provision of opportunities for public participation; and the consideration of scientific, Indigenous and community knowledge in the assessments (Canada, Government of 2019). In order to meet these objectives, the IAA sets out a comprehensive process for assessing the impact of proposed development projects constituted by several sequential stages including planning, impact statement, impact assessment, decision-making, and follow-up (Canada, Government of 2019).

Operating at the heart of the IAA – evidenced in both its primary "purpose" and through its inclusion in each of the stages – are sustainability criteria. The IAA defines sustainability as "the ability to protect the environment, contribute to the social and economic well-being of the people of Canada and preserve their health in a manner that benefits present and future generations" (Canada, Government of 2019: 41). With sustainability considered throughout the process, the final discretion on proposed projects rests with the Minister of Environment and Climate Change or the Governor in Council and must be in the "… public interest" (Canada, Government of 2019: 41).

Despite the stated intentions of the IAA, a 2021 report titled "Is Canada's Impact Assessment Act Working," found that the use of the Act in the two years following its enactment suggested that it would only be "… applied annually to a dozen or fewer projects that pose the most harmful environmental effects within federal jurisdiction" (Johnstone et al. 2021). There were 305 energy projects in operation in 2021 alone – excluding mining, forest, or other projects. The fact that the IAA was applied to only "a dozen or fewer" projects a year suggests that the government lacks the capacity or the intent to protect the country's environment (Canada, Government of 2021a).

Despite the paucity of instances of Government applications of the Act to development projects, it has garnered some support from various sources. One notes the 2024 comment that "the IAA is a good step forward, in that it mandates consideration of the effects of the proposed project on Canada's Indigenous people and of traditional ecological knowledge (TEK) provided during the environmental assessment process" (Gricius and Martel 2024: 6). The Act recognizes that Indigenous Peoples' guidance, familiarity, and experience are vital components of efforts to realize successful adaptation and mitigation plans to deal with climate change (Antwi et al. 2023: 55). It has been consistently recognized that "Local knowledge of environmental and social changes remains a crucial aspect of monitoring and navigating rapid climate changes in northern environs, and successfully adapting to climate change, and the institutional opportunities for serving the adaptation needs of Indigenous communities through co-design and cross-cultural learning are ripe" (Vogel and Bullock 2021: 2560).

However, Stephen Tsuji has voiced concern over the procedural aspects of the IAA, particularly the opacity of the final decision-making process (Tsuji 2022: 15) that Tsuji argues is both unaccountable and lacking in transparency, raising doubts about the protection of constitutionally endowed Indigenous rights (Tsuji 2022: 15). Furthermore, the Act, Tsuji states, fails to adequately address the requirements of free, prior, and informed consent of Indigenous Peoples (Tsuji 2022: 15). While the opaqueness of the decision-making process per se may not necessarily warrant skepticism, the fact that Indigenous rights are framed in a "public interest" determination gives way to the possibility that the Minister of Environment and Climate Change or the Governor in Council may unilaterally (and negatively) weigh Indigenous rights against other interests, such as economic development (Tsuji 2022: 15).

Critical as well is the fact that climate change processes in many municipalities, especially in rural areas, are not very advanced. In some places "rural plans are under-developed, lacking characteristics that comprise high-quality climate change plans, especially a strong fact base, comprehensive climate goals and policies, clear guidance on monitoring and evaluation, and a discussion of how stakeholders were engaged and contributed to the development of the plan" (Guyadeen and Henstra 2023: 130).

With respect to Indigenous knowledge systems, various representatives of the Indigenous communities in Canada raised concerns about the IAA's definition of sustainability (Tsuji 2022: 14), which they deem is insufficient due to its omission of cultural sustainability, including physical sites, landscapes, practices, and values (Tsuji 2022: 14). As Tsuji notes, "Culture is foundational to Indigenous peoples' identity and wellbeing" (Tsuji 2022: 14). In order to properly guide impact assessments, therefore, the IAA ought to elevate the importance of cultural sustainability in its framework (Tsuji 2022: 14).

National Adaptation Strategy

The National Adaptation Strategy (NAS) was announced in 2020 with the publication of Canada's climate plan, in a policy document called a "Healthy Environment and a Health Economy" (Canada, Government of 2022b). That document announced the intended creation of the NAS in partnership with the different levels of government, Indigenous Peoples, and a variety of other stakeholders. It also committed to ultimately protecting 25 percent Canada' territory through partnerships with Indigenous groups by developing "Indigenous Protected and Conserved Areas" (VanderZwaag et al. 2023: 3).

Scholars had determined that adaptation governance in the country was "weak – contributing to fragmented efforts across governments and sectors – [and that] the NAS is expected to support the establishment of mechanisms that enable effective, coordinated action within and across jurisdictions while respecting their unique needs and circumstances" (Parry et al. 2022: 2). Canada finally adopted its NAS in June 2023. The NAS sets out a common vision to address the various impacts of climate change (VanderZwaag et al. 2023), improve coordination and mainstreaming (Henstra 2022: 1), and allow a range of role players at all levels of government, working in partnership with Indigenous Peoples, to develop and implement together a range of projects to deal with climate change (Zanjani et al. 2023).

Usefully, the NAS strategy does absorb the rights of Indigenous Peoples as a foundational aspect to ensure that the Government's adaptation processes also reflected regional and local aspects. It includes a commitment to "strive for equity, as well as climate and environmental justice" (Canada, Government of 2022b). Thus, an integrated response is sought. The NAS calls for "addressing and minimizing social, gender, racial, and intergenerational inequities, which requires diverse perspectives at the table – including youth and persons with disabilities. It also includes prioritizing populations and communities at greater risk of climate change impacts, for example, due to historical and ongoing practices and policies that shape lived experiences, capacity, and access to resources" (Canada, Government of 2023b). However, some argue that while the rhetoric assumes inclusion, the reality is often otherwise. This is especially so for Indigenous Peoples generally and Indigenous women particularly (Gricius and Martel 2024).

Beyond its policy framework, the NAS could be a useful tool to increase public and government leadership awareness about climate adaption in Canada (Parry et al. 2022). In addition, the Indigenous Climate Action's 2021 Report: Decolonizing Climate Policy in Canada, discussed above, points out that "it appears that Indigenous women and 2SLGBTQIA+ folks, and Indigenous Peoples in general, and our rights, our knowledges, and our approaches to climate change are systematically excluded from the creation and implementation of climate policies and plans" (Indigenous Climate Action 2021: 9). The NAS might help further other policy priorities in addition to integrating Indigenous Peoples into climate change response actions, including advancing reconciliation with First Nations (Parry et al. 2022).

The 2020 Québec ZéN Roadmap

Unlike the PCF and IAA, the Québec ZéN Roadmap is a bottom-up, energy transition framework created by an amalgam of non-governmental organizations (Reed et al. 2021). Led by the United Front for Energy Transition (UFET), the Québec ZéN Roadmap was developed through a collaborative effort involving over 70 environmental groups, community organizations, and unions "sharing the goal of accelerating the transition towards a Québec that will be carbon neutral, more resilient and more just" (Front Commun pour la Transition Energétique 2019: 63).

In order to realize this shared vision, the Roadmap delineates three interrelated components for a net-zero emissions Quebec. Firstly, the Roadmap focuses on building *resilience* through the reclamation of environmental areas and the protection of ecosystems in Quebec (Front Commun pour la Transition Energétique 2019). Secondly, it is a *political framework of the transition to* strengthen government coherence and accountability of governments, strengthen education and social dialogue, ensure a fair transition, position human rights at the core of the transition, and ensure that the requisite finances are made available (Front Commun pour la Transition Energétique 2019). Thirdly, the Roadmap sets out sectorial and cross-sector initiatives to reduce greenhouse gas emissions (Front Commun pour la Transition Energétique 2019). The overarching goal of the Roadmap, and one that it shares with the 2021 Federal Government-led Canadian Net-Zero Emissions Accountability Act, is net zero emissions by 2050 (Front Commun pour la Transition Energétique 2019).

Similar to the PCF, the Roadmap frequently invokes the term "Indigenous," even going so far as to state that a net-zero Quebec is only attainable through "Respect[ing] the right of Indigenous Peoples to a free, prior and enlightened consent" (Front Commun pour la Transition Energétique 2019: 20).

However, despite the aspirations of Indigenous Peoples to play a central role in the implementation of the Roadmap, the UFET "was not able to reach" Indigenous communities during the drafting process (Reed et al. 2021). This is problematic and indicates a lack of commitment to include their views. As a result, the UFET proceeded with the development of the Roadmap, only consulting Indigenous Peoples in Quebec once the report had been finalized (Reed et al. 2021). The Roadmap gushes inclusion and participation for Indigenous Peoples but ultimately overlooks them in the policymaking phase, giving rise, once again, to Coulthard's notion of the "Politics of Recognition" (Coulthard 2014).

Thus again, as with the Pan-Canadian Framework, Indigenous People have in reality played a limited role in the development of the ZéN Roadmap. Some have gone as far as to argue "that this exclusion constitutes a violation of Indigenous rights to self-determination and to free, prior and informed consent" (Reed et al. 2021: 3).

More broadly, it has been argued that both the ZéN Roadmap and the Pan-Canadian Framework, despite references to Indigenous Peoples, disrespect and in some instances violate the inherent constitutional and treaty rights of Indigenous Peoples. It is clear that despite wording to the contrary, they both essentially disregard Indigenous viewpoints, Indigenous knowledge, and Indigenous approaches to dealing with climate mitigation and adaptation. Thus, these instruments are seen to conflict with Canada's pledges to promote reconciliation and improve relationships with the Indigenous Peoples of the country (Reed et al. 2021).

CONCLUSION

Climate change is going to have a devastating effect on Canada, as it will around the world in the future. What Canada will experience specifically is an increased frequency and severity of severe climate events in the country. These will significantly rise in Canada as climate change increases. There will be stronger storms, including more intense rainfall. This will result from warmer air that retains more moisture due to rising concentrations of heat-trapping gases in the atmosphere. Rising sea levels, melting snow, and shifting weather patterns will all contribute to the development of more severe and erratic storms, causing a range of effects including flooding. These will have disastrous effects on livelihoods, economies, and various communities. It will have disproportionate effects on the Indigenous Peoples in the country.

As has been shown throughout this chapter and specifically through an analysis of four environmental interventions – the 2016 Pan-Canadian Framework on Clean Growth and Climate Change, the 2019 Impact Assessment Act, the National Adaptation Strategy, and the 2020 Québec ZéN Roadmap – much needs to be done in Canada to address the legacy of the past and develop better processes to deal with climate change.

As can be seen through the evaluation of each of the four environmental interventions there has been a lack of sufficient political will to deal with the environmental injustices of Indigenous communities in Canada.

Regarding the 2016 PCF, while representing a positive step, the Framework gives a free pass to fossil fuels and simply re-states previous environmental policies. Additionally, despite its semantic inclusion of Indigenous Peoples, the PCF perpetuates colonial power dynamics by omitting Indigenous Peoples from the policymaking process. In turn, the greatest concern over the 2019 IAA's ability to tackle climate change stems from the Act's restricted application to just a dozen projects per year.

With respect to Indigenous rights and knowledge systems, there are concerns over the opaqueness in decisions are arrived at and the framing of Indigenous rights in a public interest determination. This implies that Indigenous rights can be weighed against economic development considerations. In addition, the IAA omits cultural sustainability in its definition of sustainability, potentially giving way to the repression of Indigenous knowledge systems. Lastly, the 2020 Québec ZéN Roadmap further exemplifies a case of the "Politics of Recognition" in Canada's environmental law, appearing to underline Indigenous participation but only requiring consultation with Indigenous communities after the drafting process.

6. Conclusion: Methods to integrate climate adaptative capacities and strategies with environmental justice to enhance the climate resilience of minorities and Indigenous Peoples

INTRODUCTION

This book has highlighted the significant challenges posed by climate change around the world, which are expected to intensify in the future. However, as noted, not all people experience the impacts of climate change in the same way or to the same extent. Many have already suffered its effects disproportionately. Those who will endure the worst consequences of climate change often lack the resources to address these challenges effectively. Specifically, it is poor and marginalized communities, including minorities and Indigenous Peoples, that are likely to bear a greater burden due to their geographic locations (Zhang et al. 2024) and the pre-existing environmental injustices they face. Climate change will also push more people into poverty as the effects become more apparent. It is specific minorities and Indigenous Peoples who will be most affected. These communities have already suffered the consequences of environmental injustices. The groups previously affected by environmental issues must have these matters addressed to ensure healthier and more sustainable lives for all. For instance, PFAS, or "forever chemicals," affect everyone everywhere. These chemicals, primarily ingested through food and water, have severe health implications. It is imperative to take steps to mitigate their impact everywhere.

When addressing climate change, these groups are often left behind in efforts to mitigate climate and other environmental harms. Their voices frequently go unheard. These are the very people the Sustainable Development Goals (SDGs) aim to support. However, this support is not being effectively realized, and in many places their situations are worsening. Therefore, greater effort must be directed toward assisting these communities in addressing the numerous challenges they face. Many things can be done, but safety nets are essential.

To effectively address their needs and the challenges posed by climate change, it is essential to understand and act on the interrelationship between environmental justice and climate change for these groups (da Silveira et al. 2018: 251). As discussed earlier in this book, there are profound connections between environmental justice and climate change issues. Both have separately been approached through calls for equity and justice, employing concepts such as climate justice and racial justice, among others (Walker et al. 2024). If these issues are not collectively addressed, marginalized groups will face increasing difficulty coping with climate change and the underlying environmental justice problems that frequently exacerbate its effects. Many of these environmental justice challenges are the root causes of existing problems within these communities and are likely to worsen in the future (Han and Sang 2020). The consequences are already being felt, stemming from factors such as the geographic locations of these communities, their socioeconomic conditions, and the lack of adequate support from state and local authorities.

An integrated approach is essential to addressing both climate change and environmental injustices due to the ongoing and escalating impacts these issues have on minorities and Indigenous Peoples. These communities are particularly vulnerable to the effects of climate change because they already face a disproportionate share of environmental justice challenges in the areas where they live. Without resolving the environmental justice issues they currently endure, their ability to cope with the effects of climate change will be further compromised (Pastor et al. 2024).

The ability of these groups to adapt to the changing climate is hindered by their already challenging circumstances. This stems from historical injustices that subjected them to discrimination, ostracization and marginalization. Their exclusion from economic opportunities and decision-making roles has relegated these communities to areas that are often environmentally and geographically disadvantaged. With limited resources, little political influence, and minimal representation, these communities struggle to address the issues they face. They lack the capacity to secure state resources needed to tackle the ongoing effects of past environmental injustices, which continue to impact their lives, health, housing, land, air quality, and water supply. Many of these areas are sites where waste is dumped, and even their own waste is inadequately managed. Their limited voice further reduces their ability to advocate for solutions to these problems. Therefore, an approach that recognizes the interconnected nature of climate change and environmental justice is crucial. Such an approach is already being implemented in some areas, as demonstrated by the Biden Administration's initiatives in the United States, as well as efforts in California (Pastor et al. 2024) and New York City (Foster et al. 2024).

However, it is important to note that adaptation strategies can sometimes have unintended consequences for these communities. As has been argued:

> Climate-related plans and projects create injustices. Hazard mitigation plans often emphasize personal preparedness for severe weather events while ignoring structural inequalities that create hazardous conditions and exposures for marginalized people. (Domingue 2022: 271)

Thus, it is essential to develop plans that lead to positive outcomes for all. A unified approach to addressing climate change and environmental justice is crucial, particularly one that acknowledges how adaptation processes can inadvertently worsen conditions for some communities and exacerbate existing issues. Such an integrated approach must account for current vulnerabilities and fully assess the risks involved. It is vital to consider the broad range of environmental justice challenges that many groups already endure.

The interlinkages between a range of issues were recognized at the end of 2024, when a report was published by the Intergovernmental Science-Policy Platform on Biodiversity and Ecosystem Services (IPBES). This report assessed the thematic interlinkages of biodiversity, water, food, and health. It examines the intricate and interrelated nature of problems such as food insecurity, water quality and availability, biodiversity loss, health concerns, and climate change. The report addresses these issues by offering a critical assessment of the data pertaining to the connections between the five nexus elements – biodiversity, water, food, health, and climate change. It is noted, however, that air, soil, and land are regarded as cross-cutting systems rather than independent nexus elements. Thus, the report concludes that because decisions to address the drivers and underlying causes of biodiversity, water, food, health, and climate degradation are frequently made in isolation, this can lead to misalignment, unforeseen trade-offs, and/or unintended consequences. Therefore, nexus approaches are essential. It is argued that nexus techniques recognize that issues within each component are linked to other components across various temporal and spatial scales. The results of the nexus evaluation can support holistic and synergistic management and governance by enhancing understanding of these relationships and identifying opportunities for cooperation across sectors and scales (Intergovernmental Science-Policy Platform on Biodiversity and Ecosystem Services 2024).

While the report emphasizes the need for just processes and outcomes, as well as engagement with local communities and Indigenous Peoples throughout, it does not explicitly mention environmental justice or the need to integrate those issues into an inclusive approach. However, the intrinsic need to address elements such as soil, land, and water implies that environmental justice matters must also be considered. Without this, there cannot be a genuine attempt to integrate all relevant problems and issues.

The report recognizes that, historically, a siloed approach has been used to address interrelated issues. It emphasizes the need to acknowledge their interdependencies, as well as the cascading, compounding, and synergistic effects of these issues. Accordingly, the reality of environmental injustices and their impacts on specific groups, such as minorities and Indigenous Peoples, must be recognized – particularly regarding how these injustices affect their capacity to cope with climate change. Rather than addressing various environmental problems in isolation or in a fragmented manner, a holistic approach should be adopted to collectively address all the issues while understanding their interrelationships. This would lead to better outcomes. It would also ensure that those who are more likely to be excluded have their voices heard and their longstanding problems – issues that undermine efforts to address climate change – properly addressed.

The Conclusion to the book, in these contexts, examines what progress has been made on climate change around the world. It notes that while much has been done around the world, a lot more remains to be done including by integrating environmental justice approaches. It notes that in some ways the context of dealing with climate change has evolved with the political changes that have come to the fore in the United States. It maintains that these will be impactful there, but will have an effect elsewhere as well. Critically, what needs to be done to mitigate climate change is examined, but also what adaptation steps need to be taken in conjunction with environmental justice. The Conclusion contains many recommendations on steps to be taken to mitigate and adapt to climate change, and how these need to be done with environmental justice issues in mind. Thus, a number of issues are examined to fuse adapting to climate change and environmental justice. These concern housing, water issues, air pollution, green spaces, and land and soil degradation. Also examined are the resources needed to adapt to climate change and deal with environmental justice matters.

PROGRESS ON DEALING WITH CLIMATE CHANGE GLOBALLY

As can be seen in the book, much progress has been made worldwide in addressing climate change over the last few years. However, much remains to be done, even in developed countries, as progress has not been as swift or far-reaching as necessary. This is particularly concerning given that the SDGs are now more than halfway through their timeline, from their inception in 2015 to their conclusion in 2030. Significant efforts are still required if their goals are to be achieved.

From both a climate perspective, and an environmental justice standpoint, the SDGs play a crucial role. This is because the SDGs emphasize the critical impact that sustainable management of natural resources has on social and economic development. The importance of conserving and sustainably using the world's natural resources is underscored, including in environments such as oceans, seas, freshwater ecosystems, forests, mountains, and other habitats, to protect biodiversity, ecosystems, and fauna. For example, SDG 15 aims to "[p]rotect, restore, and promote sustainable use of terrestrial ecosystems, sustainably manage forests, combat desertification, and halt and reverse land degradation and halt biodiversity loss." Similarly, SDG 13 (Lawal et al. 2022: 39) calls for urgent action to combat climate change and its impacts by strengthening resilience and adaptive capacity to climate threats and natural disasters, ensuring states incorporate climate change measures, and raising awareness on a range of related issues.

These are critical issues because, without a dramatic shift in addressing climate matters, the impact on the world's future will be even more severe. For this reason, the United Nations announced at the Hamburg Sustainability Conference in October 2024 that it would appoint an Envoy for Future Generations. This initiative builds on the outcomes of the Summit of the Future, held a month prior, where the Pact for the Future, including the Declaration on Future Generations, was drafted. These developments stem from the UN report *Our Common Agenda*. Additionally, numerous international agreements are in place, and regular international conferences are held under these agreements to address specific aspects of climate change and a wide range of interrelated issues.

While global efforts to mitigate the effects of climate change has intensified in recent years, there are a range of problems in this regard. One of these is significant pushback by those in various industries such as those in the fossil fuel and forever chemical ones. These industries have launched large campaigns to push back on the climate change agenda (Horel 2025). As *The New York Times* noted:

> For years, the fossil fuel industry and its allies have tried to overturn one of the most important federal rulings in the history of climate policy: the one that requires the government to limit greenhouse gases. They lobbied. They sued. And so far they've failed. (Zraick and Friedman 2025)

These types of campaigns will, however, be aided by Donald Trump's return to the US presidency in 2025. That is likely to have devastating consequences for the climate agenda in general. His influence will be felt both internationally and domestically. Internationally, Trump is expected to withdraw the US from various institutions that address climate issues or play interrelated roles, as well as from international agreements such as the Paris Agreement. Domestically, he is likely to reverse recent progress on climate change, as discussed in the US chapter. This includes increasing fossil fuel production, eliminating federal subsidies for clean energy and electric vehicles, and repealing numerous laws aimed at reducing greenhouse gas emissions. Such decisions are expected to lock in years of additional emissions, potentially triggering a resurgence in oil and gas production. More critically, Trump's actions will likely undo key mitigation and adaptation measures implemented in recent years, as outlined in the US chapter. His stance could embolden other nations to follow suit, or at least reduce their own commitment to addressing climate change, undermining global progress. Luckily, there appears to be a growing global consensus that action on climate change is necessary. According to a United Nations' study that surveyed 75,000 people, around 80 percent of respondents in major fossil fuel-producing economies support transitioning away from fossil fuels to clean energy. In the United States, despite the issue's deep politicization, such support stood at 66 percent. In South Africa, the support rate was even higher, at 77 percent (UNDP 2024a). The difference in these rates can, at least in part, be attributed to the political climate shaped by Donald Trump over the last decade. During this time, he has worked to undermine climate science and frame efforts to address climate change and environmental protection as being in opposition to economic development. The implications of another Trump term in office for the United States will be catastrophic for efforts to combat climate change, protect nature, and address broader environmental concerns, including biodiversity. Additionally, he is expected to reverse measures aimed at addressing the challenges faced by minorities and Indigenous Peoples. As Douglas Brinkley, a presidential historian, observed:

> Trump's the worst president for the environment in American history. A second term will be brutal. He is going to go full throttle on having the trophy of shutting down the green movement and replacing it with an older-style gasoline and coal-fired America. (Gelles 2024)

All aspects of addressing climate change will be affected under Trump's Administration. Critically, greenhouse gas emissions are likely to persist at current levels or even increase, with many of the resulting effects being irreversible. This is especially alarming, as research indicates that greenhouse gas emissions must peak no later than 2025 for effective long-term climate mitigation to occur. Therefore, significant global reductions in emissions are imperative. As the Intergovernmental Panel on Climate Change (IPCC) highlighted in its 2023 synthesis report:

> There is a rapidly closing window of opportunity to secure a liveable and sustainable future for all … "The choices and actions implemented in this decade [by the year 2030] will have impacts now and for thousands of years. (IPCC 2023)

However, the opposite is already occurring, with worldwide emissions increasing and even setting new records. The consequence of this is that with each degree of global warming, the negative impacts of climate change, along with associated losses and damages, will intensify. Conversely, every ton of carbon dioxide emissions avoided reduces the likelihood of a catastrophic disaster for the planet. Therefore, the actions taken – or not taken – in the United States will have devastating implications for the planet as a whole.

CLIMATE MITIGATION

In terms of mitigation efforts across the globe, progress is being made in certain regions, particularly to some degree in parts of the developed countries. However, developing nations face significant challenges in addressing emissions levels (Mirza et al. 2022: 64). Mitigation remains central to combating climate change. Nevertheless, some states are still unprepared to transition away from fossil fuels and industries that are environmentally harmful. While many parts of the world are advancing toward renewable energy, there remains an overreliance on fossil fuels. It is essential to establish definitive cutoffs for their continued use, including a complete end date. Without such measures, the positive steps being taken are continuously undermined by the ongoing reliance on fossil fuels.

There is, therefore, a need to reduce the emissions from fossil fuels and other gases, such as methane. This reduction must be linked to food production (Papies et al. 2024: 142) and other necessary controls to limit the release of such gases. The 2023 IPCC report finds that shifting to "sustainable diets," including reducing the consumption of red meat, could have an enormous positive effect on emissions.

Another major issue in mitigation efforts is stopping deforestation. In fact, the IPCC identifies deforestation as the second most significant mitigation tool. This is due to the critical role forests play as "carbon sinks" in absorbing carbon dioxide, among other functions. The IPCC report found that addressing deforestation could reduce carbon emissions by 4 billion tons annually by 2030. This figure is nearly double the current emissions from fossil fuels across all of Africa and South America. Additionally, there must be a focus on regenerating degraded forests, as this would enhance the world's ability to combat global warming more effectively.

There must also be a range of additional steps, including a concerted push to reduce reliance on private fossil fuel-based transport in favor of increased use of sustainable public transport (Maier et al. 2023). While there has been significant progress by states in supporting electric vehicles (Hosseini and Stefaniec 2023), other forms of transport, such as bicycles and e-bikes, also need to be encouraged. These alternatives can have substantial positive effects on emissions and on climate change more broadly.

A variety of other measures are being explored and need to be implemented in tandem to ensure more sustainable solutions. One such measure is the use of biochar to mitigate the effects of climate change. Biochar, a type of charcoal resulting from various industrial biomass processes, offers numerous benefits when used in soil. It improves soil aeration, reduces greenhouse gas emissions, decreases nutrient leaching, and lowers soil acidity (Luo et al. 2023). Additionally, biochar has the advantage of remaining in the soil for thousands of years (Khedulkar et al. 2024). Biochar also has other applications, including its use in cement (Senadheera et al. 2023). It can aid in water management (Bian et al. 2024) and boost food production, by enhancing agricultural yields, increasing nutrient availability, promoting co-composting, restoring water, and improving the overall efficiency of agricultural processes (Manikandan et al. 2023: 129725).

There are, however, significant challenges in advancing the necessary measures to play a mitigating role, including promoting greater efficiency in energy usage (Hunt and Kipouros 2023). It is also necessary for more efficient technologies to be developed and used. States ought to be regulating the energy efficiency of home appliances, and other machines, to ensure that the most efficient appliances of all types are used. While there has been progress toward adopting renewables, which are increasingly seen as the primary energy sources for the future, research highlights that wind and solar energy, in particular, are becoming more widely utilized and making the most substantial contributions in this arena. Financing remains a key issue if these processes are to be scaled up and relied upon more extensively (Arshad and Parveen 2023).

Other methods of protecting the planet are also being explored, such as the use of nuclear power (Tyler and Webler 2024: 103831) and carbon capture and storage (CCS). However, their roles are likely to remain limited, as their costs – at least in the short term – are significantly higher. Carbon pricing mechanisms, including decarbonizing efforts, carbon taxes or offsets, and emissions trading schemes, can also contribute (Dominioni 2022: 158) to ensuring that the remaining carbon budget is not exceeded (Lamboll et al. 2023: 1). In fact, "[s]topping climate change requires that emissions plummet to nearly zero rather than just dip. That is 'deep decarbonization,' and it will require bigger, more disruptive, and riskier transformations in energy and industrial systems" (Yanosek and Victor 2023). However, some question the effectiveness of these mechanisms in mitigating carbon emissions (Green 2021). Carbon offsets, in particular, remain a controversial issue in this context (West et al. 2023).

A variety of technologies are also being developed (Moriarty and Honnery 2022) to address the challenges posed by a warming planet (Chen et al. 2024). Among these, the most advanced form of geoengineering – speculative technologies designed to alter natural systems and cool the globe – appears to be the removal of carbon dioxide from the atmosphere (Zhu and Shang 2024). Other processes, such as solar radiation management (McDonald 2023: 565) and aerosol injection (Zhu and Shang 2024), also show promise. While significant investments are being made in these technologies, whether they will become sufficient components of a broader solution remains uncertain (Sovacool 2021: 100656). Concerns have also being raised about the potential for these processes to interfere with existing mitigation and adaptation goals (Ricke and Harding 2023: 828). Additionally, corporate offsets – where companies emitting large amounts of greenhouse gases take measures to counterbalance their emissions – are being scrutinized as a potential means for businesses to reduce their role in contributing to climate change.

ADAPTATION

There are significant discrepancies in adaptation strategies and processes worldwide. These differences exist not only between different regions and countries but also among communities within the same country. It is also evident that:

> Inequality shapes climate change adaptation and, if it is not accounted for in adaptation design, implementation, or evaluation, interventions may either be ineffectual or – worse – increase the vulnerability of those they seek to aid. (Eriksen et al. 2021: 2)

Inequality is having a critical effect in the world today and is a driver of problems with dealing with climate change by many around the world, including minorities and Indigenous Peoples. Inequality is in fact rampant and growing, and as the levels of inequalities expand, it is especially vulnerable groups that are affected by climate change (Oxfam 2023). This reality must be considered in domestic processes aiming to implement adaptation measures. Consequently, the concept of transformative adaptation has emerged, seeking to promote systemic change while addressing climate impacts (Fedele et al. 2019: 116). Such an approach has been adopted in various countries (Filho et al. 2022: 20).

In this context, other countries can learn from the experiences of the United States, Canada, and South Africa, as explored in this book. Despite the challenges presented by the United States, where addressing climate change is deeply politicized and intertwined with contentious topics like race and discrimination, there are valuable lessons to be drawn. Efforts to reduce discrimination, protect marginalized groups, and provide better support are essential. Additionally, fostering a deeper understanding of climate change and its impacts on vulnerable communities is critical, as is recognizing the broader societal benefits of protecting these groups. Support for these communities can take many forms, including leadership at the highest levels advocating for and implementing plans to combat climate change while addressing environmental justice issues. Governments should establish dedicated agencies to tackle these challenges, equipping them with the necessary tools and resources. Public assistance in a range of areas is going to be vital. The executive branch should be allocating resources to a range of actions, such as monitoring, policing, prosecuting, and other measures to ensure that marginalized groups are not exploited or further disadvantaged. Furthermore, mechanisms must be put in place to prevent non-state actors from taking advantage of these communities and to monitor resource utilization and extraction. Learning from comparative best practices, as highlighted in this book, can significantly enhance the effectiveness of these efforts.

In addition to a range of substantive needs related to adaptation, which will be addressed below, there is a critical need to educate the public at large about climate change (Trott et al. 2023: 1535) and, specifically, what they can do to adapt to it. There is an even greater need to educate impoverished communities, as they will bear a disproportionate burden of the impacts of climate change. They are also less likely to receive the information they need to cope with these issues. For example, early warning systems for tsunamis, cyclones, tornados, and other weather events that are easily accessible to all, including the most impoverished communities, must be established to ensure that everyone has access to that information (Udo and Naidu 2024: 31). That is relatively easy in the world today, where mobile phones proliferate. This is essential to reach all people, especially those who are marginalized, during a specific crisis. Nationwide educational initiatives are also needed, to reach as many people as possible in affordable and accessible ways on a whole range of issues, including when to escape deadly events and where to.

A range of steps also needs to be taken to avoid conflict over various environmental issues that will be exacerbated by climate change. There are already many conflicts over such matters, and these will increase as the effects of climate change are felt more acutely. For example, as climate change creates more challenges in accessing water, states and other authorities at local and other levels need to design and implement processes to address these problems. This also means that blue-green and gray infrastructure must be utilized (O'Donnell et al. 2021: 143). This type of infrastructure is defined as:

> An interconnected network of natural and designed landscape components that may include ephemeral, intermittent, and perennial water bodies, and open, green spaces which provide multiple functions such as water storage, flood control, and water purification, and many others. (Ghofrani et al. 2017: 15)

Better management processes are also needed (Olabanji et al. 2020: 102939). States must take action to address issues such as the quality of drinking water and create the necessary legislative environment (Ricalde et al. 2022: 102939). For example, PFAS or synthetic "forever" chemicals can be reduced in the water supply by legislating the allowable quantities of different chemicals in water systems. This is already happening in many places, including in the United States. Mandatory removal of these more than 15,000 chemicals above harmful levels (National Institute of Environmental Health Sciences 2024) should be required everywhere. These types of steps will benefit everyone if there is the required political will to implement them. However, resources are necessary for continuous water testing and for establishing processes to remove such contaminants. This is easier in some places, but at least some action is required everywhere.

In many parts of the world, where informal water collection occurs, it is essential to bring upstream and downstream users together to engage in dialogue resolve tensions to avoid conflict. If this does not happen, these problems could become a source of new issues that divide people, especially along identity lines, which often occur where there is disagreement over resource access and usage. This will particularly affect minorities and Indigenous Peoples. Therefore, to protect them, specific steps need to be taken to address these challenges.

To address water shortages in some areas, assistance is needed to build wells, boreholes, dams, reservoirs, and other storage systems (Elgendy et al. 2024). Existing storage systems also need to be better protected (Mi et al. 2020). Additionally, people must be encouraged to use water more sustainably. For example, in Algeria, people are compelled to reduce consumption, store, recycle, and use alternative sources of water by having intermittent water supplies (Belmeziti et al. 2025). Rainwater collection, especially in arid areas, should also be encouraged (Calderón et al. 2024).

States also have much to do to address the consequences of such issues. For example, to make water more available, especially in less well-off areas, new water systems need to be created. For this to happen, there must be a focus on building cooperation between and within communities. States, including authorities at different levels, need to help design such projects, raise financing, and assist in navigating existing regulations. In this regard, regulatory systems must be adapted to address these issues and made user-friendly. It should be noted, however, that groundwater needs to be protected, as failing to use it sustainably could lead to long-term negative effects (Elgendy et al. 2024).

People need assistance in dealing with the issues, including loans and a range of other means (Moayedi and Hayati 2023). This is because water shortages and problems in water quality are expected to be some of the greatest effects of climate change (Bartlett and Dedekorkut-Howes 2023: 651). These issues need to be anticipated, and measures to reduce their impact must be set up in advance. These problems will be exacerbated in areas where people migrate to avoid the growing effects of climate change. Many people already face a difficult choice between rebuilding and relocating. Already, substantial numbers of people have been displaced because of climate change. However, another trend is eviction of minorities and Indigenous Peoples. The Roma, for example, have had to endure such evictions for a long time all over Europe, which has forced them into areas more vulnerable to climate change (Alexandrescu et al. 2021). They have already endured marginalization, discrimination, and a range of other ills (Sequeira et al. 2024). As these vulnerable groups are displaced and move to new areas, they face a range of environmental challenges (Kuhl et al. 2014). In areas that are at greater risk of the enormous effects of climate change, people who are displaced will need to move and face new challenges. While they may leave the harshest places, they do not escape the consequences of climate change. In fact, they are likely to end up in places with environmental hazards, although perhaps not to the same extent as the places they left. This is because they may lack the resources to move to the best places. They may still end up in areas with water issues. As water challenges grow over time due to climate change, local communities may struggle to share water with newcomers, potentially causing conflict. These problems must be foreseen and militated against (Refsgaard et al. 2013: 337).

As all three country chapters indicate, much can be done to adapt to the effects of climate change. In this regard, governments can play a critical role in setting standards on a range of issues, including electric vehicles. Mitigating the effects of climate change involves regulating various practices and establishing norms for fossil fuel usage and other materials that contribute to environmental problems. States can take significant steps to address emissions and find solutions to reduce carbon dioxide, methane, and other greenhouse gases in the atmosphere.

However, when it comes to implementing some of these adaptations, because authorities must navigate a balancing act between environmental priorities and their perceived economic realities, they need to develop diverse strategies to address a range of challenges. For instance, governments must establish fair systems for accessing and sharing water resources (Ferdowsi et al. 2024). Additionally, states should assess the types of crops being planted (Abdirad et al. 2020: 1458) to ensure they can grow successfully while using water in sustainable and equitable ways (Iglesias and Garrote 2015: 113). Sustaining research and training on these issues is essential. Additionally, fair systems for allocating water between agricultural and domestic use need to be developed. Farmers must be supported in adopting more sustainable farming practices (Sabzevar et al. 2021: 224) and adapting to climate change (Vermeulen et al. 2018: 65). They require assistance in implementing various technologies, including water-efficient technologies (Alfonsi et al. 2024: 1), and need education on specific adaptation methods (Pili and Ncube 2022: 97). It is necessary to identify and address the diverse types of farming, considering the significant role the agricultural sector plays and the challenges it will face due to climate change (Soltani and Mellah 2023: 74). Small-scale farmers, in particular, require targeted assistance (Makungo and Nkuna 2023: 100361). Indigenous Peoples and minority groups rely on small-scale farming for their livelihoods (Baird 2008) but frequently encounter substantial challenges (Horst and Marion 2019: 1). Nevertheless, some have already demonstrated remarkable resilience. For example, in Africa:

> Fulani people have also used such indigenous climate-adaptation techniques, such as livestock-feed diversification, cattle stress-management techniques, and division of labour. On the other hand, the Endorois people have turned to climate-smart agroecological production systems such as the cultivation of drought-tolerant cereals, tubers, and vegetables ... Other adaptations to climate-change effects ... include livestock and crop diversification, herd adjustment by class, livestock destocking, and supplementary feeding of livestock. (Filho et al. 2021: 6224)

More broadly, measures must be developed to mitigate the worsening effects of climate change. Decisions need to be made about how to prioritize access to water during shortages. Processes should be established to facilitate water agreements, monitor water-related issues, and intervene as necessary to prevent conflict.

It is also essential to raise public awareness about climate change mitigation. An informed citizenry is less likely to fall prey to misinformation or disinformation and is better equipped to hold governments accountable for inadequate action on climate change. However, it should be noted that there is significant resistance to some of these initiatives, which must be addressed (Falla et al. 2024: 106530). While the intention here is not to delve into those issues in detail, it is important to highlight their existence. Approaches that may prove effective include community-based strategies that are inclusive, needs-driven, and avoid imposing top-down solutions (Selje et al. 2024: 155).

Adaptation for Minorities and Indigenous Peoples

Finding ways to help marginalized groups better adapt to climate change is essential. However, a significant problem is the lack of focus on addressing adaptation issues specific to these groups (Bose 2017: 575). It should also be noted that these processes do not always yield positive results for marginalized communities. This is because they often reinforce inequalities, as:

> Climate adaptation plans produce ... environmental injustices because of two processes: acts of omission and acts of commission. Acts of omission occur when planning bodies disregard the adaptation needs of vulnerable or historically marginalized groups. Acts of commission take place when planning bodies actively prioritize stakeholders who already have resources and privilege in a community. However, the claims of marginalized people can be discredited even when they are included, and this is particularly true for communities of colour. (Domingue 2022: 271; Anguelovski et al. 2016: 333)

Part of the reason for this is the limited engagement with these groups in most areas. In many places, marginalized communities, including women, are often excluded from planning processes (Sibiya et al. 2022: 7111). This exclusion is a global issue (Jayawardena 2024). As has been noted:

> Centring justice in climate adaptation processes and engaging marginalised groups can help ensure adaptation efforts address existing injustices, enhance the legitimacy of decisions, and increase the likelihood of long-term success in implementing adaptation projects. If adaptation processes neglect to engage marginalised individuals, initiatives may reinforce existing social vulnerabilities, result in negative consequences, and fail to generate support, or draw active opposition, from those that stand to be most impacted. (Brousseau et al. 2024: 29)

It is, therefore, the duty of legislatures at all levels to do more to include marginalized groups. At the national level, legislatures must ensure that laws more supportive of Indigenous and minority groups are enacted and that additional provisions are established to protect these groups and their resources from encroachment on their rights and assets. More broadly, these groups should play a central role in these processes, and their contributions must be fully recognized (Bennion and Kelly-Mundine 2021: 170). Genuine participation could be facilitated by granting these groups representation in national and regional parliaments. Participation serves as a vehicle to promote and instil a culture of good governance and as a means to prevent conflict. As Ackerman argues, "the opening up of the core activities of the state to societal participation is one of the most effective ways to improve accountability and governance" (Ackerman 2004: 447). However, participation is often limited to elites and well-resourced civil society organizations, leaving marginalized groups excluded.

Thus, it is essential to identify these groups, prioritize their needs, and engage with them. The context in which they exist must also be carefully considered (Juhola et al. 2022: 609). An intersectional analysis is necessary to assess who is most vulnerable and to make informed decisions about prioritization (Stadler et al. 2024). However, such efforts have been minimal. In fact, there is a general lack of global preparedness when it comes to climate adaptation (Malik and Ford 2024: 24).

Adaptation Strategies for Minorities and Indigenous Peoples

A variety of adaptation strategies are either currently available or can be developed to enhance the adaptive capacity of marginalized communities. Broadly, these strategies can be categorized into autonomous strategies, implemented by the affected communities themselves, and assistive strategies, supported by governments and civil society. Despite their differences, both types of strategies contribute to strengthening climate resilience and creating an enabling environment for affected poor communities to withstand and cope with climate shocks. While autonomous strategies play an important role, this brief focuses primarily on how governments and civil society can empower the poor and marginalized to become more climate-resilient. Accordingly, the emphasis will be on cooperative and assistive adaptation strategies.

There has, however, been limited recognition of how past actions involving discrimination and marginalization have left a legacy that continues to ensure certain groups are excluded from many plans and processes, including those addressing climate change. For example, while many areas focus on greening the environment, these efforts are often viewed as development initiatives rather essential measures to address lingering environmental justice issues from the past, which can undermine efforts to adapt to climate change.

Improving the adaptation strategies of the poor and marginalized – often minorities and Indigenous Peoples – requires enhancing their adaptive capacity and climate resilience. Much remains to be done to enable these groups to effectively adapt to climate change. While the rhetoric and planning are frequently present, there is little evidence, beyond a few locations, of these adaptation plans being implemented on the ground, particularly for marginalized communities. Some indications suggest that in certain large cities, such as those in the United States, Canada, South Africa, and other major economies, adaptation efforts are taking place although to different degrees. This is not taking place lineally, and some advances such as using fossil fuels are being retarded to some degree by a range of recent events, as noted. However, in many other areas, there is minimal evidence of adaptation occurring on the ground.

Rural areas are particularly vulnerable to climate change. However, even in academic circles, this issue has received little attention. Since the turn of the century, the literature has largely focused on the adaptive capacity of urban areas in developed countries (Flórez Bossio et al. 2019: 297). Over time, this focus has broadened to include urban areas in developing countries, reflecting a growing recognition that different contexts present unique dynamics affecting adaptive capacity (Flórez Bossio et al. 2019: 279). What the literature often overlooks is an analysis of how the adaptive capacity framework applies equally – if even more so – to poor populations living in rural, informal areas (Brown et al. 2024). This oversight has a particularly significant impact on Indigenous Peoples. Furthermore, there needs to be greater emphasis on impoverished urban areas, which have also been largely neglected in discussions of adaptive capacity. This is a critical issue that disproportionately affects minority communities.

Critically, as noted in the Introduction to the book, a key issue that requires much more focus is adaptive capacity. Vulnerable groups are at greater risk of adverse climatic changes (Herreros-Cantis 2024: 105474), which can also make them less resilient to climate impacts. However, just because poorer communities experience diminished adaptive capacity, lack resources, and are most at risk of climate-induced shocks or disasters, it does not mean they should be seen as passive victims of climate change without agency (Gouritin 2024). What needs to be understood is the specific challenges they face and how they can be supported. It is also essential to consider how risk reduction can be effectively achieved.

Thus, understanding that climate change disproportionately impacts poorer, marginalized communities (often specific groups) (Zhang et al. 2024) due to their structurally diminished adaptive capacity is crucial for enhancing their adaptation strategies. At a time when there is a greater push to increase the climate resilience of urban and wealthier communities, the vulnerable lived realities of those who increasingly suffer from wildfires, droughts, and arable land loss cannot be forgotten. To this end, it is important to enhance and improve both the adaptive capacities and adaptation strategies of the poor and marginalized so they are not left behind in the process of adapting to climate change and building climate resilience. Furthermore, any discussion of improving adaptive capacity and adaptation strategies inherently involves the logic of enhancement. It requires creating an enabling environment that is supportive in nature, strengthens climate resilience, and both enhances and improves adaptative capacities over time.

ENVIRONMENTAL JUSTICE AND CLIMATE CHANGE ADAPTATION, AS THEY RELATE TO EACH OTHER

There is a need for better processes to address environmental injustices alongside any adaptation steps taken, as failing to do so will undermine efforts to combat climate change. This is because both are cross-cutting issues. Addressing both will, at times, also have important climate mitigation effects. For example, waste control, as discussed below, is not only an environmental justice issue for the communities where the waste is located, but it can also have significant positive effects on methane emissions, thereby mitigating climate change. The same is true for issues such as pollution, water, green spaces, housing, etc. For this reason, states and other stakeholders need to invest significantly in adaptation capacity and developing strategies to ensure climate resilience for all. Early and sufficient investments in adaptation is essential. They must ensure that the most impoverished, including Indigenous Peoples and minorities, are reached and catered for. Without this, the suffering of these groups will increase, and the positive roles they play in mitigation efforts (as noted in the book) will be diminished.

Many issues require reform to address these matters effectively. In many parts of the world, laws, policies, and practices need extensive overhaul to tackle a range of environmental and climate challenges. Regarding environmental justice, each branch of government plays a critical role in addressing the various issues under its purview, which are discussed later. Numerous actions can be taken. For example, states can establish dedicated ministries to address the environmental concerns as well as Indigenous and minority affairs. This is exemplified by Brazil's actions in December 2022, when it appointed two separate ministers to handle these issues. While the Ministry for the Environment has existed for some time, the Minister for Indigenous Peoples was a newly created position (Günerigök 2022). Both appointees to these offices were well qualified, with one having extensive experience in environmental protection, including previous service in the role, and the other being Indigenous herself. Additionally, both appointees are women, which fosters a sense of diversity and inclusion.

However, much remains to be done to address the effects of the law and public policy in relation to environmental justice issues. Planning processes must consider the identities and needs of communities and groups, taking proactive measures to mitigate the impacts of climate change and the various environmental injustices that persist (Zoll 2022: 3). This topic will be revisited later.

A range of actions can help alleviate the challenges faced by these groups, making it easier for them to navigate administrative and bureaucratic obstacles, address their issues, and better adapt to climate change. A key step would be to ensure that the impacts of oppressive laws and practices are effectively mitigated (Vilá et al. 2022: 632). One significant limitation lies in the insufficient recognition of environmental justice and its effects (Givens et al. 2019). The reach of environmental injustice is extensive, encompassing "the lived experience of the reality of injustice on the ground, in the air, in one's food, at the workplace or school, and on the playground" (Schlosberg 2013: 37).

The following sections will address specific issues as they relate to both climate change and environmental justice.

Housing

The issue of housing highlights the significant relationship between environmental justice and climate change. Among those most vulnerable to the effects of climate change are the homeless (Kidd et al. 2023: 623), as they lack adequate shelter to cope with the challenges posed by climate-related impacts. Furthermore, poverty and homelessness disproportionately affect groups such as minorities and Indigenous Peoples. In the United States, the Department of Housing and Urban Development reported an 18 percent increase in homelessness in 2024 compared to the previous year, bringing the total to 770,000 individuals (United States Department of Housing and Urban Development 2024). However, some commentators question the accuracy of these figures, suggesting the true number may be significantly higher. Regardless, if this trend persists, more people will face heightened vulnerability to climate change (Bezgrebelna et al. 2021: 5812). While some experts emphasize prevention, others advocate for a variety of strategies to address the nexus between homelessness and climate change (Kidd et al. 2023: 623). Targeted interventions for these populations appear essential (Martin 2023: 31), especially given that climate change is likely to increase homelessness as its effects worsen over time. A 2007 warning projected that a billion people could face homelessness due to climate change by 2050 (Ashton 2007: 7). The World Bank's 2023 World Development Report similarly noted that a billion people are at risk of displacement due to rising sea levels (World Bank 2023). These individuals will eventually need to relocate and secure new housing.

Regarding housing, there are many different types, including informal housing commonly found in slums and similar areas. Despite being where the majority of housing is located, these areas often receive little attention. Supporting those living in informal housing is significantly more challenging, but given the vast number of people affected, addressing this issue must be considered an imperative. This group requires substantial support. The issue of informal housing needs will be revisited below.

While the focus here is on adaptation, it is essential to note that housing – both in terms of construction and operation – can have significant effects on climate change. Consequently, the measures taken concerning housing can yield substantial mitigation benefits (Malik et al. 2024: 406). For example, the materials used to construct formal accommodations, such as cement and steel, emit considerable greenhouse gases during manufacturing unless greener alternatives, like green hydrogen instead of coal, are used in production (Algers and Ahman 2024: 1). Therefore, changes to all aspects of housing practices can have profound climate mitigation effects, depending on the type of accommodation. Simply things such as paint color can have important implications as some colors reflect rather than absorb the sun.

Householders typically have various choices available to mitigate and adapt to climate change (Caggiano and Constantino 2004: 237). However, these choices are not equally accessible to everyone, as their availability depends on resources and other limiting factors. While the focus here is on one dimension – improving adaptation efforts – it is important to note that weatherizing accommodations can also yield significant benefits, including climate mitigation and addressing various environmental justice issues.

In this context, the repercussions of climate change on housing populations, such as minorities and Indigenous Peoples, will be particularly severe. One major issue that arises is the increasing frequency and severity of extreme weather patterns, such as hurricanes and cyclones. As the number and intensity of these storms grow, the ability of people to cope with their impacts and the resulting damages will be significantly strained. These storms are becoming more powerful, wetter, and longer-lasting, leading to increasingly deadly and destructive outcomes. There will be many of varied effects of climate change. For example, as many cities get warmer so rat populations in these places will increase, with a number of dire consequences including for people's health (Richardson et al. 2025). Much can be done to deal with such issues, but it is about political will and allocating resources through public assistance programs. Once again, these effects will be felt more acutely by some groups than others. These groups need to be assisted to deal with these issues. More broadly, many regions are experiencing rising temperatures, while others are becoming increasingly colder. The extremes currently being experienced are likely to become the norm in many areas, and new, more virulent extremes in weather patterns are expected to emerge.

The major effects of climate change will generally be felt more acutely by poor populations, including minorities and Indigenous Peoples, than by others. This disparity arises because their homes are often located in the most vulnerable areas and are less equipped to withstand extreme weather events. Additionally, these groups typically receive less assistance both during and after such events, often due to a lack of services available in their communities. As noted in the US chapter, the availability of emergency services in areas where these groups reside is significantly lower compared to other regions.

The impact of these weather events also has more profound and lasting effect on their wealth, particularly over time. If these groups are to cope with future weather patterns, these issues must be addressed. Emergency services should be expanded to ensure greater availability in affected areas, and aid should be targeted to ensure it reaches these populations. Furthermore, the effects of such events on these communities need to be monitored over time, and adjustments must be made to enhance their preparedness and resilience.

As noted in the US chapter, the ability of homeowners to obtain insurance is declining significantly in areas most affected by hurricanes, cyclones, tornadoes, and other severe storms. Consequently, states worldwide will increasingly need to shoulder the burden of crisis relief, making it imperative to identify effective strategies to address this challenge. One approach is exemplified by the actions of Vermont and New York in 2024. These states established Climate Superfunds, which require major fossil fuel companies to cover damages caused by their emissions over the past three decades. Twenty-two states have sued New York to stop that Fund. However, additional strategies must be developed to address the wide-ranging effects of climate change, as the associated costs are projected to be astronomical.

A crucial starting point is recognizing that what renders the poor – often minorities and Indigenous Peoples – particularly vulnerable to climate impacts is their lack of access to structurally sound, formal housing that can protect them from extreme heat and cold (Satterthwaite et al. 2020: 143). While the provision of adequate housing often spans national, provincial, and local jurisdictions, it is frequently city or municipal governments that impede progress due to budgetary constraints. Yet, governments have numerous mechanisms at their disposal to lower barriers and reduce the cost of entry to affordable housing. Chief among these is expanding access to key infrastructure services such as electricity, water, and sanitation (Satterthwaite et al. 2020: 143). Moreover, governments can subsidize the provision of building materials in high-risk areas to strategically enhance climate resilience. Additionally, banks can be incentivized to establish affordable loan schemes specifically designed to assist low-income communities that are often excluded from traditional loan structures due to their socioeconomic status (Satterthwaite et al. 2020: 143).

Critically, the impact of climate change will be significantly worse if efforts are not made to improve the weatherization of homes, particularly for the poor. Even with weatherization, the future effects of climate change will be devastating, especially for those living in housing that cannot withstand such impacts. Once again, Indigenous and minority groups will bear the brunt of these challenges. This vulnerability stems from their typically disadvantaged financial circumstances, which often force them to reside in areas more prone to climate change effects. These circumstances are at least partly the result of historical discrimination and marginalization, which have caused severe socioeconomic disparities, including limited access to wealth, education, and opportunities. As a result, these groups are more likely to live in locations most vulnerable to climate change and lack the resources necessary to prevent or respond to the effects of catastrophic climatic events.

Problematically, steps to improve housing conditions for these groups rarely occur, aside from isolated exceptions in certain locations. In general, their housing is often precarious (Debrunner et al. 2024: 1). To encourage investment in their housing, some degree of protection is necessary to address this precariousness. This is particularly important because these groups often rent the places where they live. Landlords must be incentivized to invest in weatherizing their properties; however, this is challenging as these properties are typically located in lower-income areas where landlords perceive little return on investment. The intersection of climate change and environmental justice is once again evident. Due to historical discrimination and marginalization, these groups often lack the resources to cope with severe climate events when they occur. Additionally, major climatic occurrences are more likely to affect less affluent areas because of their geographic vulnerability. For example, these areas are disproportionately prone to flooding. These inequities need to be calculated into adaptation processes and financing.

The challenges related to housing will be exacerbated by the significant increase in energy demands required to cope with the effects of climate change (Mastrucci et al. 2023: 761). Heating and cooling systems will have to work harder if weatherization is not implemented. Therefore, several actions must be taken to mitigate these problems. One such action involves building resilience and fostering an understanding of how to address the various challenges posed by climate change. A key priority is constructing more affordable energy-efficient houses (Panão 2024: 114583). This includes installing solar panels, smart windows, improved heating and cooling systems, and more energy-efficient appliances like electric stoves in homes. There are many ways to make buildings more sustainable (Momeni et al. 2024: 2499). A significant focus must be placed on construction due to its negative environmental impact. Consequently, environmentally friendly construction materials needed to be adopted, including phasing out the type of cement currently used, which is unsustainable and generates massive emissions. This approach should apply not only to new buildings but also to retrofitting existing structures (Motalebi et al. 2022: 105022). The critical issue is incentivizing the use of sustainable materials through rebates and other tools (Shook and Choi 2022: 100). While much of the progress in climate adaptation has occurred in developed countries, developing nations are also making strides in this area (Fowlie and Meeks 2021: 238). However, significantly more effort is required. Actions such as subsidizing equipment and assisting with installation are essential. Additionally, innovative solutions are needed to reduce the costs of these technologies.

To enhance the adaptation strategies of the poor and marginalized, it is essential to provide them with access to sustainable and efficient energy. This is critical for escaping energy poverty and ultimately strengthening their adaptive capacity and climate resilience. Solar energy appears to be the most viable immediate replacement for coal-based electricity generation. However, as the global decarbonization of electricity grids progresses and the concept of a "just transition" gains prominence, there is a legitimate concern that the poor and marginalized may be left behind due to the high costs associated with transitioning from coal to solar energy. One pressing issue is that the cost of installing solar panels, along with the space and infrastructure required to maintain solar grids, is often incompatible with the lived and material realities of people in informal settlements who lack adequate housing. A potential solution is for governments to subsidize not only access to housing but also access to solar panels, ensuring that a just transition occurs equitably. However, this assumes the existence of sufficient political will and financial resources on the part of governments to implement such measures. Unfortunately, this remains a significant obstacle.

Some groups have already found ways to address poor air quality by monitoring the air conditions and using tools such as air filters, fans, and air-conditioning, among other methods (Palinkas et al. 2023: 5393). However, these tools are inaccessible to many due to their high cost, necessitating financial assistance. Furthermore, adaptation will be less feasible if the housing and other buildings these individuals inhabit suffer from poor indoor air quality caused by the burning of various fuels for cooking and other purposes (Li et al. 2024). This situation often arises from financial constraints and other issues that exacerbate air quality problems (Sparks et al. 2024). Enabling these groups to transition to clean energy in their homes is therefore essential, wherever possible. In many locations, such as slums or squats, achieving this is particularly challenging. In more formal housing, adapting to clean energy sources such as heat pumps and solar energy will require both financial assistance and educational initiatives. Simultaneously, steps to ensure greater energy efficiency are critical (Mathias et al. 2023: 1034). Enhancing building energy efficiency – through improved insulation, better lighting, and more efficient appliances – can significantly reduce energy consumption (Olajiga et al. 2024: 1097). Once again, the feasibility of these measures depends on the type of housing and the level of support provided.

While formal homes are often the focus of sustainability efforts, slums, squats, and other areas where the poorest communities live are frequently neglected (Uddin et al. 2021: 774). However, adopting sustainable approaches is essential for all types of buildings, including large commercial structures that can have significant environmental footprints (William et al. 2020:

4549). This need also extends to public buildings, which are often overlooked (Munguia et al. 2020: 1212).

One successful model addressing these challenges is Iberdrola's "solar communities" or "solar neighborhoods." This innovative solar energy generation model relies on multistakeholder involvement to produce solar energy at reduced costs (Designit.com 2024). The power-sharing network involves schools, businesses, or wealthier neighborhoods renting their roof space to Iberdrola, a renewable energy company. Iberdrola installs solar panels on the lessor's roof space at no cost, allowing the lessor to benefit from reduced energy costs of up to 30 percent. The roof owner – be it a school, business or wealthy neighborhood – becomes a "solar host" with preferential "first access" rights to the electricity generated. Businesses, schools, and households within a defined radius become "solar neighbors," paying a subscription fee to use the solar power (World Economic Forum 2024). This scheme has achieved significant success in Spain. With advancements in battery technologies, solar energy now faces fewer limitations than in the past, when it was only available during daylight hours. For instance, improved battery systems now power up to one-fifth of California's energy needs, demonstrating the growing potential of solar energy.

For poorer communities, often located on the periphery of wealthier urban centers, a similar or identical model could be applied. Unused rooftops could serve as solar hosts, generating solar power for their less affluent solar neighbors. Governments could subsidize both the costs incurred by solar hosts and the subscription fees owed by poorer solar neighbors. Businesses could integrate similar projects into their corporate social responsibility initiatives and even contribute to the subsidization process. Costs are the largest barrier to entry for solar panels, effectively excluding poor and marginalized communities from their benefits. Initiatives like this are critical for lowering barriers to entry and expanding access to solar power. They are also vital to ensure that poorer communities are not left behind in the just transition, allowing them to receive the same adaptive benefits as their wealthier counterparts. Therefore, a range of models must be developed to play a significant role in addressing climate change and enhancing climate adaptation.

Water

In areas where minorities and Indigenous Peoples live, significant action is needed to address water-related issues. Many places experience insufficient water supply, and the water is often of poor quality.

Similarly, the growing risk of flooding in these areas requires urgent attention (Udo and Naidu 2024: 31). This issue, however, affects both poor and affluent areas (Link 2021: 174). The problem lies in the fact that poorer areas are often neglected and inadequately serviced when it comes to risk assessment. Therefore, it is essential to identify which groups are most at risk and vulnerable to these challenges (Hansson et al. 2020). Some remedial measures to address these issues are relatively straightforward (Laversa et al. 2020: 207). These include maintaining clean channels, stormwater systems, and rivers to prevent silt and other pollutants from obstructing effective runoff. Improved garbage management is another crucial step in this regard. Proper waste controls not only reduce the risk of flooding but also mitigate methane emissions from dump sites, which can contribute to climate change. Better waste management practices offer several adaptation benefits as well. Unfortunately, such measures are not implemented consistently in poorer neighborhoods, leading to persistent and preventable problems when flooding occurs. Therefore, significant efforts are needed to manage the excess water that accompanies floods. Wetlands and similar areas must be protected, and floodplains should be developed to better control water runoff. Unfortunately, there has been considerable encroachment on areas traditionally responsible for absorbing floodwaters, which has undermined flood control measures. This trend needs to be reversed, with areas once again allocated for water runoff, such as through the creation of rain gardens and the use of permeable pavements. Such measures can help prevent large volumes of water from inundating urban areas and neighborhoods.

Addressing water-related issues also necessitates dealing with hazardous waste sites. Landfills must be cleaned up to prevent hazardous runoff from contaminating groundwater and to stop the release of harmful chemicals into the air. These actions not only contribute to combating climate change but also offer significant environmental justice benefits for marginalized communities (Barhorst et al. 2024: 1).

Air Pollution

The interrelated challenges of addressing climate change and environmental justice are evident in the context of air pollution, which is an increasingly serious issue. In areas with poor air quality, governments must take bold measures to mitigate problems caused by haze and smog, which are largely the result of industrial emissions and motor vehicles. Significant adaptations can be made by reconsidering the placement of industries relative to cities and by addressing the environmental impact of vehicles (Ali et al. 2022: 5102). Developing public transport systems that are cost effective, environmentally friendly, and tailored to meet the needs of the population is a key strategy to reduce the carbon footprint of transportation. Measures such as reducing parking spaces, promoting public transit, and prioritizing pedestrian and bicycle infrastructure, along with establishing car-free zones, can effectively encourage people to opt out of personal vehicle use. Another approach involves imposing high fees on vehicles entering designated areas to discourage traffic congestion and pollution. While incentives for electric vehicles have been implemented in many regions, their impact has been less significant than anticipated due to various factors. Additionally, international trade issues have hindered the transition to electric vehicles, largely due to the high associated cost.

Another issue exacerbating air pollution is the increasing number of wildfires worldwide, which dramatically affect air quality not only in the areas where the fires occur but also in regions far removed from them. As a result, people in affected areas are sometimes unable to go outside without wearing masks (Palinkas et al. 2023). Masks need to be made available to help individuals, particularly those from lower socioeconomic backgrounds, cope with the smoke.

Addressing air pollution also requires efforts to protect the spaces where people live and work from the range of challenges caused by poor air quality. These challenges may be short term, such as those caused by wildfires, or long term, due to persistent emissions in specific areas. Limiting air pollution from entering buildings is essential, but this poses greater difficulties for minorities and Indigenous Peoples, who are often located closer to industrial sites and in buildings with poor structural quality. Studies show these groups disproportionately suffer from the adverse effects of air pollution (Fan 2024: 103042). Inequalities in their living conditions make it harder for them to adapt, as their buildings frequently have more leaks that allow greater infiltration of polluted air. Furthermore, while people may be generally aware of air quality issues, they often lack a full understanding of the extent of these problems and are unable to discern variations in air quality over time (Sisco et al. 2022). To address this, more accessible and understandable information must be disseminated regularly through appropriate channels to those in affected areas. Although air quality monitoring can be expensive, there are strategies to reduce these costs (Goodluck and Ekwutosi 2022: 89).

Green Spaces

Another issue that intersects both environmental justice and climate change is the availability of green spaces. However, there has been a widespread failure to understand the significance of green spaces for various groups and the negative implications of their absence on climate change. This is a critical issue, as green spaces are often lacking in poorer areas where marginalized groups predominantly reside. Efforts to adapt urban green spaces require a more methodical understanding of their susceptibility to climate change and their associated vulnerabilities.

One major challenge is that decisions made at the local level are primarily influenced by physical exposure, while socioeconomic vulnerability is often overlooked in initiatives to adapt urban green spaces. This oversight is partly due to insufficient research on vulnerability in this specific context (Pecharroman et al. 2024). Socioeconomic vulnerability is shaped by a combination of physical exposure and socioeconomic factors, making it more challenging to measure. As a result, it becomes exceedingly difficult for localities to develop adaptive capacity in areas where people are particularly vulnerable to climate change (Cheng et al. 2021: 100962). Green spaces can significantly benefit communities where the environment is increasingly affected by climate change. This is because:

> Urban green spaces reduce air pollution, enhance biodiversity and ecosystem services, promote health and well-being, and protect urban areas from extreme weather events or disasters such as heatwaves, droughts, and floods. (Nasir and Rahim 2023: 31)

Such places also provide opportunities for people to visit parks and other green spaces to escape the challenges they face in their living conditions. However, many of the areas where these groups reside lack green spaces. They have few parks, minimal tree coverage, little vegetation, and limited access to green spaces. This is particularly problematic as these individuals need spaces to retreat to during extreme heat, for example. Greening such areas is essential not only from an environmental justice perspective but also to facilitate adaptation to climate change. As will be discussed below, more resources must be allocated to these areas, along with individual assistance to support home weatherization. This is crucial not only to address issues related to heat but also to combat the effects of cold and poor air quality, which will be explored in greater detail later.

As is often the case, the systems least equipped to cope and most prone to causing problems are those in areas predominately inhabited by minority and Indigenous Peoples (Grabowski et al. 2022: 171). Therefore, it is essential for communities to invest in green infrastructure (Matsler et al. 2021: 104145), such as green roofs and permeable pavements, to prevent stormwater from entering sewer systems and rivers too rapidly (Rodriguez et al. 2024: 120229). Overburdened systems are more likely to result in hazardous outflows, including waste and other pollutants, which can severely impact the health of nearby residents. Significant efforts are needed to manage water and waste effectively. This includes storing and recycling wastewater, particularly in regions experiencing severe water shortages (Al-Hazmi et al. 2023: 116711). An environmental justice approach is crucial and must involve community discussion and consultation. Decisions made should be equitable and non-discriminatory.

Addressing these challenges must prioritize ensuring positive outcomes for these communities (Hammer 2019: 103). Additionally, such decisions should focus on rectifying and redressing the harm that has already occurred (McClean 2012: 339).

Land and Soil Degradation

The issue of land is a significant concern for many groups, particularly Indigenous Peoples. Their role in addressing climate change has been highlighted throughout this book. As noted, they play a critical role in protecting land and forests across the globe. Since most large-scale forests are located in rural, sparsely populated areas, these regions are often home to marginalized, poor, and vulnerable populations. Supporting these communities has the added benefit of supporting the surrounding environments. This is not to suggest that this process is always straightforward or that best practices are consistently followed. Nor does it mean that these areas are always well-managed or optimally utilized. For example, conflicts between humans and wildlife over resources can arise and may cause challenges if not effectively managed (Ahmad et al. 2016: 565). Nonetheless, local communities, including Indigenous Peoples, play an essential role as repositories of knowledge about various species and their potential benefits for humanity (Bitsch et al. 2020: 8).

With the increasing rate of deforestation, the resilience of forest ecosystems (Kelly and Kelly 2017: 10) must be supported in every possible way. Supporting Indigenous Peoples and other groups in these efforts is therefore essential. This can be achieved by implementing various measures, including providing stable and secure land tenure rights. Such measures are crucial, as the absence of secure land tenure may lead these groups to engage in projects that have detrimental effects (Fitts et al. 2022). Consequently, more must be done to strengthen the land rights of Indigenous Peoples and other groups, as land insecurity remains prevalent in many areas (Horowitz et al. 2018: 404). While many Indigenous Peoples live on land governed by customary laws, these rules are often not recognized by governments. This lack of recognition has created confusion, which has been exploited in some cases. To better protect the rights of Indigenous Peoples, it is necessary to formalize customary laws within the legal frameworks of respective countries, as has been done in certain cases. For instance, customary land titles have been recognized by the Constitutional Court of Indonesia (Jagannathan et al. 2018: 18). This can enhance the protection of groups' rights to their land. However, it can also undermine the rights of women (Whitehead and Tsikata 2003: 67). Therefore, if greater recognition is given to customary law or its equivalent, care must be taken to avoid further entrenching patriarchy (Collins 2019: 454). Currently, land acquisition – both lawful and illegal – by various entities for diverse purposes (Borras et al. 2008), is on the rise, often with negative consequences for certain groups (Moreda 2017: 698). One significant factor is that over 50–80 percent of transition minerals are estimated to be located on or near Indigenous Peoples' lands worldwide, putting these communities at severe risk of environmental degradation, land grabs, and human rights abuses (Oxfam 2023). For instance, one study estimated this figure to be 54 percent (Owen et al. 2023: 203), while another found it to be 80 percent across 700 evaluated locations (Sturman et al. 2022). Alarmingly, land grabs are sometimes facilitated by investment agreements that seek access to specific lands or resources (Künnemann and Suárez 2023: 123). Addressing this issue is crucial, as it is currently destabilizing communities (Li 2011: 281). Additionally, the gendered impacts of these practices have not been adequately studied, highlighting the need for further research in this area (Radcliffe 2014: 854). The effects of land grabs are particularly severe in rural areas (Künnemann and Suárez 2013: 123), leading to increased migration to cities (Ali et al. 2020: 189) and diminished prospects for sustainable management of the affected regions.

To address these problems, improved systems of tenure can help prevent land grabs (Künnemann and Suárez 2013: 123). There must, therefore, be a concerted push to secure land tenure, not only to redress past injustices of land dispossession but also to advance the twin objective of fostering climate resilience in poorer communities. It is essential to recognize that climate change exacerbates land tenure insecurity by driving the buying, selling, and illegal acquisition of land from poor landowners. Securing land tenure is thus a reliable means to enable autonomous climate adaptation at the local level. By pressuring governments to expedite secure land tenure regimes, Indigenous communities that have been impoverished and marginalized can begin to utilize traditional knowledge systems and revitalize local institutions to enhance communal climate resilience. This also allows these communities to strengthen their Indigenous land management systems, thereby improving their adaptive capacity (Castro and Katz 2022: 177).

It must be acknowledged, however, that Indigenous and other groups have demonstrated remarkable resilience in the face of these significant challenges. Transnational land movements have emerged in various sectors, including those involving Indigenous Peoples (Borras et al. 2008: 169). These movements have mobilized in diverse ways to counteract these issues (Cárdenas and Monterroso 2022: 338). While activism began in the 1960s, it took decades to become more effective (Niehaus and Davies 2021: 228). This is particularly evident in Latin America (Warren and Jackson 2021: 1), where, in some places such as Bolivia and Ecuador, movements (Minguet 2021: 60) have only gained significant momentum in the past two decades or so (Lupien 2020). Thus, the expansion and effectiveness of these movements has not been uniform. Some movements are composed solely of Indigenous people; however, these structures increasingly include both Indigenous people and non-Indigenous participants. There has also been significant regional variation in the issues these movements focus on, as well as in their experiences, understandings, and expectations (Hallward and Downs 2022: 183). The nature and effectiveness of these movements often depend on their geographical context, which shows considerable differentiation.

As far as land usage and soil degradation are concerned, as noted throughout this book, it is the areas where poor minorities and Indigenous Peoples live that suffer the effects of climate change and environmental injustices the most. These communities often live closer to hazardous industries and have endured the negative impacts of dumping and other harmful activities. Unfortunately, the problems they face are likely to worsen over time. This is particularly true for communities that rely heavily on farming activities for their sustenance and livelihoods.

The risks posed by climate change – such as increased heat and decreased rainfall – are nothing short of existential. These changes will have interconnected impacts on the arability and fertility of the land and soil that farmers rely on for cultivation. Despite these climate vulnerabilities, farmers have already identified several effective adaptation strategies to enhance their resilience. These strategies include: (1) crop rotation between deep – and shallow-rooted plants; (2) improved fertilization practices to increase soil nutrients; (3) supplementary or deficit irrigation, which small amounts of water are manually added to rainfed crops; and (4) intercropping, which involves cultivating two or more different crops on the same piece of land to maximize yield. It is evident that rural farmers can perceive and respond to climate impacts in ways that align with agricultural and climate research. However, as previously noted, enhancing adaptive capacity requires creating an enabling environment. Governments, at all levels and working cooperatively, can play a crucial role by administering agricultural and crop subsidies to make crops more affordable and accessible for poorer farmers. This is particularly important when market availability is limited due to scarcity or inflated prices (Williams et al. 2019: 124). Furthermore, larger, more commercial farms should be incentivized by governments to share resources, technologies, and research with small-scale rural farmers, who often lack access to these critical assets.

Resources

As highlighted throughout this Conclusion, significant efforts are required to address these pressing issues.

Funding these solutions necessitates a range of strategies, some of which have been discussed earlier. While many of the associated costs are incurred at the local level, the global challenge of holding developed nations accountable for their contributions to the world's environmental problems remains unresolved. Although an agreement was reached in 2024 to allocate $300 billion annually to the developing world by 2025, this figure falls far short of the $1.3 trillion estimated to be necessary.

Ultimately, communities will bear the majority of adaptation costs, but some will be far better equipped to manage these expense than others. Minority and Indigenous communities, in particular, are often the least able to afford adaptation measures. This is largely because these groups frequently face challenges in securing funding (Winkler and Flowers 2017: 181). This issue extends to climate financing as well (Venner et al. 2024: 1). Additionally, many human rights-focused NGOs that have taken on the responsibility of addressing environmental justice issues, especially those impacting minorities and Indigenous Peoples, have struggled to obtain the necessary resources (Zakharova and Glazkova 2024: 145). This underscores the urgent need for equitable and sustainable solutions to ensure that vulnerable communities are not left behind.

Thus, government-led social protection schemes must be a central component of strengthening the climate resilience of climate-vulnerable communities. Examples of assistive adaptation strategies include anticipatory, "climate-informed" cash transfers, as well as government subsidies for climate-resilient agricultural products and housing. Protection services must be provided for drought-, wildfire-, and famine-affected poor communities. Unique social and health insurance systems should be designed to address climate-induced health and housing vulnerabilities (McClean 2024: 4). These strategies can be adopted by national governments, with their implementation coordinated synergistically at the national, provincial, and local levels (Chaudhry 2024: 100277).

CONCLUSION

The planet is facing tremendous environmental and other challenges, which will only intensify over time without substantial mitigation efforts. The likelihood of avoiding a global temperature increase of 1.5 degrees above preindustrial levels appears remote and may already have been surpassed. Therefore, major and costly adaptations to climate change will become an unavoidable reality. Overall, much work is required to adapt to the realities of climate change. While the effects will be felt by everyone, some groups will bear the brunt of the challenges more acutely than others. Minorities and Indigenous Peoples are the most likely to be disproportionately affected and the least equipped to cope with the challenges of climate change. This is largely due to the legacy of past discrimination, marginalization, and exclusion, which continues to have significant impact on their lives. These factors have rendered them less resilient and less capable of adapting to the changing climate. To increase their adaptative capacities, a range of strategies specifically tailored for these communities – and applicable to all people in general – must be implemented as soon as possible. Simultaneously, efforts to address environmental justice harms must be established and operationalized. Importantly, these groups need to be included in the development and establishment of such processes. Critically, approaches that integrate the intersection of environmental justice with climate mitigation and adaptation efforts will have a far greater impact. Such approaches are essential to ensuring that minorities and Indigenous Peoples are not left further behind in the fight against climate change.

References

A

Abate, Randall, and Elizabeth Ann Kronk, *Climate Change and Indigenous Peoples: The Search for Legal Remedies* (Cheltenham, UK: Edward Elgar Publishing, 2013).
Abbas, Ali, et al. "Major Air Polluters in Africa Unmasked," Greepeace (2024). https://drive.google.com/file/d/1HZ4JAbS06x6vMV3V2VSNaU2JVoR0b0OB/view?_ga=2.113141316.150470983.1712238484-415848095.1712238484
Abdirad, Somayeh, Ahmad Majd, Saeed Irian, Naghmeh Hadidi, and Ghasem Hosseini Salekdeh."Differential Adaptation Strategies to Different Levels of Soil Water Deficit in Two Upland and Lowland Genotypes of Rice: A Physiological and Metabolic Approach," *Journal of the Science of Food and Agriculture* 100, no. 4 (2020): 1458–1469.
Abel, Sanderson, et al. "Carbon Tax and Environmental Quality in South Africa," *International Journal of Energy Economics and Policy* 13, no. 2 (2023): 484–488.
Abi Deivanayagam, Thilagawathi, et al. "Envisioning Environmental Equity: Climate Change, Health, and Racial Justice," *The Lancet* 402, no. 10395 (2023): 64–78.
Abraham-Dukuma, Magnus C., et al. "Improving the Climate Change Mitigation Regime of Major Emitting Countries: The Case of South Africa, China, Germany and the United States of America," *Environmental Policy and Governance* 32, no. 1 (2022): 43–55.
Abubakar, Ismaila Rimi, et al. "Environmental Sustainability Impacts of Solid Waste Management Practices in the Global South," *International Journal of Environmental Research and Public Health* 19, no. 19 (2022): 1–26.
Acharibasam, John Bosco, and Ranjan Datta. "Enhancing Community Resilience to Climate Change Disasters: Learning Experience within and from Sub-Saharan Black Immigrant Communities in Western Canada." *Sustainable Development* 32, no. 2 (2024): 1401–1411.
Ackerman, John. "Co-Governance for Accountability: Beyond 'Exit' and 'Voice'," *World Development* 32, no. 3 (2004): 447–463.
Adamson, S., and A.S. Medeiros. "The Greenlight for Government Buildings: Strategies for a Low-Carbon Building Portfolio," *FACETS* 8 (2023): 1–10.
Addaney, Michael, D.B. Jarbandhan, and William Kwadwo Dumenu, eds. *Climate Change in Africa: Adaptation, Resilience, and Policy Innovations* (New York: Springer Nature, 2023).
Adekanmbi, Toyin, et al. "Assessing Future Climate Change Impacts on Potato Yields – A Case Study for Prince Edward Island, Canada," *Foods* 12, no. 6 (2023): 1–20.
Adom, Richard Kwame, and Mulala Danny Simatele. "Analysis of Public Policies and Programmes Towards Water Security in Post-Apartheid South Africa," *Water Policy* 23, no. 3 (2021): 503–520.

Adom, Richard Kwame, Mulala Danny Simatele, and Memory Reid. "Addressing the Challenges of Water-Energy-Food Nexus Programme in the Context of Sustainable Development and Climate Change in South Africa," *Journal of Water and Climate Change* 13, no. 7 (2022): 2761–2779.

Agnew, Robert. "Dire forecast: A Theoretical Model of the Impact of Climate Change on Crime," *Theoretical Criminology* 16, no. 1 (2012): 21–42.

Ahmad, Fatima Maria. "The Road Ahead after Transformational Climate Legislation," *Fletcher Forum on World Affairs* 47 (2023): 129–145.

Ahmad, Shakeel, Shoaib Hameed, Hussain Ali, Tauheed Ullah Khan, Tahir Mehmood, and Muhammad Ali Nawaz. "Carnivores' Diversity and Conflicts with Humans in Musk Deer National Park, Azad Jammu and Kashmir, Pakistan," *European Journal of Wildlife* Research 62, no. 5 (2016): 565–576.

Ahren, Mattias. "The Relevance of the UN Declaration on the Rights of Indigenous Peoples to Vibrant, Viable and Sustainable Sámi Communities," in *The Significance of Sámi Rights*, eds. Dorothée Cambou and Oyvind Ravna (London: Routledge, 2024), 5–21.

Air Quality Life Index. "The Index," University of Chicago (2021). https://aqli.epic.uchicago.edu/the-index/

Ajibo, Kenneth I. "Transboundary Hazardous Wastes and Environmental Justice: Implications for Economically Developing Countries," *Environmental Law Review* 18, no. 4 (2016): 267–283.

Al Shamsi, Hilal, et al. "Implications of Language Barriers for Healthcare: A systematic Review," *Oman Medical Journal* 35, no. 2 (2020): 1–7.

Al-Breiki, Mohammed, and Yusuf Bicer. "Liquified Hydrogen vs. Liquified Renewable Methane: Evaluating Energy Consumption and Infrastructure for Sustainable Fuels," *Fuel* 350 (2023): 1–15.

Al-Hazmi, Hussein E., Ali Mohammadi, Aleksander Hejna, Joanna Majtacz, Amin Esmaeili, Sajjad Habibzadeh, Mohammad Reza Saeb, Michael Badawi, Eder C. Lima, and Jacek Mąkinia. "Wastewater Treatment for Reuse in Agriculture: Prospects and Challenges," *Environmental Research* 236 (2023): 116711.

Al-Kohlani, Sumaia A., and Heather E. Campbell. "Extending Environmental Justice Research to Religious Minorities," *Review of Policy Research* 39, no. 1 (2022): 90–112.

Alarcon, Vladimir J., et al. "Potential Challenges for the Restoration of Biscayne Bay (Florida, USA) in the Face of Climate Change Effects Revealed with Predictive Models," *Ocean & Coastal Management* 247 (2024): 1–33.

Alava, Juan José, and Gerald G. Singh. "Changing Air Pollution and CO2 Emissions during the COVID-19 Pandemic: Lesson Learned and Future Equity Concerns of Post-COVID Recovery," *Environmental Science & Policy* 130 (2022): 1–8.

Albertyn, Catherine. "Gendered Transformation in South African Jurisprudence: Poor Women and the Constitutional Court," *Stellenbosch Law Review* 22, no. 3 (2011): 591–613.

Albertyn, Cathi. "Radical Connectedness: Reproductive Rights, Climate Justice and Gender Equality," in *Feminist Frontiers in Climate Justice: Gender Equality, Climate Change and Rights*, eds. Cathi Albertyn, et al. (Cheltenham, UK: Edward Elgar Publishing, 2023), 138–164.

Alexandrescu, Filip, Ionuț-Marian Anghel, Júlia Adorjáni, Lucrina Ştefănescu, Alina Pop, and Anca Mihai. "On the Path of Evictions and Invisibilization: Poor Roma Facing Climate Vulnerability," *Cities* 114 (2021): 103201–103211.

Alfonsi, Robert Massimo, Merle Naidoo, and Alexandros Gasparatos. "Stakeholder Perspectives for Information and Communication Technologies (ICTs) for Sustainable Food Systems in South Africa: Challenges, Opportunities and a Proposed ICT Framework," *Environment, Development and Sustainability* (2024): 1–34.

Algers, Jonas, and Max Åhman. "Phase-In and Phase-Out Policies in the Global Steel Transition," *Climate Policy* (2024): 1–14.

Ali, Mifrah, Iffat Siddique, and Sawaid Abbas. "Characterizing Air Pollution and Its Association with Emission Sources in Lahore: A Guide to Adaptation Action Plans to Control Pollution and Smog," *Applied Sciences* 12, no. 10 (2022): 5102–5119.

Ali, Mudssar, et al. "Assessing Indigenous and Local Knowledge of Farmers about Pollination Services in Cucurbit Agro-Ecosystem of Punjab, Pakistan," *Saudi Journal of Biological Sciences* 27, no. 1 (2020): 189–194.

Ali, Muhammad Imran. "Comparative Legal Research – Building a Legal Attitude for a Transnational World," *Journal of Legal Studies "Vasile Goldiş"* 26, no. 40 (2020): 66–80.

Alibudbud, Rowalt. "Gender in Climate Change: Safeguarding LGBTQ+ Mental Health in the Philippine Climate Change Response from a Minority Stress Perspective," *Journal of Preventive Medicine and Public Health* 56, no. 2 (2023): 196–199.

Allan, Jennifer et al. "State of global environmental governance 2023," International Institute for Sustainable Development (2024). https://orca.cardiff.ac.uk/id/eprint/167251/1/state-global-environmental-governance-2023.pdf

Allen, Davis, et al. "The Fraud of Plastic Recycling: How Big Oil and the Plastics Industry Deceived the Public for Decades and Caused the Plastic Waste Crisis," The Center for Climate Integrity (2024). https://climateintegrity.org/uploads/media/Fraud-of-Plastic-Recycling-2024.pdf

Alvarez, Camila H. "Structural Racism as an Environmental Justice Issue: A Multilevel Analysis of the State Racism Index and Environmental Health Risk from Air Toxics," *Journal of Racial and Ethnic Health Disparities* 10, no. 1 (2023): 244–258.

American Lung Association. "Disparities in the Impact of Air Pollution," American Lung Association (2023). https://www.lung.org/clean-air/outdoors/who-is-at-risk/disparities

American Lung Association. "State of the AIR 2024 Report," American Lung Association (2024). https://www.lung.org/getmedia/dabac59e-963b-4e9b-bf0f-73615b07bfd8/State-of-the-Air-2024.pdf

American Public Health Association. "Eliminating Exposure to Asbestos," American Public Health Association, 5 November (2019). https://www.apha.org/policies-and-advocacy/public-health-policy-statements/policy-database/2020/01/10/eliminating-exposure-to-asbestos

Amorim-Maia, Ana T., et al. "Intersectional Climate Justice: A Conceptual Pathway for Bridging Adaptation Planning, Transformative Action, and Social Equity," *Urban Climate* 41 (2022): 1–18.

Amprako, Jessica Lucinda. "The United Nations World Water Development Report 2015: Water for a Sustainable World," *Future of Food: Journal on Food, Agriculture and Society* 4, no. 2 (2016): 64–65.

Anderson, Sarah, et al. "Inequality in Agency Responsiveness: Evidence from Salient Wildfire Events," *Resources for the Future Series* 20–22 (2020): 1–36.

Anderson, Sarah E., Andrew J. Plantinga, and Matthew Wibbenmeyer. "Inequality in Agency Response: Evidence from Salient Wildfire Events," *The Journal of Politics* 85, no. 2 (2023): 625–639.
Andrijevic, Marina, et al. "Future Cooling Gap in Shared Socioeconomic Pathways," *Environmental Research Letters* 16, no. 9 (2021): 1–11.
Anekwe, Ifeanyi Michael Smarte, et al. "Addressing Climate Change Challenges in South Africa: A Study in KwaZulu Natal Province," *Climate Crisis: Adaptive Approaches and Sustainability* (2024): 475–496.
Anguelovski, Isabelle, and Joan Martinez-Alier. "The 'Environmentalism of the Poor' Revisited: Territory and Place in Disconnected Global Struggles," *Ecological Economics* 102 (2014): 167–176.
Anguelovski, Isabelle, Linda Shi, Eric Chu, Daniel Gallagher, Kian Goh, Zachary Lamb, Kara Reeve, and Hannah Teicher. "Equity Impacts of Urban Land Use Planning for Climate Adaptation: Critical Perspectives from the Global North and South," *Journal of Planning Education and Research* 36, no. 3 (2016): 333–348.
Annecke, Wendy, and Peta Wolpe. "What Role for Social Policies in the Framework of the Just Transition in South Africa?" *AFD Research Papers* 230 (2022): 1–33.
Anser, Muhammad Khalid, et al. "Beyond Climate Change: Examining the Role of Environmental Justice, Agricultural Mechanization, and Social Expenditures in Alleviating Rural Poverty," *Sustainable Futures* 6 (2023): 1–17.
Antwi, Effah Kwabena, et al. "Review of Climate Change Adaptation and Mitigation Implementation in Canada's Forest Ecosystem's part II: Successes and Barriers to Effective Implementation," *Environmental Reviews* 32, no. 1 (2023): 42–67.
Anyanwu, Chima, et al. "The Impact of Environmental Conditions on Non-Communicable Diseases in Sub-Saharan Africa: A Scoping Review of Epidemiologic Evidence," *Journal of Global Health* 14 (2024): 1–9.
Aon. "Climate and Catastrophe Insight 2024," Aon (2024). https://assets.aon.com/-/media/files/aon/reports/2024/climate-and-catastrophe-insights-report.pdf
Arkush, David, and Donald Braman. "Climate Homicide: Prosecuting Big Oil for Climate Deaths," *Harvard Environmental Law Review* 48, no. 1 (2024): 1–70.
Arowolo, Ayoyemi Lawal, and Olalekan Moyosore Lalude. "'Their Color is a Diabolic Die,' Colonialism and the State of Environmental Justice in Africa," *Society & Sustainability* 4, no. 2 (2022): 1–8.
Arshad, Ameena, and Shagufta Parveen. "Exploring the Impact of Financial Inclusion on Greenhouse Gas Emissions (CO2) and Energy Efficiency: Evidence from Developing Countries," *Environment, Development and Sustainability* (2023): 27723–27738.
Ashton, J. "A Billion People to Be Homeless by 2050 Due to Climate Change," *The Birmingham Post* (Newspaper, England), 14 May (2007), 7.
Associated Press and *Guardian* Staff. "Trump Campaign Attacks Biden Over Recognition of Transgender Day of Visibility," *The Guardian*, 31 March (2024). https://www.theguardian.com/us-news/2024/mar/31/biden-transgender-day-of-visibility-trump-republicans
Atapattu, Sumudu A., et al. "Intersections of Environmental Justice and Sustainable Development: Framing the Issues," in *The Cambridge Handbook of Environmental Justice and Sustainable Development*, eds. Sumudu A. Atapattu, Carmen G. Gonzalez, and Sara L. Seck (Cambridge: Cambridge University Press, 2021), 1–23.
Atencio, Mario, et al. "Federal Statutes and Environmental Justice in the Navajo Nation: The Case of Fracking in the Greater Chaco Region," *American Journal of Public Health* 112, no. 1 (2022): 116–123.

Austen, Ian and Vjosa Isai. "Canada's Logging Industry Devours Forests Crucial to Fighting Climate Change" *The New York Times*, 4 January (2024). https://www.nytimes.com/2024/01/04/world/canada/canada-boreal-forest-logging.html?campaign_id=2&emc=edit_th_20240103&instance_id=111645&nl=todaysheadlines&regi_id=69606629&segment_id=154215&user_id=c3d27092138c7c10d85add87f2186ba3

Australian Centre of Advanced Photovoltaics. "Scoping Study: Solar Panel End-of-life Management in Australia," Australian Centre of Advanced Photovoltaics (2024). https://apvi.org.au/scoping-study-solar-panel-end-of-life-management-in-australia/

Avenancio-León, Carlos F., and Howard Troup. "The Assessment Gap: Racial Inequalities in Property Taxation," *The Quarterly Journal of Economics* 137, no. 3 (2022): 1383–1434.

Averchenkova, Alina., Kate Gannon, and Patrick Curran. *Governance of Climate Change Policy: A Case Study of South Africa* (London, Grantham Research Institute on Climate Change and the Environment and Centre for Climate Change Economic and Policy, London School of Economics and Political Science, 2019). https://www.lse.ac.uk/granthaminstitute/wp-content/uploads/2019/06/GRI_Governance-of-climate-change-policy_SA-case-study_policy-report_40pp.pdf

Aygün, Okan, et al. "Landscape and Climate Conditions Influence the Hydrological Sensitivity to Climate Change in Eastern Canada," *Journal of Hydrology* 615 (2022): 128595.

B

Baba, Sofiane, Shoeb Mohammad, and Carson Young. "Managing Project Sustainability in the Extractive Industries: Towards a Reciprocity Framework for Community Engagement," *International Journal of Project Management* 39, no. 8 (2021): 887–901.

Bachtel, Lauren, Kerry McGrath, Andrew Turner, and John Bobka. "Navigating Environmental Justice Issues in Federal Permitting," *Natural Resources & Environment* 36, no. 2 (2021): 4–8.

Baciu, Alina, et al. "The State of Health Disparities in the United States," in *Communities in Action: Pathways to Health Equity*, eds. Alina Baciu, et al. (US: National Academies Press, 2017).

Bae, Junghwan, Soyoung Kang, and Michael J. Lynch. "Drinking Water Injustice: Racial Disparity in Regulatory Enforcement of Safe Drinking Water Act Violations," *Race and Justice* (2023): 1–21. https://journals.sagepub.com/doi/pdf/10.1177/21533687231189854

Bahls, Christine. "Achieving Equity in Health," *Health Affairs* 34, no. 4 (2011): 1–6.

Baicker-McKee, Steven. "Celebrating Ten Years of Joule," *Joule: Duquesne Energy and Environmental Law Journal* 10 (2022): 1–108. https://sites.law.duq.edu/joule/wp-content/uploads/2022/04/Masthead_FINAL-combined-1.pdf

Bailey, Conner, Robert Gramling, and Shirley B. Laska. "Complexities of Resilience: Adaptation and Change Within Human Communities of Coastal Louisiana," in *Perspectives on the Restoration of the Mississippi Delta*, eds. John W. Day et al. (Dordrecht: Springer, 2014), 125–140.

Baird, Kim, et al. "Insights from First Nations, Government and Industry Leaders on Criteria for Successful Impact Benefit Agreements," *Environmental Management* 14 (2024): 1–15.

Baird, Rachel. "The Impact of Climate Change on Minorities and Indigenous Peoples," Minority Rights Group (2008). https://minorityrights.org/app/uploads/2024/01/download-524-the-impact-of-climate-change-on-minorities-and-indigenous-peoples.pdf

Balasubramanian, Srinidhi, Nina G.G. Domingo, Natalie D. Hunt, Madisen Gittlin, Kimberly K. Colgan, Julian D. Marshall, Allen L. Robinson et al. "The Food We Eat, the Air We Breathe: A Review of the Fine Particulate Matter-Induced Air Quality Health Impacts of the Global Food System," *Environmental Research Letters* 16, no. 10 (2021): 103004.

Balci, Gökcay, et al. "A Roadmap to Alternative Fuels for Decarbonising Shipping: The Case of Green Ammonia," *Research in Transportation Business & Management* 53 (2024): 1–15. https://doi.org/10.1016/j.rtbm.2024.101100

Bang, Guri. "The United States: Conditions for Accelerating Decarbonisation in a Politically Divided Country," *International Environmental Agreements: Politics, Law and Economics* 21, no. 1 (2021): 43–58.

Barankin, Ram A., et al. "Evidence-Driven Approach for Assessing Social Vulnerability and Equality during Extreme Climatic Events," *Frontiers in Water* 2 (2021): 1–10. https://www.frontiersin.org/journals/water/articles/10.3389/frwa.2020.544141/full

Barcalow, Kate Monti, and Jeremy Spoon. "Traditional Cultural Properties or Places, Consultation, and the Restoration of Native American Relationships with Aboriginal Lands in the Western United States," *Human Organization* 77, no. 4 (2018): 291–301.

Barclay, Jenni, et al. "Historical Trajectories of Disaster Risk in Dominica," *International Journal of Disaster Risk Science* 10 (2019): 149–165.

Barhorst, Charles, Abigail Mitchell, Sydney Smith, Lauren Soileau, and Ruth Vanhaverbeke."Stormwater Management for Environmental and Climate Justice Community Resilience," in *2024 Waste-management Education Research Conference* (WERC), 5 (IEEE, 2024), 1–6.

Barnwell, Garret, and Nick Wood. "Climate Justice is Central to Addressing the Climate Emergency's Psychological Consequences in the Global South: A Narrative Review," *South African Journal of Psychology* 52, no. 4 (2022): 486–497.

Barr, Stephanie L., Brendon M.H. Larson, Thomas J. Beechey, and Daniel J. Scott. "Assessing climate change adaptation progress in Canada's protected areas," *The Canadian Geographer/Le Géographe canadien* 65, no. 2 (2021): 152–165.

Barrage, Lint. "The Nobel Memorial Prize for William D. Nordhaus," *The Scandinavian Journal of Economics* 121, no. 3 (2019): 884–924.

Barrage, Lint, and William Nordhaus. "Policies, Projections, and the Social Cost of Carbon: Results from the DICE-2023 model," *Proceedings of the National Academy of Sciences* 121, no. 13 (2024): 1–8.

Barsa, Michael, and David Dana. "The Major Questions Doctrine's Upside for Combatting Climate Change," *New York University Environmental Law Journal* 32 (2024): 1–14.

Bartlett, Jenelle A., and Ayşın Dedekorkut-Howes. "Adaptation Strategies for Climate Change Impacts on Water Quality: A Systematic Review of the Literature," *Journal of Water and Climate Change* 14, no. 3 (2023): 651–675.

Basso, Andrew R. *Destroy Them Gradually: Displacement as Atrocity* (New Brunswick: Rutgers University Press, 2024).

Basson, Gideon. "Poverty Discrimination under the Promotion of Equality and Prevention of Unfair Discrimination Act: A Transformative Substantive Equality Approach," *South African Journal on Human Rights* 39, no. 1 (2023): 26–51.

Beard, Sharon, et al. "Government Program Celebrates 25 Years of Commitment to Environmental Justice Movement," *New Solutions: A Journal of Environmental and Occupational Health Policy* 32, no. 4 (2023): 277–287.

Beaudoin, Jean-Michel, et al. "What does 'First Nation Deep Roots in the Forests' Mean? Identification of Principles and Objectives for Promoting Forest-Based Development," *Canadian Journal of Forest Research* 46, no. 4 (2016): 508–519.

Becares, Laia. "Which Ethnic Groups Have the Poorest Health?" in *Ethnic Identity and Inequalities in Britain*, eds. Stephen Jivraj and Ludi Simpson (Bristol: Policy Press, 2015), 123–140.

Bélanger, Julie, and Dafydd Pilling. "The State of the World's Biodiversity for Food and Agriculture," Food and Agriculture Organization of the United Nations (2019). https://openknowledge.fao.org/server/api/core/bitstreams/50b79369-9249-4486-ac07-9098d07df60a/content

Bellamy, Paige. "Free, Prior Informed Consent and Extractive Industry: Indigenous Action Is the Past, Present, and Future of Global Environmental Justice," *Barry Law Environmental & Earth Law Journal* 13 (2023): 105–136.

Belmeziti, Ali, Adlane M'hammedi Bouzina, and Bernard De Gouvello. "Emergence of a New Paradigm of Water Supply through Adaptation Strategies of the Intermittent System in Algerian Residential Buildings," *Cities* 158 (2025): 105657–105665.

Benesh, Melanie. "This is a Multi-Administration Failure to Take Action on PFOA and PFOS and on the Broader Class of PFAS Chemicals That May Pose Health Effects," *Chemical & Engineering News* 98, no. 27 (2020): 20–20.

Benevolenza, Mia A., and LeaAnne DeRigne. "The Impact of Climate Change and Natural Disasters on Vulnerable Populations: A Systematic Review of Literature," *Journal of Human Behavior in the Social Environment* 29, no. 2 (2019): 266–281.

Bennion, Lily and Juanita Kelly-Mundine. "Clashes in Conservation: First Nations Sites, Communities and Culture in Australian Cultural Heritage Management," *Journal of the Institute of Conservation* 44, no. 3 (2021): 170–182.

Benson, Krista L. "Erosions of Settler State Recognition of Sovereignty and Reproductive Justice," *Adoption & Culture* 10, no. 2 (2022): 225–230.

Bergenas, Johan, and Ariella Knight. "Green Terror: Environmental Crime and Illicit Financing," *The SAIS Review of International Affairs* 35, no. 1 (2015): 119–131.

Bergh, Andreas, and Anders Kärnä. "Economic Freedom and Populism," in *Handbook of Research on Economic Freedom*, ed. Niclas Berggren (Cheltenham, UK: Edward Elgar Publishing, 2024), 297–314.

Berry, Peter, and Rebekka Schnitter. "Health of Canadians in a changing climate: advancing our knowledge for action." Ottawa, Ontario: Government of Canada (2022). https://changingclimate.ca/health-in-a-changing-climate/

Betts, Richard et al. *Assessing Impacts of Earth System Tipping Points on Human Societies*, no. EGU24-6874 (Göttingen, Germany: Copernicus Meetings, 2024).

Bezgrebelna, Mariya, Kwame McKenzie, Samantha Wells, Arun Ravindran, Michael Kral, Julia Christensen, Vicky Stergiopoulos, Stephen Gaetz, and Sean A. Kidd. "Climate Change, Weather, Housing Precarity, and Homelessness: A Systematic Review of Reviews," *International Journal of Environmental Research and Public Health*18, no. 11 (2021): 5812.

Bhatti, Ahmad Zeeshan, Aitazaz Ahsan Farooque, Nicholas Krouglicof, Wayne Peters, Bishnu Acharya, Qing Li, and M. Sheraz Ahsan. "Climate Change Impacts on Precipitation and Temperature in Prince Edward Island, Canada," *World Water Policy* 7, no. 1 (2021): 9–29.

Bian, Xia, Zhilin Ren, Lingling Zeng, Feiyang Zhao, Yikang Yao, and Xiaozhao Li. "Effects of Biochar on the Compressibility of Soil with High Water Content," *Journal of Cleaner Production* 434 (2024): 1–12.

Biddlestone, Mikey, Flavio Azevedo, and Sander van der Linden. "Climate of Conspiracy: A Meta-Analysis of the Consequences of Belief in Conspiracy Theories about Climate Change," *Current Opinion in Psychology* 46 (2022): 1–5.

Biermann, Frank, and Agni Kalfagianni. "Planetary Justice: A Research Framework," *Earth System Governance* 6 (2020): 1–14.

Bilal, Adrien, and Diego R. Känzig. "*The Macroeconomic Impact of Climate Change: Global vs. Local Temperature*," National Bureau of Economic Research, no. w32450 (2024). https://dkaenzig.github.io/diegokaenzig.com/Papers/bk_micc.pdf

Billiot, Shanondora and Jessica Parfait. "Reclaiming Land: Adaptation Activities and Global Environmental Change Challenges Within Indigenous Communities," in *People and Climate Change: Vulnerability, Adaptation, and Social Justice*, eds. Lisa Reyes Mason and Jonathan Riggs (Oxford: Oxford University Press, 2019), 108–121.

Birchall, S. Jeff, Seghan MacDonald, and Nicklas N. Baran. "An Assessment of Systems, Agents, and Institutions in Building Community Resilience to Climate Change: A Case Study of Charlottetown, Canada," *Urban Climate* 41 (2022): 101062.

Bitsch, Jessica, Matthias Jurek, Lønning Gjerdi, Björn Alfthan, Tiina Kurvits, Anna Sinisalo, Davnah Payne, Eva Spehn, Carolina Adler, and Aino Lyydia Kulonen, "Policy Brief: Elevating Mountains in the Post-2020 Global Biodiversity Framework 2.0," (2020): 1–8. https://lib.icimod.org/records/d4vsb-ndy62

Blodgett, Abigail D. "An Analysis of Pollution and Community Advocacy in 'Cancer Alley': Setting an Example for the Environmental Justice Movement in St. James Parish, Louisiana," *Local Environment* 11, no. 6 (2006): 647–661.

Boerner, Christopher, and Thomas Lambert. "Environmental Injustice," *Public Interest* 118 (1995): 61–82.

Böhler, Heike, Marcel Hanegraaff, and Kai Schulze. "Does Climate Advocacy Matter? The Importance of Competing Interest Groups for National Climate Policies," *Climate Policy* 22, no. 8 (2022): 961–975.

Bolibol, Andrew, et al. "Health Insurance Coverage and Access to Care Among LGBT Adults, 2013–19: Study Examines Health Insurance and Access to Care Among LGBT Adults," *Health Affairs* 42, no. 6 (2023): 858–865.

Bonnett, Nicole L., and S. Jeff Birchall. "The Influence of Regional Strategic Policy on Municipal Climate Adaptation Planning," *Regional Studies* 57, no. 1 (2023): 141–152.

Boone, Christopher G. "Environmental Justice, Sustainability and Vulnerability," *International Journal of Urban Sustainable Development* 2, no. 1–2 (2010): 135–140.

Booth, Annie L., and Norman W. Skelton. "'There's a Conflict Right There': Integrating Indigenous Community Values Into Commercial Forestry in the Tl'azt'en First Nation," *Society and Natural Resources* 24, no. 4 (2011): 368–383.

Borras Jr, Saturnino M., Marc Edelman, and Cristóbal Kay. "Transnational Agrarian Movements: Origins and Politics, Campaigns and Impact," *Journal of Agrarian Change* 8, no. 2–3 (2008): 169–204.

Bose, Purabi. "Climate Adaptation: Marginal Populations in the Vulnerable Regions," *Climate and Development* 9, no. 6 (2017): 575–578.

Bosi, Lorenzo, and Gianluca Fazio, *The Troubles in Northern Ireland and Theories of Social Movements* (The Netherlands: Amsterdam University Press, 2017).

Boulanger, Yan, et al. "Multi-Model Projections of Tree Species Performance in Quebec, Canada under Future Climate Change," *Global Change Biology* 28, no. 5 (2022): 1884–1902.

Bouzarovski, Stefan. "Just Transitions: A Political Ecology Critique," *Antipode* 54, no. 4 (2022): 1003–1020.

Boyce, Daniel G., et al. "A Prospective Framework to Support Climate-Adaptive Fisheries in Canada," *FACETS* 8, no. 1 (2023): 1–15.

Boyd, David. *2024 Report of the Special Rapporteur on the Issue of Human Rights Obligations Relating to the Enjoyment of a Safe, Clean, Healthy, and Sustainable Environment, David E. Boyd, "Business, Planetary Boundaries, and the Right to a Clean, Healthy, and Sustainable Environment"* UN Doc. A/HRC/55/43, 2 January (2024). https://documents.un.org/doc/undoc/gen/n21/240/69/pdf/n2124069.pdf?token=QlT3CwM8Je1YrRf2Sa&fe=true

Bradt, Jacob T., et al. "Voluntary Purchases and Adverse Selection in the Market for Flood Insurance," *Journal of Environmental Economics and Management* 110 (2021): 102515.

Bratspies, Rebecca. "Administering Environmental Justice: How New York's Environmental Rights Amendment Could Transform Business as Usual," *Pace Environmental Law Review* 41, no. 2 (2024): 100–126.

Bravo, Mercedes A., et al. "Racial Isolation and Exposure to Airborne Particulate Matter and Ozone in Understudied US Populations: Environmental Justice Applications of Downscaled Numerical Model Output," *Environment International* 92 (2016): 247–255.

Bray, Laura A. "Settler Colonialism and Rural Environmental Injustice: Water Inequality on The Navajo Nation," *Rural Sociology* 86, no. 3 (2021): 586–610.

Bresson, É., et al. "Climate Change Risks and Vulnerabilities during Mining Exploration, Operations, and Reclamation: A Regional Approach for the Mining Sector in Québec, Canada," *CIM Journal* 13, no. 2 (2022): 77–96.

Bridges, Khiara M. "The Dysgenic State: Environmental Injustice and Disability-Selective Abortion Bans," *California. Law Review* 110 (2022): 297–369.

Brink, Ebba, et al. "Weapons of the Vulnerable? A Review of Popular Resistance to Climate Adaptation," *Global Environmental Change* 80 (2023): 1–19.

Brink, Ebba, et al. "Quiet Resistance Speaks: A Global Literature Review of the Politics of Popular Resistance to Climate Adaptation Interventions," *World Development* 177 (2024): 106530, 1–11.

Brinkmann, Robert. "Our Great Sustainability Challenge," in *Practical Sustainability: A Guide to a More Sustainable Life*, ed. Robert Brinkmann (Cham, Switzerland: Palgrave Macmillan, 2021), 37–65.

Brisman, Avi, and Nigel South. "Consumer Technologies, Crime and Environmental Implications," in *The Routledge Handbook of Technology, Crime and Justice*, eds. M.R. McGuire and Thomas Holt (Abingdon, Oxon: Routledge, 2017), 310–324.

Bromley, Benjamin C., Sameer H. Khan, and Scott J. Kenyon. "Dust as a Solar Shield," *PLOS Climate* 2, no. 2 (2023): 1–22.

Brousseau, Jennifer J., Marc J. Stern, Malia Pownall, and Lara J. Hansen. "Understanding How Justice is Considered in Climate Adaptation Approaches: A Qualitative Review of Climate Adaptation Plans," *Local Environment* (2024): 1644–1663.

Browde, Pippa. "Sacrificing Sovereignty: How Tribal-State Tax Compacts Impact Economic Development in Indian Country," *Hastings Law Journal* 74 (2022): 1–44.

Brown, David, Benjamin Bégou, Floriane Clement, Brendan Coolsaet, Lisa Darmet, Mathilde Gingembre, Zuzana V. Harmáčková, Adrian Martin, Barbora Nohlová, and Cécile Barnaud. "Conceptualising Rural Environmental Justice in Europe in an Age of Climate-Influenced Landscape Transformations," *Journal of Rural Studies* 110 (2024): 1–12.

Brown, Julia. "Assuming Too Much? Participatory Water Resource Governance in South Africa," *The Geographical Journal* 177, no. 2 (2011): 171–185.

Brunnée, Jutta. "Climate Change, Global Environmental Justice and International Environmental Law," in *Environmental Law and Justice in Context,* eds. Jonas Ebbesson and Phoebe Okowa (Cambridge: Cambridge University Press, 2009), 316–332.

Buettner, Amanda. "Land Claims Cases of Indigenous Nations in New York," Library of Congress, 21 July (2022). https://blogs.loc.gov/law/2022/07/land-claims-cases-of-indigenous-nations-in-new-york/

Buiquy, Myranda. "The Digital Isolation of Indigenous Communities," *American Indian Law Journal* 11, no. 1 (2022): 1–27.

Bullard, Robert D. "Environmental Racism and Invisible Communities," *West Virginia Law Review* 96 (1993a): 1037–1050.

Bullard, Robert D. "The Threat of Environmental Racism," *Natural Resources & Environment* 7, no. 3 (1993b): 23–56.

Bullard, Robert D. "Environmental Justice: It's More Than Waste Facility Siting," *Social Science Quarterly* 77, no. 3 (1996): 493–499.

Bullard, Robert D. "Environmental Justice-Once a Footnote, Now a Headline," *Harvard Environmental Law Review* 45 (2021): 243–248.

Bullard, Robert D., Maaz Gardezi, Carrie Chennault, and Hannah Dankbar. "Climate Change and Environmental Justice: A Conversation with Dr. Robert Bullard," *Journal of Critical Thought and Praxis* 5, no. 2 (2016).

Büntgen, Ulf, et al. "Recent Summer Warming over the Western Mediterranean Region is Unprecedented Since Medieval Times," *Global and Planetary Change* 232 (2024): 1–13.

Busayo, Emmanual et al. "Spatial Planning and Climate Change Adaptation Assessment: Perspectives from Mdantsane Township Dwellers in South Africa," *Habitat International* 90 (2019): 1–9.

Bush, Elizabeth, and Donald Stanley Lemmen, eds. *Canada's Changing Climate Report. Government of Canada* (2019). https://natural-resources.canada.ca/sites/www.nrcan.gc.ca/files/energy/Climate-change/pdf/CCCR_FULLREPORT-EN-FINAL.pdf

Byskov, Morten Fibieger, and Keith Hyams. "Epistemic Injustice in Climate Adaptation." *Ethical Theory and Moral Practice* 25, no. 4 (2022): 613–634.

C

Cadieux, Philippe, et al. "Projected Effects of Climate Change on Boreal Bird Community Accentuated by Anthropogenic Disturbances in Western Boreal Forest, Canada," *Diversity and Distributions* 26, no. 6 (2020): 668–682.

Caesar, L., et al. *Planetary Health Check Report 2024* (Potsdam: Potsdam Institute for Climate Impact Research, 2024), 1–96. https://www.planetaryhealthcheck.org/storyblok-cdn/f/301438/x/a4efc3f6d5/planetaryhealthcheck2024_report.pdf

Caggiano, Holly, and Sara Constantino. "Household Decision-Making in the Climate Crisis," in *Routledge Handbook of Climate Change and Society* (Routledge, 2024), 237–257.

Calderón Archina, Aldana, Diego Escolar, Guillermo Heider, Marcos J. Niborski, Esteban G. Jobbágy, and Patricio N. Magliano. "Rainwater Harvesting Technologies in Arid Plains of Argentina: Small Local Strategies vs. Large Centralized Projects," *Frontiers in Environmental Science* 12 (2024): 1486798–148680

Calzadilla, Paola Villavicencio. "Human Rights and the New Sustainable Mechanism of the Paris Agreement: A New Opportunity to Promote Climate Justice," *Potchefstroom Electronic Law Journal/Potchefstroomse Elektroniese Regsblad* 21, no. 1 (2018) 1–39, 2.

Campbell, Fiona. "Canada's Carbon Tax Increase: What You Need to Know," *Forbes*, 3 April (2024). https://www.forbes.com/advisor/ca/taxes/canada-carbon-tax-increase/

Campbell-Duruflé, Christopher. "The Canadian Net-Zero Emissions Accountability Act: A Tepid Response to the Paris Agreement," *UBC Law Review* 56, no. 2 (2023): 340–396.

Canada, Government of. "Pan-Canadian Framework on Clean Growth and Climate Change," (2016). https://publications.gc.ca/collections/collection_2017/eccc/En4-294-2016-eng.pdf#page=5.10

Canada, Government of. "Impact Assessment Act," (2019), accessed 22 May 2024. https://cdn.climatepolicyradar.org/navigator/CAN/2019/impact-assessment-act_e2af5712ac4694bc078a75121d385045.pdf

Canada, Government of. "Crown-Indigenous Relations and Northern Affairs Canada, Information Bulletin, Climate Change Preparedness in the North Program," 10 July 2020 (2020). https://www.rcaanc-cirnac.gc.ca/eng/1481305554936/1594738066665

Canada, Government of. "Canada's Natural Resources: Major Projects Inventory," 2021 (2021a). https://natural-resources.canada.ca/sites/nrcan/files/emmc/pdf/EN_version-2021_MPI_Report.pdf#page=14.07

Canada, Government of. "Healthy Environment, Healthy Economy: Canada's Strengthened Climate Plan," (2021b). https://www.canada.ca/en/services/environment/weather/climatechange/climate-plan/climate-plan-overview/healthy-environment-healthy-economy.html

Canada, Government of. "Climate Change Adaptation Plans and Actions," 02 February 2022 (2022a). https://www.canada.ca/en/environment-climate-change/services/climate-change/adapting/plans.html

Canada, Government of. "Crown-Indigenous Relations and Northern Affairs Canada, Information Bulletin, The First Nation Adapt Program," 02 November 2022 (2022b). https://www.rcaanc-cirnac.gc.ca/eng/1481305681144/1594738692193

Canada, Government of. Bill S-5, Strengthening Environmental Protection for a Healthier Canada Act – Summary of Amendments, (2023a). https://www.canada.ca/en/services/environment/pollution-waste-management/strengthening-canadian-environmental-protection-act-1999/bill-c-28-strengthening-environmental-protection-healthier-canada-act-summary-amendments.html

Canada, Government of. "Canada's National Adaptation Strategy: Full Strategy," (2023b). https://www.canada.ca/en/services/environment/weather/climatechange/climate-plan/national-adaptation-strategy/full-strategy.html

Canada, Government of. Crown-Indigenous Relations and Northern Affairs Canada, Information Bulletin, Indigenous Community-Based Climate Monitoring Program, 01 December 2023 (2023c). https://www.rcaanc-cirnac.gc.ca/eng/1509728370447/1594738205979#sec1

Canada, Government of. "Output-Based Pricing System," 09 February 2024 (2024a). https://www.canada.ca/en/environment-climate-change/services/climate-change/pricing-pollution-how-it-will-work/output-based-pricing-system.html

Canada, Government of. "Government of Canada Announces Support for Indigenous-Led Climate Solutions in Remote Indigenous Communities," 06 October 2024 (2024b). https://www.canada.ca/en/natural-resources-canada/news/2024/10/government-of-canada-announces-support-for-indigenous-led-climate-solutions-in-remote-indigenous-communities.html

Canada, Government of. "Indigenous Guardians," 09 February 2024 (2024c). https://www.canada.ca/en/environment-climate-change/services/environmental-funding/indigenous-guardians.html

Canada, Office of the Auditor General of Canada. "Canada's Commitments and Actions on Climate Change," (2024). https://www.oag-bvg.gc.ca/internet/English/att__e_43947.html

Canada Parliament. Legis*info*, (2023). https://www.parl.ca/legisinfo/en/bill/44-1/c-226. https://www.parl.ca/documentviewer/en/44-1/bill/C-226/royal-assent

Canadian Human Rights Commission. "Fast Talk on Environmental Racism in Canada," (2024). https://www.chrc-ccdp.gc.ca/sites/default/files/2024-02/Fast%20talk%20on%20environmental%20racism%20in%20Canada%20%E2%80%93%20Summary%20report.pdf

Canadian Taxpayers Federation. "Carbon Tax Support Survey," 27 February (2024). https://www.taxpayer.com/media/Poll-carbon-tax-hike-CTF-Leger.pdf

Cannon, Clare, et al. "Translating and Embedding Equity-Thinking into Climate Adaptation: An Analysis of US Cities," *Regional Environmental Change* 23, no. 1 (2023): 1–12.

Cantoni, Roberto, Federico Caprotti, and Jiska de Groot. "Solar Energy at the Peri-Urban Frontier: An Energy Justice Study of Urban Peripheries from Burkina Faso and South Africa," *Energy Research & Social Science* 94 (2022): 1–11.

Cantzler, Julia Miller, and Megan Huynh. "Native American Environmental Justice as Decolonization," *American Behavioral Scientist* 60, no. 2 (2016): 203–223.

Cappuccio, Laura. "The Protection of the Political Rights of Minorities in the ACHR," in *The Right to Political Participation: A Study of the Judgments of the European and Inter-American Courts of Human Rights*, eds. Gabriella Citroni, Irene Spigno, and Palmina Tanzarella (Abingdon, Oxon: Routledge, 2022), 253–275.

Carbon Majors Database. "The Carbon Majors Database: Launch Report," Carbon Majors (2024). https://carbonmajors.org/briefing/The-Carbon-Majors-Database-26913

Cárdenas, Omaira Bolaños, and Iliana Monterroso. "Achieving Political Rights, Enhancing Forest Livelihoods: Latin American Indigenous and Afrodescendant Women's Views," in *Routledge Handbook of Community Forestry*, eds. Janette Bulkan, John Palmer, Anne M. Larson, and Mary Hobley (New York: Routledge, 2022), 38–354.

Caretta, Martina Angela, and Ruth A. Morgan. "Special Issue on Indigenous Knowledge for Water-Related Climate Adaptation," *Climate and Development* 13, no. 9 (2021): 761–765.

Carlson, Kirsten Matoy. "Congress and Indians," *University of Colorado Law Review* 86 (2015): 77–180.
Carman, Jennifer, et al. "Exploring Support for Climate Justice Policies in the United States," Yale Program on Climate Change Communication, 4 August (2022). https://climatecommunication.yale.edu/publications/exploring-support-for-climate-justice-policies-in-the-united-states/
Carrington, Damian. "We Asked 380 Top Climate Scientists What They Felt About the Future … They Are Terrified, but Determined to Keep Fighting. Here's What They Said," *The Guardian*, 8 May (2024a). https://www.theguardian.com/environment/ng-interactive/2024/may/08/hopeless-and-broken-why-the-worlds-top-climate-scientists-are-in-despair?CMP=share_btn_url&campaign_id=54&emc=edit_clim_20240509&instance_id=122871&nl=climate-forward&regi_id=73699248&segment_id=166157&te=1&user_id=3cc6620316c0a93e52f5fac3913b2a16
Carrington, Damian. "'Bewildering' to Omit Meat-Eating Reduction from UN Climate Plan," *The Guardian*, 18 March (2024b). https://www.theguardian.com/environment/2024/mar/18/bewildering-to-omit-meat-eating-reduction-from-un-climate-plan
Carrington, Damian, and Seán Clarke. "Revealed: The 1,200 Big Menthane Leaks from Waste Dumps Trashing the Planet," *The Guardian*, 12 February (2024). https://www.theguardian.com/environment/2024/feb/12/revealed-the-1200-big-methane-leaks-from-waste-dumps-trashing-the-planet
Carson, Kelly. "The Water Is Coming: How Policies for Internally Displaced Persons Can Shape the US Response to Sea Level Rise and the Redistribution of the American Population," *Hastings Law Journal* 72 (2020): 1279–1312.
Case, David W. "The Role of Information in Environmental Justice," *Mississippi Law Journal* 81 (2011): 701–746.
Casey, Joan A., et al. "Race/Ethnicity, Socioeconomic Status, Residential Segregation, and Spatial Variation in Noise Exposure in the Contiguous United States," *Environmental Health Perspectives* 125, no. 7 (2017): 1–10.
Castellón, Idna G. "Cancer Alley and the Fight Against Environmental Racism," *Villanova Environmental Law Journal* 32 (2021): 15–43.
Castro, Brianna, and Christina Katz. "Land Tenure Insecurity and Climate Adaptation: Environmental Realities in Colombia and Implications for Integrated Environmental Rights and Participatory Policy," in *Land Tenure Security and Sustainable Development*, eds. Holland, Margaret, Yuta Masuda, and Brian Robinson (Cham: Switzerland: Palgrave Macmillan, 2022), 177–199, 183.
CBS News. "EPA Proposes Rule to Replace All Lead Water Pipes in U.S. Within 10 Years: 'Trying to Right a Longstanding Wrong,'" CBS News, 30 November (2023). https://www.cbsnews.com/news/epa-rule-replace-lead-water-pipes-us-within-10-years/
Ceballos, Gerardo, and Paul R. Ehrlich. "The Misunderstood Sixth Mass Extinction," *Science* 360, no. 6393 (2018): 1080–1081.
Cecco, Leyland. "Mercury Exposure Linked to High Youth Suicides in Canada First Nation," *The Guardian*, 20 July (2023). https://www.theguardian.com/world/2023/jul/20/canada-mercury-poisoning-first-nations-indigenous-youth-suicides
Center for Health, Environment & Justice. "The Intersection of Climate Change, Environmental Injustice and Racism – the Center for Health, Environment & Justice." The Center for Health, Environment & Justice, 9 June (2020). https://chej.org/the-intersection-of-climate-change-environmental-injustice-and-racism

Central Intelligence Agency. *The World Factbook: Canada*, last updated 15 May (2024a). https://www.cia.gov/the-world-factbook/countries/canada/#people-and-society

Central Intelligence Agency. "The World Factbook: South Africa," CIA, last updated 15 May (2024b). https://www.cia.gov/the-world-factbook/countries/south-africa/

Centre for Environmental Rights. "LAC Statement on Minister Creecy's Decision to Grant Sasol's MES Appeal," Centre for Environmental Rights, 10 May (2024a). https://cer.org.za/news/lac-statement-on-minister-creecys-decision-to-grant-sasols-mes-appeal

Centre for Environmental Rights. "Climate Change Act Signed into Law – But Not Yet Operational," 1 August (2024b). https://cer.org.za/news/climate-change-act-signed-into-law

Cha, J. Mijin, Caroline Farrell, and Dimitris Stevis. "Climate and Environmental Justice Policies in the First Year of the Biden Administration," *Publius: The Journal of Federalism* 52, no. 3 (2022): 408–427.

Chakraborty, Jayajit, et al. "Hurricane Harvey and People with Disabilities: Disproportionate Exposure to Flooding in Houston, Texas," *Social Science & Medicine* 226 (2019): 176–181.

Chamberlain, Lisa, and Melissa Fourie. "Using Climate Litigation to Strengthen Advocacy Strategies: The Life after Coal Campaign in South Africa," *Journal of Human Rights Practice* 16, no. 1 (2024): 248–257.

Chambers, Lori. "Boil-Water Advisories and Federal (in) Action: The Politics of Potable Water in Pikangikum First Nation," *Journal of Canadian Studies* 51, no. 2 (2018): 289–310.

Chan, Emily Ying Yang, *Essentials for Health Protection: Four Key Components* (New York: Oxford University Press, 2019).

Chanza, Nelson, Walter Musakwa, and Anton de Wit. "Prospects for Strengthening Adaptation Governance Through Indigenous Knowledge Systems," in *Indigenous Knowledge and Climate Governance: A Sub-Saharan African Perspective*, eds. Eromose E. Ebhuoma and Llewellyn Leonard (Cham, Switzerland: Springer International Publishing, 2022), 141–152.

Chaudhry, Divya. "Climate Change and Health of the Urban Poor: The Role of Environmental Justice," *The Journal of Climate Change and Health* 15 (2024): 100277–100282.

Chavardès, Raphaël D., et al. "Regional Drought Synchronised Historical Fires in Dry Forests of the Montane Cordillera Ecozone, Canada," *International Journal of Wildland Fire* 31, no. 1 (2021): 67–80.

Chawla, Neelam, and Basanta Kumar. "Does Indian Real Estate Regulation Protect Urban Homebuyers? Policy Implications," *Cogent Business & Management* 9, no. 1 (2022): 2117164.

Chen, Sibo. *Energy Politics and Discourse in Canada: Probing Progressive Extractivism* (New York: Taylor & Francis, 2023).

Chen, Yangxin, John C. Moore, and Duoying Ji. "Simulated Responses and Feedbacks of Permafrost Carbon under Future Emissions Pathways and Idealized Solar Geoengineering Scenarios," *Environmental Research Letters* 19, no. 2 (2024): 24050.

Cheng, Yuan Daniel, James R. Farmer, Stephanie L. Dickinson, Scott M. Robeson, Burnell C. Fischer, and Heather L. Reynolds. "Climate Change Impacts and Urban Green Space Adaptation Efforts: Evidence from US Municipal Parks and Recreation Departments," *Urban Climate* 39 (2021): 100962–100978.

Chongtham, Nirmala, and Madho Singh Bisht, *Bamboo Shoot: Superfood for Nutrition, Health and Medicine* (Abingdon, Oxon: CRC Press, 2020).

Chu, Eric K., and Clare E.B. Cannon. "Equity, Inclusion, and Justice as Criteria for Decision-Making on Climate Adaptation in Cities," *Current Opinion in Environmental Sustainability* 51 (2021): 85–94.

Chukwudum, Queensley C. "Climate Finance Across Sub-Saharan Africa: Decision Trees and Network Flows," *South African Journal of Science* 120, no. 3–4 (2024): 1–12.

Cirkovi, Elena. "Self-Determination and Indigenous Peoples in International Law," *American Indian Law Review* 31, no. 2 (2007): 375–400.

Citizen's Climate Lobby. "Perception Gap Plagues Canada's Carbon Price," 19 January (2024). https://citizensclimatelobby.org/blog/policy/perception-gap-plagues-canadas-carbon-price/

Clark, Janine Natalya. "Re-thinking Memory and Transitional Justice: A Novel Application of Ecological Memory," *Memory Studies* 14, no. 4 (2021): 695–712.

Clark, W.C. "Sustainability Science: An Emerging Interdisciplinary Frontier. The Rachel Carson Distinguished Lecture Series," (2007), quoted in Klein, Julie Thompson, "Sustainability and Collaboration: Crossdisciplinary and Cross-Sector Horizons," *Sustainability* 12, no. 4 (2020): 1515–1528, 1516.

Climate Action Tracker. "South Africa: Overview," Climate Action Tracker (2024). https://climateactiontracker.org/countries/south-africa/

Climate Central. "A Decade of Growth in Solar and Wind Power: Trends Across the U.S.," Climate Central, 3 April (2024). https://www.climatecentral.org/report/solar-and-wind-power-2024

Climate Trace. "Climate TRACE Unveils Open Emissions Database of More Than 352 Million Assets," Climate Trace (2023). https://climatetrace.org/news/climate-trace-unveils-open-emissions-database-of-more-than

Climate Transparency. "Climate Transparency Report 2022: South Africa," Climate Transparency (2022): https://www.climate-transparency.org/wp-content/uploads/2022/10/CT2022-South-Africa-Web.pdf

Climate Watch. "Historical GHG Emissions," Climate Watch (1990). https://www.climatewatchdata.org/ghg-emissions?chartType=percentage&end_year=2016&start_year=1990

Cochran, Patricia, et al. "Indigenous Frameworks for Observing and Responding to Climate Change in Alaska," in *Climate Change and Indigenous Peoples in the United States: Impacts, Experiences and Actions,* eds. Julie Koppel Maldonado, Benedict Colombi, and Rajul Pandya (Cham, Switzerland: Springer, 2014), 49–59.

Cock, Jacklyn. "South Africa Must End its Coal Habit. But it's at Odds about When and How," *The Conversation,* 27 February (2019). https://theconversation.com/south-africa-must-end-its-coal-habit-but-its-at-odds-about-when-and-how-112016

Cohen, Daniel Aldana, et al. "Securing Climate Justice Federally: A Political Economy Approach to Targeted Investments," *Environmental Justice* 16, no. 5 (2023): 351–359.

Colbourne, Rick, et al. "Indigenous Works and Two Eyed Seeing: Mapping the Case for Indigenous-Led Research," *Qualitative Research in Organizations and Management: An International Journal* 15, no. 1 (2020): 68–86.

Collins, Andrea M. "Financialization, Resistance, and the Question of Women's Land Rights," *International Feminist Journal of Politics* 21, no. 3 (2019): 454–476.

Collins, Michael. "Biden Designates Castner Range in Texas and Avi Kwa Ame in Nevada as National Monuments," *USA Today,* 21 March (2023). https://eu.usatoday

.com/story/news/politics/2023/03/21/joe-biden-declare-castner-range-avi-kwa-ame-national-monuments/11511671002/

Conklin, Michael. "Good For the Goose but Not for The Gander: Biden's Promise to Appoint a Black Female to The Supreme Court and Title VII Principles," *Texas A&M Law Review* 9, no. 2 (2022): 1–22.

Consumer Federation of America. "EXPOSED: A Report on 1.6 Trillion Dollars of Uninsured American Homes," Consumer Federation of America, 11 March (2024). https://consumerfed.org/press_release/millions-of-consumers-lack-vital-homeowners-insurance-resulting-in-1-6-trillion-in-unprotected-market-value/#:~:text=Using%20data%20from%20the%20American,compared%20to%20homeowners%20in%20general

Contat, Justin, et al. "When Climate Meets Real Estate: A Survey of the Literature," *Real Estate Economics* 52, no. 3 (2024): 618–659.

Contreras, Jesse. "The Impact of Sanitation on Child Health in Vulnerable Populations: a Multicountry Analysis across Income Levels" (PhD dissertation, University of Michigan, 2020).

Convention on the Conservation of Migratory Species of Wild Animals. "State of the World's Migratory Species," Convention on the Conservation of Migratory Species of Wild Animals (2024). https://www.cms.int/sites/default/files/publication/State%20of%20the%20Worlds%20Migratory%20Species%20report_E.pdf

Cooper, Evlondo. "How Broadcast TV News Covered Environmental Justice in 2023" *Media Matters*, 3 April 2024. https://www.mediamatters.org/broadcast-networks/how-broadcast-tv-news-covered-environmental-justice-2023

Cormany, Lauren. "Standing in the Way of Environmental Justice," *Utah Law Review* 2024, no. 1 (2024): 167–187.

Corntassel, Jeff J. "Who Is Indigenous? 'Peoplehood' and Ethnonationalist Approaches to Rearticulating Indigenous Identity," *Nationalism and Ethnic Politics* 9, no. 1 (2003): 75–100.

Corntassel, J. and Holder, C. "Who's Sorry Now? Government Apologies, Truth Commissions, and Indigenous Self-Determination in Australia, Canada, Guatemala, and Peru," *Human Rights Review* 9, no. 4 (2008): 465–489.

Coulthard, Glen. *Red Skin, White Masks: Rejecting the Colonial Politics of Recognition* (Minneapolis: University of Minnesota Press, 2014).

Cowie, Robert H., Philippe Bouchet, and Benoît Fontaine. "The Sixth Mass Extinction: Fact, Fiction or Speculation?" *Biological Reviews* 97, no. 2 (2022): 640–663.

Coyle, Marcia, et al. "Unequal Protection: The Racial Divide in Environmental Law. A Special Investigation," *National Law Journal* 15 (1992): 52–58.

Crago, Christine L., et al. "Income and Racial Disparities in Financial Returns From Solar PV Deployment," *Energy Economics* 117 (2023): 1–36.

Craig, Robin Kundis. "Climate Adaptation Law and Policy in the United States," *Frontiers in Marine Science* 9 (2022): 1059734, 04.

Creamer, John, et al. "Poverty in the United States: 2021," *US Census Bureau* (2022). https://www.census.gov/content/dam/Census/library/publications/2022/demo/p60-277.pdf

Crepelle, Adam. "The United States First Climate Relocation: Recognition, Relocation, and Indigenous Rights at the Isle de Jean Charles," *Belmont Law Review* 6 (2018): 1–40.

Crepelle, Adam. "Decolonizing Reservation Economies: Returning to Private Enterprise and Trade," *The Journal of Business, Entrepreneurship & the Law* 12 (2019a): 413–470.

Crepelle, Adam. "Lies, Damn Lies, and Federal Indian Law: The Ethics of Citing Racist Precedent in Contemporary Federal Indian Law," *NYU Review of Law & Social Change* 44 (2019b): 529–579.

Crepelle, Adam. "Tribal Courts, The Violence Against Women Act, and Supplemental Jurisdiction: Expanding Tribal Court Jurisdiction to Improve Public Safety in Indian Country," *Montana Law Review* 81 (2020a): 59–94.

Crepelle, Adam. "White Tape and Indian Wards: Removing the Federal Bureaucracy to Empower Tribal Economies and Self-Government," *University of Michigan Journal of Law Reform* 54 (2020b): 563–609.

Crepelle, Adam. "Tribes, Vaccines, and COVID-19: A Look at Tribal Responses to the Pandemic," *Fordham Urban Law Journal* 49 (2021a): 31–64.

Crepelle, Adam. "The Time Trap: Addressing the Stereotypes That Undermine Tribal Sovereignty," *Columbia Human Rights Law Review* 53 (2021b): 189–246.

Crepelle, Adam. "Tribal Law's Indian Law Problem: How Supreme Court Jurisprudence Undermines the Development of Tribal Law and Tribal Economies," *Virginia Journal of Social Policy & the Law* 29 (2022): 93–127.

Crepelle, Adam. "Making Red Lives Matter: Public Choice Theory and Indian Country Crime," *Lewis & Clark Law Review* 27 (2023a): 769–823.

Crepelle, Adam. "An Intertribal Business Court," *American Business Law Journal* 60, no. 1 (2023b): 61–109.

Crowe, Nicola. "Pollution and Religion: India's Environmental Dilemma," STEAR (2021): https://www.stearthinktank.com/post/india-s-environmental-dilemma

Crowley, Ryan, et al. "Environmental Health: A Position Paper from the American College of Physicians," *Annals of Internal Medicine* 175, no. 11 (2022): 1591–1593.

Crutzen, Paul J. "Geology of Mankind," in *A Pioneer on Atmospheric Chemistry and Climate Change in the Anthropocene*, eds. Paul J. Crutzen and Hans Günter Brauch (Cham, Switzerland: Springer International Publishing, 2016), 211–230.

Cullen, Mark R., et al. "Disease-Specific Health Disparities: A Targeted Review Focusing on Race and Ethnicity," *Healthcare* vol. 10, no. 4 (2022): 1–16.

Cunningham, Calum X., et al. "Increasing Frequency and Intensity of the Most Extreme Wildfires on Earth," *Nature Ecology & Evolution* (2024): 1–6.

Curtice, Kaitlin, and Esther Choo. "Indigenous Populations: Left Behind in the COVID-19 Response," *The Lancet* 395, no. 10239 (2020): 1753.

D

da Silveira, Maria Eduarda Medeiros, et al. "Environmental Justice and Climate Change Adaptation in the Context of Risk Society," *Theory and Practice of Climate Adaptation* (2018): 251–268.

Daly, Erin, Maria Antonia Tigre, and Natalia Urzola. "Common but Differentiated Constitutionalisms: Does 'Environmental Constitutionalism' Offer Realistic Policy Options for Improving UN Environmental Law and Governance? US and Latin American Perspectives," *World Trade Institute Advanced Studies* (2024): 172–205.

Damico, James S., Mark Baildon, and Alexandra Panos. "Climate Justice Literacy: Stories-We-Live-By, Ecolinguistics, and Classroom Practice," *Journal of Adolescent & Adult Literacy* 63, no. 6 (2020): 683–691.

Dana, David. "Climate Change Adaptation Through the Prism of Individual Rights," in *Climate Liberalism: Perspectives on Liberty, Property and Pollution*, ed. Jonathan H. Adler (Berlin: Springer International Publishing, 2023), 79–102.

Dana, David A., and Deborah Tuerkheimer. "After Flint: Environmental Justice as Equal Protection," *Northwestern University Law Review Online* 111 (2016): 93–103.
Dasgupta, Shouro, Johannes Emmerling, and Soheil Shayegh. "Inequality and Growth Impacts of Climate Change – Insights from South Africa," *Environmental Research Letters* 18, no. 12 (2023): 1–22.
Datta, Amber W., et al. "Imagining Reef Futures after Mass Coral Bleaching Events," *Environmental Science & Policy* 151 (2024a): 1–12.
Datta, Ranjan, et al. "Decolonizing Meanings Climate Crisis and Land-based Adaptions: From Indigenous Women's Perspectives in Western Canada." In *Women's Studies International Forum* 104, no. 102913 (2024b): 1–9.
Daughton, Christian G., and Thomas A. Ternes. "Pharmaceuticals and Personal Care Products in the Environment: Agents of Subtle Change?" *Environmental Health Perspectives* 107, no. 6 (1999): 907–938.
Daum, Thomas, et al. "Addressing Agricultural Labour Issues is Key to Biodiversity-Smart Farming," *Biological Conservation* 284 (2023): 1–14.
Davenport, Coral. "How Abrupt U-Turns Are Defining U.S. Environmental Regulations," *The New York Times*, 26 April (2024). https://www.nytimes.com/2024/04/26/climate/biden-trump-environmental-regulations.html?campaign_id=34&emc=edit_sc_20240430&instance_id=121921&nl=science-times&regi_id=73699248&segment_id=165189&te=1&user_id=3cc6620316c0a93e52f5fac3913b2a16
Davenport, Coral, and Lisa Friedman. "With Ready Orders and an Energy Czar, Trump Plots Pivot to Fossil Fuels," *The New York Times*, 8 November (2024). https://www.nytimes.com/2024/11/08/climate/trump-transition-epa-interior-energy.html
Davidson, Osha. "Why Biden's Pause on New LNG Export Terminals is a BFD," Yale Climate Connections, 14 February (2024): https://yaleclimateconnections.org/2024/02/why-bidens-pause-on-new-lng-export-terminals-is-a-bfd/?utm_source=substack&utm_medium=email
Davis, Lacey J., et al. "Environmental Injustice and Escherichia Coli in Urban Streams: Potential for Community-Led Response," *Wiley Interdisciplinary Reviews: Water* 9, no. 3 (2022): e1583.
Davis-Castro, Carla Y. "Indigenous Peoples in Latin America: Statistical Information," Congressional Research Services (2020). https://sgp.fas.org/crs/row/R46225.pdf
de Belmont, Lucas. "Mass Atrocities against Indigenous Peoples: Atrocity Structure and the Brazilian Amazon under Bolsonaro," *Global Responsibility to Protect* 1, no. aop (2024): 1–30.
de Jong, Irthe J.M. "Beyond the Turn to Human Rights: A Call for an Intersectional Climate Justice Approach," *The International Journal of Human Rights* 28, no. 5 (2024): 738–758.
de Sousa Santos, Boaventura, *The End of the Cognitive Empire: The Coming of Age of Epistemologies of the South* (Durham, NC: Duke University Press, 2018).
Deb, Nikhil and Genshaft-Volz, Nadia. "In Addressing Climate Change, Business as Usual Is Climate Injustice," Just Security (2023). https://www.justsecurity.org/84663/in-addressing-climate-change-business-as-usual-is-climate-injustice/
Debrunner, Gabriela, Katrin Hofer, Michael Wicki, Fiona Kauer, and David Kaufmann. "Housing Precarity in Six European and North American Cities: Threatened by the Loss of a Safe, Stable, and Affordable Home," *Journal of the American Planning Association* (2024): 1–17.

Deep South Center for Environmental Justice and Bullard Center for Environmental and Climate Justice. "Gulf South Environmental Justice Leaders React to EPA Clean Power Plant Rules," 26 April (2024). https://www.dscej.org/the-latest

Deloria Jr, Vine, *God is Red: A Native View of Religion* (Colorado, USA: Fulcrum Publishing, 2003).

Deloria, Philip J., et al. "Unfolding Futures: Indigenous Ways of Knowing for the Twenty-First Century," *Daedalus* 147, no. 2 (2018): 6–16.

Democratic Party. "DNC Releases 2024 Party Platform Draft, Outlining Historic Record and Bold Agenda for President Biden and Vice President Harris to Finish the Job," 13 July (2024). https://democrats.org/news/dnc-releases-2024-party-platform-draft-outlining-historic-record-and-bold-agenda-for-president-biden-and-vice-president-harris-to-finish-the-job/

DePillis, Lydia. "White House Aims to Reflect the Environment in Economic Data," *The New York Times*, 20 January (2023). https://www.nytimes.com/2023/01/20/business/economy/economic-statistics-climate-nature.html?action=click&algo=bandit-all-surfaces-shadow-lda-unique-diversify-time-cutoff-30&alpha=0.05&block=trending_recirc&fellback=false&imp_id=682345842&impression_id=409c1f85-98a6-11ed-83c6-45de77831f29&index=0&pgtype=Article&pool=pool%2F91fcf81c-4fb0-49ff-bd57-a24647c85ea1&region=footer&req_id=882042332&shadow_vec_sim=0.08094519188569906&surface=eos-most-popular-story&variant=0_bandit-eng30s-diversify-shadow-lda-unique

Depledge, Duncan. "Low-Carbon Warfare: Climate Change, Net Zero and Military Operations," *International Affairs* 99, no. 2 (2023): 667–685.

Deranger, Eriel. *Climate Emergency: A Yellowhead Brief.* 1–6. Yellowhead Institute, July (2021). https://yellowheadinstitute.org/wp-content/uploads/2021/07/eriel-deranger-climateemergency-yi-brief.pdf

Derenoncourt, Ellora, et al. "Wealth of Two Nations: The US Racial Wealth Gap, 1860–2020," *The Quarterly Journal of Economics* 139, no. 2 (2024): 693–750.

Derman, Brandon Barclay, *Struggles for Climate Justice: Uneven Geographies and the Politics of Connection* (Berlin: Springer Nature, 2020).

Designit.com. "Revolutionising Energy Services: Iberdrola's Solar Communities," (2024). https://www.designit.com/work/solar-communities

Devisscher, Tahia, Jillian Spies, and Verena C. Griess. "Time for Change: Learning from Community Forests to Enhance the Resilience of Multi-value Forestry in British Columbia, Canada," *Land Use Policy* 103 (2021): 105317.

Dialesandro, John, et al. "Dimensions of Thermal Inequity: Neighborhood Social Demographics and Urban Heat in the Southwestern US," *International Journal of Environmental Research and Public Health* 18, no. 3 (2021): 941–955.

Diamond, James D. "An Uncomfortable Truth: Law as a Weapon of Oppression of the Indigenous Peoples of Southern New England," *Roger Williams University Law Review*: Vol. 27(2) (2022): 255–287.

Dienemann, Wolfgang. "Heidelberg Materials' CO2 Roadmap: On the Way to Carbon Neutrality," *ce/papers* 6, no. 6 (2023): 110–113.

Dietz, Thomas, et al. "Climate Change and Society," *Annual Review of Sociology* 46, no. 1 (2020): 135–158.

Diezmartínez, Claudia V., and Anne G. Short Gianotti. "US Cities Increasingly Integrate Justice into Climate Planning and Create Policy Tools for Climate Justice," *Nature Communications* 13, no. 1 (2022): 1–10.

Diffenbaugh, Noah S., and Marshall Burke. "Global Warming has Increased Global Economic Inequality," *Proceedings of the National Academy of Sciences* 116, no. 20 (2019): 9808–9813.

Dike, Samuel C., and Nyemachi Okparaibea. "An Appraisal of the Fiscal Laws of Oil and Gas Industry Operation in Nigeria," *African Journal of International Energy and Environmental Law* 7, no. 1 (2023): 1–13.

Dobbs, G. Rebecca, and Cassandra Johnson Gaither. "How Much Heirs' Property Is There? Using LightBox Data to Estimate Heirs' Property Extent in the US," *Journal of Rural Social Sciences* (2023): 1–24.

Dobson, Geoffrey P. "Wired to Doubt: Why People Fear Vaccines and Climate Change and Mistrust Science," *Frontiers in Medicine* 8 (2022): 1–9.

Doelle, Meinhard, and Roman Dremliuga. "Comparing Russian and Canadian Climate Policy," *Arctic Review on Law and Politics* 13 (2022): 258–285.

Dombey, Maia. "Environmental Racism: How Governments Are Systematically Poisoning Indigenous Communities & the UN's Role," *University of Miami International & Comparative Law Review* 27 (2019): 131–154.

Domingue, Simone Justine. "The (In)Dispensability of Environmental Justice Communities: A Case Study of Climate Adaptation Injustices in Coastal Louisiana and Narratives of Resistance," *Environmental Justice* 15, no. 4 (2022): 271–278.

Dominioni, Goran. "Motivated Reasoning and Implicit Carbon Prices: Overcoming Public Opposition to Carbon Taxes and Emissions Trading schemes," *European Journal of Risk Regulation* 13, no. 1 (2022): 158–173.

Donoghue, Sophie, and Ryan Katz-Rosene. "Evaluating the Comprehensiveness of Municipal Climate Change Adaptation Plans in Ontario, Canada," *Regional Environmental Change* 23, no. 44 (2023): 1–12.

Dornan, Keith, and Jamie Hearn. "Crossing the Dark and Fearful River: Monsanto, PCBs, and Emerging Tort Theories," *American Indian Law Journal* 10, no. 2 (2022): 1–66.

Doucette, Sonya Remington, et al. "Teaching STEM Through Climate Justice and Civic Engagement," *Science Education and Civic Engagement* 15, no. 1 (2023): 6–16.

Douglas, Karen M., and Robbie M. Sutton. "Climate Change: Why the Conspiracy Theories are Dangerous," *Bulletin of the Atomic Scientists* 71, no. 2 (2015): 98–106.

Doyle, Timothy, Doug McEachern, and Sherilyn MacGregor, *Environment and Politics* (London: Routledge, 2015).

Duarte, Mafalda and Razan Al Mubarak. "Indigenous Peoples Guard Our Nature. We Must Empower Them," *Context*, 9 August (2024). https://www.context.news/nature/opinion/indigenous-peoples-guard-our-nature-we-must-empower-them

Duku, Confidence, and Lars Hein. "Assessing the Impacts of Past and Ongoing Deforestation on Rainfall Patterns in South America," *Global Change Biology* 29, no. 18 (2023): 5292–5303.

Duncan Jr, John C. "Multicultural Participation in the Public Hearing Process: Some Theoretical, Pragmatical, and Analeptical Considerations," *Columbia. Journal of Environmental Law* 24 (1999): 169–248.

Durand-Moreau, Quentin, Jesse Lafontaine, and Jennifer Ward. "Work and Health Challenges of Indigenous People in Canada," *The Lancet Global Health* 10, no. 8 (2022): 1189–1197.

E

Earthjustice. "Biden Administration Says Yes to the Willow Project – But the Fight Is Just Heating Up," (2023). https://earthjustice.org/brief/2023/biden-administration-says-yes-to-the-willow-project-but-the-fight-is-just-heating-up

Earthjustice. "Earthjustice Applauds New Rules Limiting Future Oil Drilling in Western Arctic," Earthjustice, 19 April (2024). https://earthjustice.org/press/2024/earthjustice-applauds-new-rules-limiting-future-oil-drilling-in-western-arctic

Eckert, Lauren E., et al. "Indigenous Knowledge and Federal Environmental Assessments in Canada: Applying Past Lessons to the 2019 Impact Assessment Act." *Facets* 5, no. 1 (2020): 67–90.

Edirisinghe, Asanka, and Sandie Suchet-Pearson. "Nature as a sentient being: Can rivers be legal persons?" *Review of European, Comparative & International Environmental Law* 33, no. 2 (2024): 224–235.

Edwards, Morgan R. et al. "Assessing Inequities in Electrification Via Heat Pumps Across the US," *Joule* (2024): 1–49.

Eghdami, Sadegh, et al. "Gap Analysis of Climate Adaptation Policymaking in Coastal Virginia," *Frontiers in Climate* 5 (2023): 1–14.

Ekhator, Eghosa O., and Edward O. Okumagba. "Climate Change and Multinationals in Nigeria: A Case for Climate Justice," in *Climate Litigation and Justice in Africa*, eds. Kim Bouwer, et al. (Bristol: Bristol University Press, 2024), 247–273.

Ekins, Paul. *Stopping Climate Change: Policies for Real Zero* (New York: Taylor & Francis, 2023).

Eldemire, Ashleigh, Kimberly Luchtenberg, and Matthew Wynter. "Black Tax: Evidence of Racial Discrimination in Municipal Borrowing Costs," *Hutchins Center Working Papers* (2022). https://www.brookings.edu/wp-content/uploads/2022/11/WP81-Eldemire-Poindexter-et-al.pdf

Elgendy, Mohamed, Sonia Hassini, and Paulin Coulibaly. "Review of Climate Change Adaptation Strategies in Water Management," *Journal of Hydrologic Engineering* 29, no. 1 (2024): 3123001–3123015.

Elmore, Skylar, et al. "Time Well Spent: Exploring the Role of Attitude and Topic Importance on Selective Exposure to Agricultural Messages," *Journal of Applied Communications* 107, no. 1 (2023): 1–23.

Emmer, Christine, Julia Dorn, and Jutta Mata. "The Immediate Effect of Discrimination on Mental Health: A Meta-Analytic Review of the Causal Evidence," *Psychological Bulletin* 150(3) (2024): 215–252.

Epstein, Paul R., et al. "Full Cost Accounting for the Life Cycle of Coal," *Annals of the New York Academy of Sciences* 1219, no. 1 (2011): 73–98.

Equalshope. "Open Letter to the United Nations Secretary-General and President of the World Bank-Setting Serious Goals to Combat Inequality," Equalshope (2023). https://equalshope.org/index.php/2023/07/17/setting-serious-goals-to-combat-inequality/

Eriksen, Siri, et al. "When Not Every Response to Climate Change is a Good One: Identifying Principles for Sustainable Adaptation," *Climate and Development* 3, no. 1 (2011): 7–20.

Eriksen, Siri, E. Lisa F. Schipper, Morgan Scoville-Simonds, Katharine Vincent, Hans Nicolai Adam, Nick Brooks, Brian Harding et al. "Adaptation Interventions and Their Effect on Vulnerability in Developing Countries: Help, Hindrance or Irrelevance?," *World Development* 141 (2021): 1–16.

Ernst, Yolandi, et al. "The African Regional Greenhouse Gases Budget (2010–2019)," *Global Biogeochemical Cycles* 38, no. 4 (2024): 1–33.
Estrada, Alejandro, et al. "Global Importance of Indigenous Peoples, Their Lands, and Knowledge Systems for Saving the World's Primates from Extinction," *Science Advances* 8, no. 31 (2022): 1–19.
Etieyibo, Edwin. *A Case for Environmental Justice* (Maryland: Rowman & Littlefield, 2022).
Etzel, Ruth A., et al. "Climate Change Impacts on Health Across the Life Course," *Journal of Global Health* 14 (2024): 3–18.
European Commission Joint Research Centre and United Nations Convention to Combat Desertification, *World Drought Atlas 2024* (2024). https://www.unccd.int/sites/default/files/2024-12/World%20Drought%20Atlas%202024.pdf
European Parliament. "MEPs Adopt New Law Banning Greenwashing and Misleading Product Information," European Parliament (2024). https://www.europarl.europa.eu/news/en/press-room/20240112IPR16772/meps-adopt-new-law-banning-greenwashing-and-misleading-product-information
Evans, Geoff, and Liam Phelan. "Transition to a Post-Carbon Society: Linking Environmental Justice and Just Transition Discourses," *Energy Policy* 99 (2016): 329–339.
Evans-Agnew, Robin A., and Juan Aguilera. "Climate Justice is Environmental Justice: System change for Promoting Planetary Health and a Just Transition from Extractive to Regenerative Action," *Health Promotion Practice* 24, no. 4 (2023): 597–602.

F

Fa, Julia E., et al. "Importance of Indigenous Peoples' Lands for the Conservation of Intact Forest Landscapes," *Frontiers in Ecology and the Environment* 18 no. 3 (2020): 135–140.
Falla, Ana Maria Vargas, Ebba Brink, and Emily Boyd. "Quiet Resistance Speaks: A Global Literature Review of the Politics of Popular Resistance to Climate Adaptation Interventions," *World Development* 177 (2024): 106530–106540.
Fan, Yichun. "Social Cost of Lifestyle Adaptation: Air Pollution and Outdoor Physical Exercise," *Journal of Environmental Economics and Management* 127 (2024): 103042–103059
FAO. "Hunger Hotspots: FAO–WFP Early Warnings on Acute Food Insecurity: November 2024 to May 2025 Outlook" (Rome: FAO, 2024). https://openknowledge.fao.org/items/21824350-cfc8-4ac6-87ef-ce925df137e3-promotion-and-protection-human-rights-and-fundamental-freedoms
FAO, IFAD, UNICEF, WFP, and WHO. "The State of Food Security and Nutrition in the World 2023. Urbanization, Agrifood Systems, Transformation, and Healthy Diets Across the Rural-Urban Continuum," United Nations (2023): https://www.fao.org/documents/card/en/c/cc3017en
Fayazi, Mahmood, Isabelle-Anne Bisson, and Eugene Nicholas. "Barriers to Climate Change Adaptation in Indigenous Communities: A Case Study on the Mohawk Community of Kanesatake, Canada," *International Journal of Disaster Risk Reduction* 49 (2020): 1–9.

Fedele, Giacomo, Camila I. Donatti, Celia A. Harvey, Lee Hannah, and David G. Hole. "Transformative Adaptation to Climate Change for Sustainable Social-Ecological Systems," *Environmental Science & Policy* 101 (2019): 116–125.

Ferdowsi, Ahmad, Farzad Piadeh, Kourosh Behzadian, Sayed-Farhad Mousavi, and Mohammad Ehteram. "Urban Water Infrastructure: A Critical Review on Climate Change Impacts and Adaptation Strategies," *Urban Climate* 58 (2024): 102132–102148.

Fernandes-Jesus, Maria, Barnes Brendon, and Raquel F. Diniz. "Communities Reclaiming Power and Social Justice in the Face of Climate Change," *Community Psychology in Global Perspective* 6, no. 2/2 (2020): 1–21.

Fiack, Duran, et al. "Sustainable Adaptation: Social Equity and Local Climate Adaptation Planning in US Cities," *Cities* 115 (2021): 103235.

Figueroa, Robert Melchior. "Environmental Justice," in *The Routledge Companion to Environmental Ethics*, eds. Benjamin Hale, Andrew Light, and Lydia Lawhon (London: Routledge, 2022), 767–782.

Filho, Walter L., Newton R. Matandirotya, Johannes M. Lütz, Esubalew Abate Alemu, Francis Q. Brearley, Anastasia Ago Baidoo, Adolphine Kateka et al. "Impacts of Climate Change to African Indigenous Communities and Examples of Adaptation Responses," *Nature Communications* 12, no. 1 (2021): 6224–6227.

Filho, Walter Leal, Franziska Wolf, Stefano Moncada, Amanda Lange Salvia, Abdul-Lateef Babatunde Balogun, Constantina Skanavis, Aristea Kounani, and Patrick D. Nunn. "Transformative Adaptation as a Sustainable Response to Climate Change: Insights from Large-Scale Case Studies," *Mitigation and Adaptation Strategies for Global Change* 27, no. 3 (2022): 1–20.

Fine, Michael. "Climate Adaptation and the Need for a National Land Use Policy," *Boston University Law Review* 103, no. 6 (2023): 1843–1877.

Finn, Arden. "A National Minimum Wage in the Context of the South African Labour Market," A Southern Africa Labour and Development Research Unit Working Paper Number 153 (2015). Cape Town: SALDRU, University of Cape Town. http://nationalminimumwage.co.za/wp-content/uploads/2015/09/NMW-RI-Descriptive-StatisticsFinal.pdf

Fisher, Dana R., Oscar Berglund, and Colin J. Davis. "How Effective are Climate Protests at Swaying Policy – and what Could Make a Difference?" *Nature* 623, no. 7989 (2023): 910–913.

Fitts, Lucia A., Zoila A. Cruz-Burga, Hannah Legatzke, and María de los Ángeles La Torre. "Is timber management a realistic conservation alternative for indigenous Amazonian communities?," *Ethnobiology and Conservation* 11 (2022): 1–31.

Fjellheim, Eva Maria. "'You Can Kill Us with Dialogue:' Critical Perspectives on Wind Energy Development in a Nordic-Saami Green Colonial Context," *Human Rights Review* 24, no. 1 (2023): 25–51.

Flavelle, Christopher. "The Climate Future: The Risk to Homes," *The New York Times*, Last accessed 13 May (2024). https://www.nytimes.com/interactive/2024/05/13/climate/insurance-homes-climate-change-weather.html?campaign_id=34&emc=edit_sc_20240514&instance_id=123287&nl=science-times&regi_id=73699248&segment_id=166578&te=1&user_id=3cc6620316c0a93e52f5fac3913b2a16

Fletcher, Matthew L.M. "In 5–4 Ruling, Court Dramatically Expands the Power of States to Prosecute Crimes on Reservations," SCOTUSblog, 29 June (2022). https://www.scotusblog.com/2022/06/in-5-4-ruling-court-dramatically-expands-the-power-of-states-to-prosecute-crimes-on-reservations/

Fletcher, Matthew L.M., and Wenona T. Singel. "Lawyering the Indian Child Welfare Act," *Michigan Law Review* 120 (2021): 1755–1798.

Fletcher, William J., et al. "Introduction to the Holocene Climate," in *European Glacial Landscapes*, eds. David Palacios, et al. (Amsterdam: Elsevier, 2024), 65–72.

Florey, Katherine. "Beyond Uniqueness: Reimagining Tribal Courts' Jurisdiction," *California Law Review* 101 (2013): 1499–1607.

Flórez Bossio, Camila, James Ford, and Danielle Labbé. "Adaptive Capacity in Urban Areas of Developing Countries," *Climatic Change* 157 (2019): 27–29.

Fløttum, Kjersti and Gierstad Øyvind. "The Role of Social Justice and Poverty in South Africa's National Climate Change Response White Paper," *South African Journal of Human Rights* 29, no. 1 (2013): 61–90.

Fombad, Madeleine C. "Knowledge Management for Climate Change in South Africa: A Proposed Strategy," *International Federation of Library Associations and Institutions Journal*, no. 2 (2024): 292–309.

Forchtner, Bernhard, and Balša Lubarda. "Scepticisms and Beyond? A Comprehensive Portrait of Climate Change Communication by the Far Right in the European Parliament," *Environmental Politics* 32, no. 1 (2023): 43–68.

Ford, James D., et al. "Including Indigenous Knowledge and Experience in IPCC Assessment Reports," *Nature Climate Change* 6, no. 4 (2016): 349–353.

Fornalé, Elisa. "Vulnerability, Intertemporality, and Climate Litigation," *Nordic Journal of Human Rights* 41, no. 4 (2023): 357–377.

Forsyth, Miranda, et al. "A Future Agenda for Environmental Restorative Justice?" *The International Journal of Restorative Justice* 4, no. 1 (2021): 17–40.

Forsyth, Tim, and Constance L. McDermott. "When Climate Justice Goes Wrong: Maladaptation and Deep Co-Production in Transformative Environmental Science and Policy," *Political Geography* 98 (2022): 1–39.

Fos, Peter John, Peggy Ann Honore, and Russel L. Honore. "Air Pollution and Covid-19: A Comparison of Europe and the United States," *European Journal of Environment and Public Health* 5, no. 2 (2021): 1–7.

Foster, Sheila R. et al. "NPCC4: Advancing Climate Justice in Climate Adaptation Strategies for New York City," *Annals of the New York Academy of Sciences* 1539, no. 1 (2024): 77–126.

Fosu, Augustin Kwasi. "Growth, Inequality, and Poverty Reduction in Developing Countries: Recent Global Evidence," *Research in Economics* 71, no. 2 (2017): 306–336.

Fowlie, Meredith, and Robyn Meeks. "The Economics of Energy Efficiency in Developing Countries," *Review of Environmental Economics and Policy* 15, no. 2 (2021): 238–260.

Fowler, K.N. "Designing, Sustaining, and Evaluating Environmental Justice Projects: The Critical Community Science Tool," *Connected Science Learning* 5, no. 1 (2023): 1–7. https://www.tandfonline.com/doi/full/10.1080/24758779.2023.12318621

Fox, Laura. "Big AG is the Next Big Oil: How Liability Insurers and Reinsurers Should Prepare for Climate Change Litigation," Available at SSRN 4782669 (2024): 1–47. https://ssrn.com/abstract=4782669

Fox, A., et al. "On Track or Off Course? Assessing Progress toward the 30X30 Target for the Ocean," *Metabolic Consulting*, accessed 22 October (2024). https://for-the-ocean.org/wp-content/uploads/2024/10/Together-for-the-Ocean-Report-Updated.pdf

Francis, David, and Edward Webster. "Poverty and Inequality in South Africa: Critical Reflections," *Development Southern Africa* 36, no. 6 (2019): 788–802.

Francis, David, Adam Habib, and Imraan Valodia. "South Africa's Post-Apartheid Economic Development Trajector," in *The Oxford Handbook of the South African Economy*, eds. Arkebe Oqubay, Fiona Tregenna, and Imraan Valodia (Oxford: Oxford University Press, 2021): 91–110.

Frank, T. "Disasters Displaced More Than 3 Million Americans in 2022," *Scientific American* 6 (2023). https://www.scientificamerican.com/article/disasters-displaced-more-than-3-million-americans-in-2022/

Franklin, Janet, and Glen M. MacDonald. "Climate Change and California Sustainability – Challenges and Solutions," *Proceedings of the National Academy of Sciences* 121, no. 32 (2024): 1–7. https://www.pnas.org/doi/pdf/10.1073/pnas.2405458121

Fransen, Taryn. et al. "Taking Stock of the Implementation Gap in Climate Policy," *Nature Climate Change* 13 (2023): 752–755.

Frederick, Bryan and McCulloch, Caitlin. "Under the Weather – The National Security Risks from Climate Change Could Go Well Beyond What the U.S. Government Thinks," *Just Security* (2024). https://www.justsecurity.org/93111/under-the-weather-the-national-security-risks-from-climate-change-could-go-well-beyond-what-the-u-s-government-thinks/

Fregoso, Jimenez P. "Intersectionality as a Tool to Adjudicate Human Rights Law; A Case Study on the Inter-American System of Human Rights (PhD dissertation, University of Essex, 2022).

Freiin von Bibra, Maria-Letizia. "Manifest Mythmaking: The Role of US 'Manifest Destiny' in Nineteenth and Twenty-First Century Indigenous Dispossession," *Webster Review of International History* 2, no. 2 (2022): 15–31.

Friedman, Lisa. "A War Against Climate Science, Waged by Washington's Rank and File," *The New York Times*, 15 June (2020). https://www.nytimes.com/2020/06/15/climate/climate-science-trump.html

Friedman, Lisa. "White House Takes Aim at Environmental Racism, but Won't Mention Race," *The New York Times*, 15 February (2022). https://www.nytimes.com/2022/02/15/climate/biden-environment-race-pollution.html?smid=nytcore-ios-share&referringSource=articleShare

Front Commun pour la Transition Energétique. *Road Map for Québec's Transition Toward Carbon Neutrality, Version 1.0.* (Quebec: Front Commun pour la Transition Energétique, 2019).

Frost, Natasha. "Thursday Briefing," *The New York Times*, 31 October (2024). https://www.nytimes.com/2024/10/31/briefing/correspondents-election-britain-budget.html?smid=nytcore-ios-share&referringSource=articleShare

Fuentes-George, Kemi. "The Legacy of Colonialism on Contemporary Climate Governance," *Georgetown Journal of International Affairs* 24, no. 1 (2023): 91–98.

Fyson, Claire, et al. "When Will Global Greenhouse Gas Emissions Peak?" Climate Analytics (2023). https://climateanalytics.org/publications/2023/2030-targets-aligned-to-15cevidence-from-the-latest-global-pathways/

G

Gadson, Arnita, and Rochelle H. Holm. "Environmental Health Justice Across the Globe," *Journal of Epidemiology & Community Health* 77, no. 1 (2023): 63–64.

Gambhir, Ajay, Fergus Green, and Peter J.G. Pearson. "Towards a Just and Equitable Low-Carbon Energy Transition," *Grantham Institute Briefing Paper* 26 (2018): 1–18.
Ganguly, Geetanjali. "Breaking New Ground in Private Climate Litigation," in *Research Handbook on Climate Change Litigation*, eds. Francesco Sindico, et al. (Cheltenham, UK: Edward Elgar Publishing, 2024), 123–144.
Garcia, Beatriz, and Lucas Lixinski. "Beyond Culture: Reimagining the Adjudication of Indigenous Peoples' Rights in International Law," *Intercultural Human Rights Law Review* 15 (2020): 127–169.
Gardner-Frolick, Rivkah, David Boyd, and Amanda Giang. "Selecting Data Analytic and Modeling Methods to Support Air Pollution and Environmental Justice Investigations: A Critical Review and Guidance Framework," *Environmental Science & Technology* 56, no. 5 (2022): 2843–2860.
Garnett, Stephen T., et al. "A Spatial Overview of the Global Importance of Indigenous Lands for Conservation," *Nature Sustainability* 1, no. 7 (2018): 369–374.
Garschagen, Matthias, et al. "The Consideration of Future Risk Trends in National Adaptation Planning: Conceptual Gaps and Empirical Lessons," *Climate Risk Management* 34 (2021): 1–14.
Gauci, Vincent, et al. "Global Atmospheric Methane Uptake by Upland Tree Woody Surfaces," *Nature* 631 (2024): 796–800.
Gayle, Damien. "Millions Suffering in Deadly Pollution 'Sacrifice Zones', Warns UN Expert," *The Guardian*, 10 March (2022). https://www.theguardian.com/environment/2022/mar/10/millions-suffering-in-deadly-pollution-sacrifice-zones-warns-un-expert
Gayle, Damien, Matthew Taylor and Ajit Niranjan. "Human Rights Experts Warn Against European Crackdown on Climate Protesters," *The Guardian*, 12 October (2023). https://www.theguardian.com/environment/2023/oct/12/human-rights-experts-warn-against-european-crackdown-on-climate-protesters
Geisler, Charles, and Ben Currens. "Impediments to Inland Resettlement under Conditions of Accelerated Sea Level Rise," *Land Use Policy* 66 (2017): 322–330.
Gelles, David. "A Climate-Focused Guide to Election Day," *The New York Times*, 5 November (2024). https://www.nytimes.com/2024/11/05/climate/2024-election-climate-change.html
Gelles, David, and Manuela Andreoni. "A Seismic Supreme Court Decision," *The New York Times*, 2 July (2024). https://www.nytimes.com/2024/07/02/climate/supreme-court-climate-chevron.html
Gençay, Gökçe, and Üstüner Birben. "The Effectiveness of Law on Forest Protection," in *Recent Researches in Interdisciplinary Sciences* (2016): 722–734.
Gerbatsch, Kirsten D. "Treaty-Based Climate Change Claims: Litigation Pathways in the Face of Cultural Devastation." *Public Land & Resources Law Review* 45 (2022): 155–195.
Gerrard, Michael B., and Edward McTiernan. "New York Adopts Nation's Strongest Environmental Justice Law," *New York Law Journal* (2023): 1–5.
Gerrard, Michael, et al. "West Virginia v. Environmental Protection Agency: The Agency's Climate Authority," *Environmental Law Reports.* 52 (2022): 1–13.
Gewin, Virginia. "How to Include Indigenous Researchers and Their Knowledge," *Nature* 589, no. 7841 (2021): 315–318.
Ghofrani, Zahra, Victor Sposito, and Robert Faggian. "A Comprehensive Review of Blue-Green Infrastructure Concepts," *International Journal of Environment and Sustainability* 6, no. 1 (2017): 15–36, 15.

Giacomini, Giada, *Indigenous Peoples and Climate Justice: A Critical Analysis of International Human Rights Law and Governance* (London: Springer Nature, 2022).

Giang, Amanda, et al. "Exposure, Access, and Inequities: Central Themes, Emerging Trends, and Key Gaps in Canadian Environmental Justice Literature from 2006 to 2017." *The Canadian Geographer/Le Géographe canadien* 66, no. 3 (2022): 434–449.

Gilles, Cate. "No One Ever Told Us: Native Americans and the Great Uranium Experiment," in *Governing the Atom*, eds. John Byrne and Steven M. Hoffman (Routledge, 2020), 103–125.

Gilliam, Christa C., et al. "Climate Change: Environmental Justice, Human Rights, and Peaceful Practices," in *Perspectives on Justice, Indigeneity, Gender, and Security in Human Rights Research*, eds. Laura E. Reimer and Katerina Standish (Singapore: Springer Nature Singapore, 2023), 301–351.

Giordano, Christian. "Affiliation, Exclusion and the National State: 'Ethnic Discourses' and Minorities in East Central Europe," in *Rethinking Nationalism and Ethnicity*, ed. Hans-Rudolf Wicker (Routledge, 2020), 175–192.

Giovannoni, Elena, and Giacomo Fabietti. "What Is Sustainability? A Review of the Concept and Its Applications," in *Integrated Reporting*, eds. Cristiano Busco, et al. (Cham, Switzerland: Springer International Publishing, 2013), 21–40.

Givens, Jennifer E., Xiaorui Huang, and Andrew K. Jorgenson. "Ecologically Unequal Exchange: A Theory of Global Environmental Injustice," *Sociology Compass* 13, no. 5 (2019): 199–205.

Glauner, Lindsay. "The Need for Accountability and Reparations: 1830–1976 the United States Government's Role in the Promotion, Implementation, and Execution of the Crime of Genocide against Native Americans," *DePaul Law Review* 51 (2001): 911–962.

Global Carbon Project. "Fossil Fuel CO2 Emissions Increase Again in 2024," *Global Carbon Budget* (2024). https://globalcarbonbudget.org/fossil-fuel-co2-emissions-increase-again-in-2024/

Global Energy Monitor. "Boom and Bust Coal 2024: Tracking the Global Coal Pipeline," Global Energy Monitor (2024). https://globalenergymonitor.org/report/boom-and-bust-coal-2024/

Global Methane Pledge. "Fast Action on Methane to Keep a 1.5 °C Future Within Reach," Global Methane Pledge (nd). https://www.globalmethanepledge.org

Global Monitoring Laboratory. "Trends in CO2, CH4, N2O, SF6," Global Monitoring Laboratory (nd). https://gml.noaa.gov/ccgg/trends/gl_gr.html?campaign_id=54&emc=edit_clim_20240409&instance_id=119740&nl=climate-forward&regi_id=73699248&segment_id=163030&te=1&user_id=3cc6620316c0a93e52f5fac3913b2a16

Godsey, Mark A. "Educational Inequalities, the Myth of Meritocracy, and the Silencing of Minority Voices: The Need for Diversity on America's Law Reviews," *Harvard BlackLetter Law Journal* 12 (1995): 59–91.

Goga, Sumayya, and Pamela Mondliwa. "Structural Transformation, Economic Power, and Inequality in South Africa," in *Structural Transformation in South Africa: The Challenges of Inclusive Industrial Development in a Middle-Income Country*, eds. Antonio Andreoni et al. (Oxford: Oxford University Press, 2021), 165–188.

Goldstein, Bernard D. "The Role of Populism and Private Property Rights in President Trump's Decision to Withdraw from the Paris Agreement on Climate Change," in

The Perils of Populism: The End of the American Century?, ed. Adebowale Akande (New York: Springer Nature Switzerland, 2023), 469–493.

Gomes, Marcus, and Steffen Böhm. "Right-Wing Populism versus Climate Capitalism: Climate Change Governance under Scrutiny," in *Business Elites and Populism: The Odd Couple?*, eds. Magnus Feldmann and Glenn Morgan (Oxford: Oxford University Press, 2023), 245–262.

Gone, Joseph P. "Indigenous Historical Trauma: Alter-Native Explanations for Mental Health Inequities," *Dædalus* 152, no. 4 (2023): 130–150.

Gonzalez, Alicia Lorena. *The Bolsonaro Effect: State Violence on Indigenous Rights in Brazil* (Indiana: Indiana University, 2023).

Gonzalez, Carmen G. "Environmental Justice and International Environmental Law," in *Routledge Handbook of International Environmental Law*, eds. Erika J. Techera, et al. (London: Routledge, 2012), 107–128.

Gonzalez, Teddy, and Jillian Knox. "In the Dark: The Scapegoating of Renewables after Grid Failures," *National Resources Journal* 63 (2023): 30–69.

Gonzalez, David J.X., et al. "Historic Redlining and the Siting of Oil and Gas Wells in the United States," *Journal of Exposure Science & Environmental Epidemiology* 33, no. 1 (2023): 76–83.

Goodluck, Okoro, and Umeuduji Joel Ekwutosi. "Monitoring and Adaptation to Air Pollution: The Case of Oil Producing Communities in the Niger Delta Region," American *Journal of Bioscience and Bioengineering* 10, no. 4 (2022): 89–104.

Goodrick, Wade, and Nola Redelinghuys. "Hydraulic Fracturing as an Environmental and Social Justice Issue in South Africa," in *Environment, Climate, and Social Justice: Perspectives and Practices from the Global South*, eds. Devendraraj Madganagopal, et al. (Singapore: Springer Nature Singapore, 2022), 297–322.

Gouritin, Armelle. "Climate Displacement in Mexico and Critical Environmental Justice:Towards a New Line of Research," *Perfiles Latinoamericanos* 32, no. 64 (2024): 1–24.

Grabowski, Zbigniew Jakub, Katinka Wijsman, Claudia Tomateo, and Timon McPhearson. "How Deep Does Justice Go? Addressing Ecological, Indigenous, and Infrastructural Justice Through Nature-Based Solutions in New York City," *Environmental Science & Policy* 138 (2022): 171–181

Graff, Michelle et al. "Opportunities to Advance Research on Energy Insecurity." *Nature Energy* 8, no. 6 (2023): 550–553.

Grahn-Farley, Maria. "The Human Rights Claim in Climate Justice: An Argument for Reintroducing the Principle of Anti-Discrimination and to Strengthening the Anti-Domination Principle When Children Go to Court," *Journal of Gender Race & Justice* 25 (2022): 439–498.

Greaves, Wilfrid. "Climate Change and Security in Canada." *International Journal* 76, no. 2 (2021): 183–203.

Green, Jessica F. "Does Carbon Pricing Reduce Emissions? A Review of Ex-Post Analyses," *Environmental Research Letters* 16, no. 4 (2021): 43004–43021.

Greenspoon, Lior, et al. "The Global Biomass of Wild Mammals," *Proceedings of the National Academy of Sciences* 120, no. 10 (2023): 1–7.

Gricius, Gabriella, and Annie Martel. "Integrating Indigenous Women's Traditional Knowledge for Climate Change in Canada." *AlterNative: An International Journal of Indigenous Peoples* 20, no.3 (2024): 377–387.

Günerigök, Servet. "Brazil's Lula Names Ministers for Environment and Indigenous Peoples," Anadolu Agency, 31 December (2022). https://www.aa.com.tr/en/americas/brazils-lula-names-ministers-for-environment-and-indigenous-peoples-/2775811

Gupta, Joyeeta, and Courtney Vegelin. "Inclusive Development, Leaving No One Behind, Justice and the Sustainable Development Goals," *International Environmental Agreements: Politics, Law and Economics* 23, no. 2 (2023): 115–121.

Gustafsson, Maria-Therese, and Almut Schilling-Vacaflor. "Indigenous Peoples and Multiscalar Environmental Governance: The Opening and Closure of Participatory Spaces," *Global Environmental Politics* 22, no. 2 (2022): 70–94.

Guyadeen, Dave, and Daniel Henstra. "Evaluating Rural Municipal Climate Change Plans in Ontario, Canada," *Journal of Rural Studies* 97 (2023): 123–139.

H

Hagstrom, Anna L., et al. "Yale School of Public Health Symposium: An Overview of the Challenges and Opportunities Associated with Per-and Polyfluoroalkyl Substances (PFAS)," *Science of The Total Environment* 778 (2021): 146–192.

Hajizadeh, Mohammad, Min Hu, Yukiko Asada, and Amy Bombay. "Explaining the Gaps in Psychological Distress and Suicidal Behaviours Between Non-Indigenous and Indigenous Adults Living Off-Reserve in Canada: A Cross-Sectional Study." *Canadian Medical Association Open Access Journal* 9, no. 1 (2021): E215–E223.

Hale, Rebecca L., et al. "Social Dimensions of Urban Flood Experience, Exposure, and Concern," *Journal of the American Water Resources Association* 54, no. 5 (2018): 1137–1150.

Hall, Nina, and Lucy Crosby. "Climate Change Impacts on Health in Remote Indigenous Communities in Australia," *International Journal of Environmental Health Research* 32, no. 3 (2020): 487–502.

Hallward, Maia, and Jonathan Taylor Downs. "Differences in African Indigenous Rights Messaging in International Advocacy Coalitions," *Human Rights Review* 23, no. 2 (2022): 183–204.

Hamann, Ralph and O'Riordan, Timothy. "South African Environmental Policy on the Move," *South African Geographic Journal* 82, no. 2 (2000): 11–22.

Hammer, P.J. "The Flint Water Crisis, the Karegnondi Water Authority and Strategic–structural Racism," *Critical Sociology* 45, no. 1 (2019): 103–119.

Han, Heejin, and Sang Wuk Ahn. "Youth Mobilization to Stop Global Climate Change: Narratives and Impact," *Sustainability* 12, no. 10 (2020): 1–23.

Hannon, Mary T. "Beyond a Sliver of a Full Moon: Acknowledging and Abolishing White Bias to Restore Safety & Sovereignty to Indian Country," *American Indian Law Journal* 9, no. 2 (2021): 257–290.

Hansen, James E., et al. "Global Warming in the Pipeline," *Oxford Open Climate Change* 3, no. 1 (2023): kgad008.

Hansson, Sten, Kati Orru, Andra Siibak, Asta Bäck, Marco Krüger, Friedrich Gabel, and Claudia Morsut. "Communication-Related Vulnerability to Disasters: A Heuristic Framework," *International Journal of Disaster Risk Reduction* 51 (2020): 1–9.

Hanto, Jonathan, et al. "South Africa's Energy Transition – Unraveling Its Political Economy," *Energy for Sustainable Development* 69 (2022): 164–178.

Harp, Ryan D., and Kristopher B. Karnauskas. "Global Warming to Increase Violent Crime in the United States," *Environmental Research Letters* 15, no. 3 (2020): 1–9.

Harper, Sherilee L., et al. "Estimating Climate Change and Mental Health Impacts in Canada: A Cross-Sectional Survey Protocol," *Plos One* 18, no. 10 (2023): 1–12.

Harrington, A. *The Future of Peace: Incorporation of Intergenerational Equity and Justice in Peace Treaties and Reconciliation Agreements* (Cheltenham, UK: Edward Elgar Publishing, 2023).

Harrison, Paula A., et al. "Selecting Methods for Ecosystem Service Assessment: A Decision Tree Approach," *Ecosystem Services* 29 (2018): 481–498.

Harvey, Jeffrey A., et al. "Scientists' Warning on Climate Change and Insects," *Ecological Monographs* 93, no. 1 (2023): 2–37.

Harwatt, Helen, et al. "Options for a Paris-Compliant Livestock Sector. Timeframes, Targets and Trajectories for Livestock Sector Emissions from a Survey of Climate Scientists," Harvard Law School (2024). https:// animal.law.harvard.edu/wp-content/uploads/Paris-compliant-livestock-report.pdf

Hausman, Lauren M. "Bridging the Chasmic Gap Between Two Methods of Constitutional Interpretation," *California Western Law Review* 60, no. 2 (2024): 315–361.

Health Canada. "Health Canada Releases Assessment Report on Effects of Climate Change on Health," News release, February (2022). https://www.canada.ca/en/health-canada/news/2022/02/health-canada-releases-assessment-report-on-effects-of-climate-change-on-health.html

Health Effects Institute. "State of Global Air 2022," Health Effects Institute (2022). https://www.stateofglobalair.org/resources/report/state-global-air-report-2024

Hedge, Paul, Elizabeth Ingrid Van Putten, Cass Hunter, and Mibu Fischer. "Perceptions, Motivations and Practices for Indigenous Engagement in Marine Science in Australia," *Frontiers in Marine Science* 7 (2020): 522, 1–12.

Heidegger, Patrizia, and Katy Wiese. "Pushed to the Wastelands: Environmental Racism Against Roma Communities in Central and Eastern Europe." *Brussels: European Environmental Bureau* (2020). https://eeb.org/library/pushed-to-the-wastelands-environmental-racism-against-roma-communities-in-central-and-eastern-europe/

Heikka-Huber, Lisa. "Challenges to Reindeer, Reciprocity, and Indigenous Sami Sovereignty Amidst the Impact of Green Energy Developments," *IdeaFest: Interdisciplinary Journal of Creative Works and Research from Cal Poly Humboldt* 7, no. 1 (2023): 52–64.

Henricksen, Wes. "Fossil Fuel Fraud," *Case Western Reserve Journal of International Law* 56 (2024): 409–438.

Henry, Matthew S. *Hydronarratives: Water, Environmental Justice, and a Just Transition* (University of Nebraska Press, 2023).

Henry, Matthew S. "Transition Obstructionism and 'Embodied Energy Injustice': A Wyoming Case Study," *Contemporary Social Science* (2024): 1–18.

Henstra, Daniel. "A Whole-of-Government Approach to Climate Adaptation." *Canadian Climate Institute Report* (2022), 1. https://climateinstitute.ca/wp-content/uploads/2022/07/A-whole-of-government-approach-to-climate-adaptation.pdf

Herman, Jody L., Andrew R. Flores, and Kathryn K. O'Neill. "How Many Adults and Youth Identify as Transgender in the United States?" *William Institute UCLA Law School* (2022). https://williamsinstitute.law.ucla.edu/wp-content/uploads/Trans-Pop-Update-Jun-2022.pdf

Hermansen, Erlend A.T., et al. "Climate Action Post-Paris: How Can the IPCC Stay Relevant?" *NPJ Climate Action* 2, no. 30 (2023): 1–8.

Hernández, Diana. "Sacrifice Along the Energy Continuum: A Call For Energy Justice," *Environmental Justice* 8, no. 4 (2015): 151–156.

Hernández-Truyol, Berta E. "Who's Afraid of Being WOKE? – Critical Theory as Awakening to Erascism and Other Injustices," *Journal of Critical Race and Ethnic Studies* 1, no. 5 (2023): 1–36.
Hernes, Hans-Kristian, et al. "Indigenous Rights and Governance Theory," in *Indigenous Peoples, Natural Resources and Governance*, eds. Monica Tennberg, Hernes, Hans-Kristian, Else Grete Broderstad, and Monica Tennberg (New York: Routledge, 2022), 1–217.
Herreros-Cantis, Pablo, et al. "Shifting Landscapes of Coastal Flood Risk: Environmental (In) justice of Urban Change, Sea Level Rise, and Differential Vulnerability in New York City." *Urban Transformations* 2, no. 1 (2020): 1–28.
Herreros-Cantis, Pablo, Lawrence Hoffman, Christopher Kennedy, Young Kim, Joel Charles, Victoria Gillet, Anne Getzin et al. "Co-Producing Research and Data Visualization for Environmental Justice Advocacy in Climate Change Adaptation: The Milwaukee Flood-Health Vulnerability Assessment," *Cities* 15, no. 5 (2024): 105474–105490.
Hillary, Luke S., et al. "Wastewater and Public Health: The Potential of Wastewater Surveillance for Monitoring COVID-19," *Current Opinion in Environmental Science & Health* 17 (2020): 14–20.
Höhne, Chris, Christian Kahmann, and Mathis Lohaus. "Continuity and Change in Norm Translations after the Paris Agreement: From First to Second Nationally Determined Contributions," *Global Environmental Politics* 24, no. 2 (2024): 69–97.
Hoicka, Christina E., Katarina Savic, and Alicia Campney. "Reconciliation Through Renewable Energy? A Survey of Indigenous Communities, Involvement, and Peoples in Canada," *Energy Research & Social Science* 74 (April 2021): 101897.
Holifield, Ryan, Michael Porter, and Gordon Walker. "Spaces of Environmental Justice: Frameworks for Critical Engagement," *Antipode* 41, no. 4 (2009): 591–612.
Holroyd, Carin. "Corporate Social Responsibility, Indigenous Peoples and Mining in Scandinavia," in *Local Communities and the Mining Industry*, eds. Nicholas D. Brunet and Sheri Longboat (New York: Routledge, 2023), 103–122.
Horel, Stéphane. "PFAS: How the Chemical Industry Is Derailing a Ban on 'Forever Chemicals,'" *Le Monde.fr*, Le Monde, 15 January 2025. https://www.lemonde.fr/en/les-decodeurs/article/2025/01/15/pfas-how-the-chemical-industry-is-derailing-a-ban-on-forever-chemicals_6737079_8.html
Horowitz, Leah S. "'It Shocks Me, The Place of Women': Intersectionality and Mining Companies' Retrogradation of Indigenous Women in New Caledonia," *Gender, Place & Culture* 24, no. 10 (2017): 1419–1440.
Horowitz, Leah S., Arn Keeling, Francis Lévesque, Thierry Rodon, Stephan Schott, and Sophie Thériault. "Indigenous Peoples' Relationships to Large-Scale Mining in Post/Colonial Contexts: Toward Multidisciplinary Comparative Perspectives," *The Extractive Industries and Society* 5, no. 3 (2018): 404–414.
Horst, Megan, and Amy Marion. "Racial, Ethnic and Gender Inequities in Farmland Ownership and Farming in the US," *Agriculture and Human Values* 36 (2019): 1–16.
Hosseini, Keyvan, and Agnieszka Stefaniec. "A Wolf in Sheep's Clothing: Exposing the Structural Violence of Private Electric Automobility," *Energy Research & Social Science* 99, (2023): 1–6.
Hou, Kun, et al. "High Ambient Temperatures are Associated with Urban Crime Risk in Chicago," *Science of The Total Environment* 856 (2023): 1–9.
Howell, Junia, and James R. Elliott. "Damages Done: The Longitudinal Impacts of Natural Hazards on Wealth Inequality in the United States," *Social Problems* 66, no. 3 (2019): 448–467.

Howitt, Richard. “Decolonizing People, Place and Country: Nurturing Resilience across Time and Space,” *Sustainability* 12, no. 15 (2020): 58–82.
Hughes, Hannah. *The IPCC and the Politics of Writing Climate Change* (Cambridge: Cambridge University Press, 2024).
Huizenga, Daniel. “Governing Territory in Conditions of Legal Pluralism: Living Law and Free, Prior, and Informed Consent (FPIC) in Xolobeni, South Africa,” *The Extractive Industries and Society* 6, no. 3 (2019): 711–721.
Hull, Eric V. “Environmental Injustice and COVID-19: Addressing the Link between Pandemics and Pollution in Racial and Ethnic Minority Communities under the Clean Air Act,” *Georgetown Environmental Law Review* 35 (2022): 113–154.
Humby, Tracy-Lynn. “Environmental Justice and Human Rights on the Mining Wastelands of the Witwatersrand Gold Fields,” *Revue Générale de Droit* 43 (2013): 67–112.
Hunt, Lester C., and Paraskevas Kipouros. “Energy Demand and Energy Efficiency in Developing Countries,” *Energies 16*, no. 3 (2023): 1056–1081.
Hurlbert, Margot, Tanushree Das, and Charisse Vitto. “Climate Change Energy Futures in Business, Industry and Mining in Saskatchewan, Canada,” *International Journal of Climate Change Strategies and Management* 16, no. 1 (2024): 44–62.

I

Idiazabal Gorrotxategi, Itziar, and Manel Pérez-Caurel, *Linguistic Diversity, Minority Languages and Sustainable Development.* Servicio Editorial de la Universidad del País Vasco/Euskal Herriko Unibertsitatearen Argitalpen Zerbitzua, *addi.ehu.es* (2019). https://doi.org/10/34728
Iglesias, Ana, and Luis Garrote. “Adaptation Strategies for Agricultural Water Management Under Climate Change in Europe,” *Agricultural Water Management* 155 (2015): 113–124.
Illsley, Barbara M. “Good Neighbour Agreements: The First Step to Environmental Justice?” *Local Environment* 7, no. 1 (2002): 69–79.
Indigenous Climate Action. “Decolonizing Climate Policy in Canada,” (2021) 9. https://cpcil.ca/wp-content/uploads/youzer/ICA-Decolonizing-Climate-Policy-in-Canada-report-from-phase-one.pdf
Industriouslabs. “New Report Unveils Alarming Health Costs of U.S. Coal-Based Steelmaking,” (2024). https://industriouslabs.org/archive/report-dirty-steel-dangerous-air
Intergovernmental Science-Policy Platform on Biodiversity and Ecosystem Services “Summary for Policymakers of the Thematic Assessment of the Interlinkages Among Biodiversity, Water, Food and Health (Nexus Assessment),” 16 December (2024). https://ipbes.canto.de/pdfviewer/viewer/viewer.html?v=IPBES11Media&portalType=v/IPBES11Media&column=document&id=cj0uc5396d1ed5418tsuic2r45&suffix=pdf&print=1
International Bar Association. “Achieving Justice and Human Rights in an Era of Climate Disruption: International Bar Association Climate Change Justice and Human Rights Task Force Report,” International Bar Association (2014). https://www.ibanet.org/MediaHandler?id=0f8cee12-ee56-4452-bf43-cfcab196cc04
International Energy Agency. “Coal 2023: Executive Summary,” International Energy Agency (2023). https://www.iea.org/reports/coal-2023/executive-summary

International Energy Agency. "Accelerating Just Transitions for the Coal Sector: Strategies for Rapid, Secure and People-Centred Change," International Energy Agency (2024a), 7. https://iea.blob.core.windows.net/assets/b12ee835-3448-4b7d-a83f-5566e8c222f9/AcceleratingJustTransitionsfortheCoalSector-WEOSpecialReport.pdf

International Energy Agency. "Major Growth of Clean Energy Limited the Rise in Global Emissions in 2023," International Energy Agency (2024b). https://www.iea.org/news/major-growth-of-clean-energy-limited-the-rise-in-global-emissions-in2023

International Labour Organization (ILO). "Discrimination at Work on the Basis of Sexual Orientation and Gender Identity: Results of the ILO's PRIDE Project," International Labor Organization (nd). https://www.ilo.org/resource/brief/discrimination-work-basis-sexual-orientation-and-gender-identity-results

International Labour Organization (ILO). "Eliminating Discrimination Against Indigenous and Tribal Peoples in Employment and Occupation: A Guide to ILO Convention No. 11," International Labor Organization (2007). https://www.ilo.org/publications/eliminating-discrimination-against-indigenous-and-tribal-peoples-employment-0

International Labour Organization (ILO). "Sustainable Development Goals Indigenous Peoples in Focus 2016," International Labour Organization (2016). https://www.ilo.org/publications/sustainable-development-goals-indigenous-peoples-focus

International Maritime Organization. "IMO Agrees Possible Outline for Maritime 'Net-Zero Framework'," International Maritime Organization (2024). https://www.imo.org/en/MediaCentre/PressBriefings/pages/IMO-agrees-possible-outline-for-net-zero-framework.aspx

IPCC. "Sixth Assessment Report — IPCC." IPCC (2023). https://www.ipcc.ch/assessment-report/ar6/

Ismail, Faizel. "Beyond the Just Energy Transition Narrative: How South Africa Can Support the AfCFTA to Advance Climate Resilient Development," *South African Journal of International Affairs* 30, no. 2 (2023): 245–262.

J

Jagannathan, Ravi, Ashwin Ravikumar, and Marco Sammon. "Environmental, Social, and Governance Criteria: Why Investors Should Care," *Journal of Investment Management* 16 (2018): 18–31.

Jaureguiberry, Pedro, et al. "The Direct Drivers of Recent Global Anthropogenic Biodiversity Loss," *Science Advances* 8, no. 45 (2022): 1–11.

Jayaram, Dhanasree, and Radhika Ajayan. "Human Security in Times of Climate Change: Climate Diplomacy for Integrated Action," in *Disaster Risk Reduction for Resilience: Disaster Risk Management Strategies*, eds. Saeid Eslamian and Faezeh Eslamian (New York: Springer International Publishing, 2022), 261–293.

Jayawardena, Iresh. "Indigenous, Ethnic, and Racial Diversity and Climate Justice: New Zealand's Climate Adaptation Policies and Policy Approaches," *Environmental Research: Climate* (2024): 1–18.

Jbaily, Abdulrahman, et al. "Air Pollution Exposure Disparities across US Population and Income Groups," *Nature* 601, no. 7892 (2022): 228–233.

Jimoh, Musa Yusuf, Peter Bikam, and Hector Chikoore. "The Influence of Socioeconomic Factors on Households' Vulnerability to Climate Change in Semiarid Towns of Mopani, South Africa," *Climate* 9, no. 1 (2021): 1–20.

Jodoin, Sébastien, Annalisa Savaresi, and Margaretha Wewerinke-Singh. "Rights-Based Approaches to Climate Decision-Making," *Current Opinion in Environmental Sustainability* 52 (2021): 45–53.

Jodoin, Sébastien, et al. "Nothing About Us Without Us: The Urgent Need for Disability-Inclusive Climate Research," *PLOS Climate* 2, no. 3 (2023): 1–3.

Johansen, Bruce E. "Mining: Tearing at Mother's Breast," in *Resource Devastation on Native American Lands: Toxic Earth, Poisoned People*, ed. Bruce E. Johansen (Cham, Switzerland: Springer International Publishing, 2023), 191–227.

Johnson, Gbemende. "Gender, Diversity, and the United States Judiciary," *SAIS Review of International Affairs* 41, no. 1 (2021): 61–71.

Johnson, Danielle Emma, Meg Parsons, and Karen Fisher. "Indigenous Climate Change Adaptation: New Directions for Emerging scholarship," *Environment and Planning E: Nature and Space* 5, no. 3 (2022): 1541–1578.

Johnson-Jennings, Michelle, Shanondora Billiot, and Karina Walters. "Returning to our Roots: Tribal Health and Wellness Through Land-Based Healing," *Genealogy* 4, no. 3 (2020): 91–111.

Johnston, P. et al. "Climate Change Impacts in South Africa: What Climate Change Means for a Country and its People," *University of Cape Town* (2024). https://web.csag.uct.ac.za/%7Ecjack/South%20Africa_FINAL_22%20Jan_ONLINE.pdf

Johnstone, Anna., et al. "Is Canada's Impact Assessment Act Working?" (2021). https://www.wcel.org/sites/default/files/publications/2021-impact-assessment-act-report-en-web.pdf

Jones, Jack. "Defending Race-Conscious Policy: New York State's Criteria for Identifying Disadvantaged Communities," *Columbia Journal of Environmental Law* 49, no. 2 (2024): 425–478.

Jones, Matthew W., Douglas I. Kelley, Chantelle A. Burton, Francesca Di Giuseppe, Maria Lucia F. Barbosa, Esther Brambleby, Andrew J. Hartley et al. "State of Wildfires 2023–2024," *Earth System Science Data* 16, no. 8 (2024): 3601–3685.

Josey, Kevin P., et al. "Air Pollution and Mortality at the Intersection of Race and Social Class," *New England Journal of Medicine* 388, no. 15 (2023): 1396–1404.

Juhola, Sirkku, et al. "Connecting Climate Justice and Adaptation Planning: An Adaptation Justice Index," *Environmental Science & Policy* 136 (2022): 609–619.

Jung, Meen Chel, Michael G. Yost, Andrew L. Dannenberg, Karen Dyson, and Marina Alberti. "Legacies of Redlining Lead to Unequal Cooling Effects of Urban Tree Canopy," *Landscape and Urban Planning* 246 (2024): 105028.

Juon, Hee-Soon, et al. "Racial Disparities in Occupational Risks and Lung Cancer Incidence: Analysis of the National Lung Screening Trial," *Preventive Medicine* 143 (2021): 1–15.

K

Kaiho, Kunio. "Extinction Magnitude of Animals in the Near Future," *Scientific Reports* 12, no. 1 (2022): 1–8.

Kalaba, Mmatlou. "South Africa's Carbon Tax: Balancing Climate Action and Economic Development," *South African Institute of International Affairs* 81 (2020): 1–21.

Kallis, Giorgos, Erik Gómez-Baggethun, and Christos Zografos. "To Value or Not To Value? That is Not the Question," *Ecological Economics* 94 (2013): 97–105.
Kameri-Mbote, Patricia, and Philippe Cullet. "Environmental Justice and Sustainable Development Integrating Local Communities in Environmental Management," *International Environmental Law Research Centre* (1996). https://www.ielrc.org/content/w9601.pdf
Karali, Nihan, Nina Khanna, and Nihar Shah. "Climate Impact of Primary Plastic Production," *Lawrence Berkeley National Laboratory* (2024). https://escholarship.org/uc/item/6cc1g99q
Kashwan, Prakash. "Globalization of Environmental Justice," in *The Oxford Handbook of Comparative Environmental Politics*, eds. Jeannie Lynn Sowers, Stacy D. VanDeveer, and Erika Weinthal (Oxford: Oxford University Press, 2023).
Kato-Huerta, Jarumi, and Davide Geneletti. "Analysing the Treatment of Environmental Justice and Nature-Based Solutions in the Urban Climate Action Plans of Latin American Metropolitan Areas," *Local Environment* 28, no. 11 (2023): 1388–1409.
Kaufman, Joel D. and Anjum Hajat. "Confronting Environmental Racism," *Environmental Health Perspectives* 129, no. 5, 20 May (2021): 1–2.
Kaza, Silpa, et al. "What a Waste 2.0: A Global Snapshot of Solid Waste Management to 2050," The World Bank (2018). https://openknowledge.worldbank.org/entities/publication/d3f9d45e-115f-559b-b14f-28552410e90a
Ke, Piyu, et al. "Low Latency Carbon Budget Analysis Reveals a Large Decline of the Land Carbon Sink in 2023," *arXiv* preprint arXiv:2407.12447 (2024): 1–35. https://arxiv.org/abs/2407.12447
Keary, Michael. "The New Prometheans: Technological Optimism in Climate Change Mitigation Modelling," *Environmental Values* 25, no. 1 (2016): 7–28.
Kelly, Ute, and Rhys Kelly. "Resilience, Solidarity, Agency–Grounded Reflections on Challenges and Synergies," *Resilience* 5, no. 1 (2017): 10–28.
Kermani, Amir, and Francis Wong. "Racial Disparities in Housing Returns," National Bureau of Economic Research (2021). https://www.nber.org/system/files/working_papers/w29306/w29306.pdf
Khanyile, Samkelisiwe, and Christina Culwick Fatti. "Interrogating Park Access and Equity in Johannesburg, South Africa," *Environment and Urbanization* 34, no. 1 (2022): 10–31.
Khedulkar, Akhil Pradiprao, Annadurai Thamilselvan, Ruey-an Doong, and Bidhan Pandit. "Sustainable High-Energy Supercapacitors: Metal Oxide-Agricultural Waste Biochar Composites Paving the Way for a Greener Future," *Journal of Energy Storage* 77, (2024): 1–17.
Khine, Myo Myo, and Uma Langkulsen. "The Implications of Climate Change on Health Among Vulnerable Populations in South Africa: A Systematic Review," *International Journal of Environmental Research and Public Health* 20, no. 4 (2023): 1–19. https://www.mdpi.com/journal/ijerph
Kidd, Sean A., Mariya Bezgrebelna, Shakoor Hajat, Lynne Keevers, Arun Ravindran, Vicky Stergiopoulos, Samantha Wells et al. "A Response Framework for Addressing the Risks of Climate Change for Homeless Populations," *Climate Policy* 23, no. 5 (2023): 623–636.
Killean, Rachel, and Lauren Dempster. "Mass Violence, Environmental Harm and the Limits of Transitional Justice," *Genocide Studies and Prevention* 16, no. 1 (2022): 11–39.
Kim, Jina J. "Leave No One Behind: Realizing Environmental Justice through Climate Litigation Remedies," *Ecology Law Quarterly* 48 (2021): 409–432.

Kim, Ji-Su, et al. "Global Pattern of Microplastics (MPs) in Commercial Food-Grade Salts: Sea Salt as an Indicator of Seawater MP Pollution," *Environmental Science and Technology* 52, no. 21 (2018): 12819–12828.

Kim, Junsik, Reid Ewing, and Alessandro Rigolon. "Does Green Infrastructure Affect Housing Prices Via Extreme Heat and Air Pollution Mitigation? A Focus on Green and Climate Gentrification in Los Angeles County, 2000–2021," *Sustainable Cities and Society* 102 (2024): 105225.

Kinay, Pelin, et al. "Reporting Evidence on the Environmental and health Impacts of Climate Change on Indigenous Peoples of Atlantic Canada: A Systematic Review," *Environmental Research: Climate* 2, no. 2 (2023): 1–24.

King, Malcolm, Alexandra Smith, and Michael Gracey. "Indigenous Health Part 2: The Underlying Causes of the Health Gap," *The Lancet* 374, no. 9683 (2009): 76–85.

Kinol, Alaina, et al. "Climate Justice in Higher Education: A Proposed Paradigm Shift Towards a Transformative Role for Colleges and Universities," *Climatic Change* 176, no. 2 (2023): 1–29.

Klass, Alexandra B. "Energy Transitions in the Trump Administration and Beyond," *Environmental Law* 51 (2021): 241–266.

Klaudt, Dustin W. "Can Canada's 'Living Tree' Constitution and Lessons from Foreign Climate Litigation Seed Climate Justice and Remedy Climate Change?" *Journal of Environmental Law and Practice* 31, no. 3 (2018): 185–243.

Klein, Julie Thompson. "Sustainability and Collaboration: Crossdisciplinary and Cross-Sector Horizons," *Sustainability* 12, no. 4 (2020a): 1515–1528, 1516.

Klein, Kenneth S. "Ashes to Ashes: A Way Home for Climate Change Survivors," *Arizona Law Review* 63 (2021): 679–730.

Klein, Naomi. "Let them Drown: The Violence of Othering in a Warming World," *London Review of Books* 38, no. 11 (2016): 11–13.

Klein, Naomi, *On Fire: The (Burning) Case for a Green New Deal* (New York: Simon & Schuster, 2020b).

Klinsky, Sonja, and Jasmina Brankovic. *The Global Climate Regime and Transitional Justice* (New York: Routledge, 2018).

Knox, John H. "Human Rights, Environmental Protection, and the Sustainable Development Goals," *Washington International Law Journal* 24 (2015): 517–528.

Kojola, Erik, and David N. Pellow. "New Directions in Environmental Justice Studies: Examining the State and Violence," *Environmental Politics* 30, no. 1–2 (2021): 100–118.

Kom, Zongho, et al. "Indigenous Knowledge Indicators Employed by Farmers for Adaptation to Climate Change in Rural South Africa," *Journal of Environmental Planning and Management* 66, no. 13 (2023): 2778–2793.

Konisky, David M. "Inequities in enforcement? Environmental Justice and Government Performance," *Journal of Policy Analysis and Management: The Journal of the Association for Public Policy Analysis and Management* 28, no. 1 (2009): 102–121.

Kortleve, Anniek J., et al. "Over 80% of the European Union's Common Agricultural Policy Supports Emissions-Intensive Animal Products," *Nature Food* 5, no. 4 (2024): 288–292.

Kotz, Maximilian, et al. "Global Warming and Heat Extremes to Enhance Inflationary Pressures," *Communications Earth & Environment* 5, no. 1 (2024a): 1–13.

Kotz, Maximilian, et al. "The Economic Commitment of Climate Change," *Nature* 628, no. 8008 (2024b): 551–557.

Kranz, Johanna, et al. "The (UN) Political Perspective on Climate Change in Education – A Systematic Review," *Sustainability* 14, no. 7 (2022): 1–44.

Krauss, Clifford. "Surging U.S. Oil Production Brings Down Prices and Raises Climate Fears," *The New York Times*, 1 December (2023). https://www.nytimes.com/2023/12/01/business/energy-environment/us-oil-production-record-climate.html

Krieg, Eric J., and Daniel R. Faber. "Not So Black and White: Environmental Justice and Cumulative Impact Assessments," *Environmental Impact Assessment Review* 24, no. 7–8 (2004): 667–694.

Krondorfer, Björn. *Unsettling Empathy: Working with Groups in Conflict* (Maryland: Rowman & Littlefield Publishers, 2020).

Kruczkiewicz, Andrew, et al. "Are Underserved Populations Left Out of National Flood Mitigation Efforts and Facing Greater Impact? A Method to Assess Racial Inequality at the Census Tract Level," *Journal of Climate Resilience and Climate Justice* 1 (2023): 78–92.

Kubik, Zaneta, Alisher Mirzabaev, and Julian May. "Climate Change, Food and Nutrition Security, and Human Capital," in *Handbook of Labor, Human Resources and Population Economics*, ed. Klaus F. Zimmermann (Cham, Switzerland: Springer International Publishing, 2023), 1–37.

Kuh, Katrina Fischer, and Michael Charles Leach. "Climate Change Mitigation Law and Policy in the United States and Canada." In *Research Handbook on Climate Change Mitigation Law*, eds Leonie Reins and Jonathan Verschuuren (Cheltenham, UK: Edward Elgar Publishing, 2022), 103–137.

Kuhl, Laura, Paul H. Kirshen, Matthias Ruth, and Ellen M. Douglas. "Evacuation as a Climate Adaptation Strategy for Environmental Justice Communities," *Climatic Change* 127, no. 3 (2014): 493–504.

Künnemann, Rolf, and Sofía Monsalve Suárez. "International Human Rights and Governing Land Grabbing: A view from Global Civil Society," *Globalizations* 10, no. 1 (2013): 123–139.

Kwame, Adom Richard, Simatele Mulala Danny, and Reid Memory. "The Threats of Climate Change on Water and Food Security in South Africa," *American Journal of Environment and Climate* 1, no. 2 (2022): 73–91.

Kwon, Minji, et al. "Forgone Summertime Comfort as a Function of Avoided Electricity Use," *Energy Policy* 183 (2023): 1–10.

L

Laboucan, Melina. "Indigenous Clean Energy Leadership for Canada's Clean, Electric Future Waves of Change," Canadian Climate Institute (2022). https://climatechoices.ca/wp-content/uploads/2022/02/ICE-report-ENGLISH-FINAL.pdf

LaDuke, Winona. *All Our Relations: Native Struggles for Land and Life* (Boston, Massachusetts: South End Press, 1999).

Lajoie-O'Malley, Alana. "Discursive Infrastructures of Settler Colonialism: The Coastal Gaslink Pipeline, Indigenous Workers, and the Ongoing Promise of Integration," *Journal of Canadian Studies* 58, no. 2 (2024): 366–391.

Lamboll, Robin D., Zebedee R.J. Nicholls, Christopher J. Smith, Jarmo S. Kikstra, Edward Byers, and Joeri Rogelj. "Assessing the Size and Uncertainty of Remaining Carbon Budgets," *Nature Climate Change* (2023): 1–8.

Landrigan, Philip J., et al. "Pollution and Children's Health," *Science of the Total Environment* 650 (2019): 2389–2394.

Lane, Haley M., et al. "Historical Redlining is Associated with Present-Day Air Pollution Disparities in US Cities," *Environmental Science & Technology Letters* 9, no. 4 (2022): 345–350.

Langmack, Fin-Jasper. "Remedies for Climate Change – A Decisive Push Towards Paris?" in *Netherlands Yearbook of International Law 2021: A Greener International Law – International Legal Responses to the Global Environmental Crisis*, eds. Daniëlla Dam-de Jong and Fabian Amtenbrink (The Hague: TMC Asser Press, 2023), 19–45.

Lantushenko, Viktoriya, and Carolin Schellhorn. "The Rising Risks of Fossil Fuel Lobbying," *Global Finance Journal* 56 (2023): 955–977.

Laversa, Tom, Ian Berryb, and Colin A. Boothc. "From Umbrellas to Sandbags—An Integration of Flood Risk Management, Engineering and Social Insights," Sustainable Water Engineering (2020): 207–228.

Lawal, Musediq Olufemi et al. "Influence of Indigenous Spiritual Beliefs in Natural Resources Management and Climate Change Mitigation and Among the Yorùbás in Nigeria," in *Indigenous Knowledge and Climate Governance*, eds. E. Ebhuoma and L. Leanard (Cham, Switzerland: Springer, 2022), 39–51.

Lawson, Max, and David Jacobs. "Inequality Kills Methodology Note," *Oxfam* (2022), 2–22.

LeClair, Jessica, Theresa Watts, and Susan Zahner. "Nursing Strategies for Environmental Justice: A Scoping Review," *Public Health Nursing* 38, no. 2 (2021): 296–308.

Lee, Charles. "Toxic Waste and Race in the United States," in *Race and the Incidence of Environmental Hazards,* eds. Bunyan Bryant and Paul Mohai (Abingdon, Oxon: Routledge, 2019), 10–27.

Lee, De-Chih, Hailun Liang, and Leiyu Shi. "The Convergence of Racial and Income Disparities in Health Insurance Coverage in the United States," *International Journal for Equity in Health* 20 (2021): 1–8.

Lee, Eun Kyung, et al. "Health Outcomes in Redlined Versus Non-Redlined Neighborhoods: A Systematic Review and Meta-Analysis," *Social Science & Medicine* 294 (2022): 114696.

Lee, Marc. "A Critical Guide to the Pan-Canadian Framework on Clean Growth and Climate Change," Canadian Centre for Policy Alternatives 14 December (2016). https://www.policynote.ca/a-critical-guide-to-the-pan-canadian-framework-on-clean-growth-and-climate-change/

Lee, Renae Hee Eun. "Rising Tides, Rising Solutions: A Maryland Program Proposal to Navigate the Flood Insurance Frontier," *Maryland Law Review Online* 83 (2024). https://digitalcommons.law.umaryland.edu/endnotes/99

Legagneux, Pierre, et al. "Our House is Burning: Discrepancy in Climate Change vs. Biodiversity Coverage in the Media as Compared to Scientific Literature," *Frontiers in Ecology and Evolution* 5 (2018): 1–6.

Lehtonen, Markku. "The Environmental–Social Interface of Sustainable Development: Capabilities, Social Capital, Institutions," *Ecological Economics* 49, no. 2 (2004): 199–214.

Leiserowitz, Anthony, et al. "Global Warming's Six Americas, Fall 2023," Yale Program on Climate Change Communication, 14 December (2023). https://climatecommunication.yale.edu/publications/global-warmings-six-americas-fall-2023/

Leiserowitz A., et al. *Climate Change in the American Mind: Politics & Policy, Spring 2024* (New Haven, CT: Yale Program on Climate Change Communication, 2024).

https://climatecommunication.yale.edu/wp-content/uploads/2024/06/climate-change-american-mind-politics-policy-spring-2024b.pdf
Lekunze, Angwe Rachel. "Rethinking the (Un) just Transition: A Review of the Impact of Neoliberal Approaches to Energy Governance in South Africa," *Journal of Law, Society and Development* (2024): 1–18.
Lelechenko, Anzhela, et al. "Critical Analysis of the Achievements of the Global Millennium Development Goals and New Approaches to Sustainable Development Management," *E3S Web of Conferences*, vol. 211 (2020): 1–18.
Lenton, Timothy M., et al. "Quantifying the Human Cost of Global Warming," *Nature Sustainability* 6, no. 10 (2023): 1237–1247.
Leocadia, Zhou. "An Analysis of the Differences in Vulnerability to Climate Change: A Review of Rural and Urban Areas in South Africa," *Climate* 10, no. 118 (2022): 1–22.
Levin, Matthew. "Power to the People? True Justice under New Jersey's Environmental Justice Law," *Seton Hall Journal of Legislation and Public Policy* 48, no. 1 (2023): 250–277.
Levy, Jonathan I. "2022 Inflation Reduction Act: Climate Investments Are Public Health Investments," *American Journal of Public Health* 112, no. 11 (2022): 1525–1525.
Levy, Sheri R., et al. "A Human Rights-Based Approach to Climates Injustices at the Local, National, and International Levels: Program and Policy Recommendations," *Social Issues and Policy Review* 18, no. 1 (2024): 3–30.
Li, Buyun. "Scientific Meaning and Significance of the Right to Development," *Journal of Human Rights* (2015): 307–311.
Li, Jiajun, Shichen Li, Yingqi Zeng, Xiang Zhou, Lingjie Zeng, Meng Liu, Changsheng Cao, Yunfei Xia, and Jun Gao. "Cooking-Related Thermal Comfort and Carbon Emissions Assessment: Comparison Between Electric and Gas Cooking in Air-Conditioned Kitchens," *Building and Environment* 265 (2024): 1–15.
Li, Tania Murray. "Centering Labor in the Land Grab Debate," *The Journal of Peasant Studies* 38, no. 2 (2011): 281–298.
Li, Xiangdong, Ling Jin, and Haidong Kan. "Air Pollution: A Global Problem Needs Local Fixes," *Nature* 570, no. 7762 (2019): 437–439.
Li, Yuze, et al. "The Climate Impact of High Seas Shipping," *National Science Review* 10, no. 3 (2023): 1–15.
Liboiron, Max. *Pollution is Colonialism* (Durham, NC: Duke University Press, 2021).
Liddell, Jessica L., Catherine E. McKinley, and Jennifer M. Lilly. "Historic and Contemporary Environmental Justice Issues Among Native Americans in the Gulf Coast Region of the United States," *Studies in Social Justice* 15, no. 1 (2021): 1–24.
Lidskog, Rolf, and Göran Sundqvist. *Governing the Air: The Dynamics of Science, Policy, and Citizen Interaction* (Cambridge, MA: MIT Press, 2011).
Lidskog, Rolf, and Göran Sundqvist. "When Does Science Matter? International Relations Meets Science and Technology Studies," *Global Environmental Politics* 15, no. 1 (2015): 1–20.
Lightfoot, Sheryl, and Sarah Maddison. "Introduction," in *Handbook of Indigenous Public Policy,* eds. Sheryl Lightfoot and Sarah Maddison (Cheltenham, UK: Edward Elgar Publishing, 2024).
Lindsay, Sally, et al. "The Impact of Climate Change Related Extreme Weather Events on People with Pre-Existing Disabilities and Chronic Conditions: A Scoping Review," *Disability and Rehabilitation* 45, no. 25 (2023): 4338–4358.

Link, Lewis E. "How Hurricane Katrina Influenced the Design of Hurricane Protection and Risk Reduction Systems and National Approaches to Risk and Resilience. Part 2: Designing the Hurricane and Storm Damage Risk Reduction System and Resulting Long-Term Engineering Guidance and Practice Changes," *Water Policy* 23, no. S1 (2021): 174–187.

Lioubimtseva, Elena. "The Role of Inclusion in Climate Vulnerability Assessment and Equitable Adaptation Goals in Small American Municipalities," *Discover Sustainability* 3, no. 1 (2022): 1–16.

Liu, Tong, and Chao Fan. "Impacts of Disaster Exposure on Climate Adaptation Injustice Across US Cities," *Sustainable Cities and Society* 89 (2023): 104371.

Lively, Cathy. "COVID-19 in the Navajo Nation Without Access to Running Water: The Lasting Effects of Settler Colonialism," *Voices in Bioethics* 7 (2021): 374–378.

Lohmann, Maggie. "Law of the Land: The Continuing Legacy of Indian Law's Racist Roots and Its Impact on Native American Land Rights," *West Virginia Law Review* 125, no. 1 (2022): 329–370.

Lombard, A. "The Human-Nature Nexus: A Sustainability Framework for Social Work?" in *Environment, Climate, and Social Justice: Perspectives and Practices from the Global South*, eds. Devendraraj Madhanagopal, et al. (Cham, Switzerland: Springer, 2022), 45–60.

Lomeo, Davide, et al. "Remote Sensing and Spatial Analysis Reveal Unprecedented Cyanobacteria Bloom Dynamics Associated with Elephant Mass Mortality," *Science of the Total Environment* 957 (2024): 1–13.

Loppie, Samantha, Charlotte Reading, and Sarah de Leeuw. "Indigenous Experiences with Racism and Its Impacts," National Collaborating Centre for Aboriginal Health (2020). https://www.nccih.ca/docs/determinants/FS-Racism2-Racism-Impacts-EN.pdf

Loser, Nicole. "Litigation against Coal-Fired Power in South Africa: Lessons from and for Global Climate Litigation to Reduce Greenhouse Gas Emissions," in *Climate Litigation and Justice in Africa*, eds. Kim Bouwer, et al. (Bristol: Bristol University Press, 2024), 68–101.

Lowitt, Sandy. "A Just Transition Finance Roadmap For South Africa: A First Iteration," *Trade & Industrial Policy Strategies* (2021). https://justtransitionforall.com/wp-content/uploads/2022/12/TIPS_UK_PACT_A_Just_Transition_Finance_Roadmap_for_South_Africa_First_Iteration_December_2021.pdf

Loyn, David. "Indigenous People 'Worst-Off World Over'," *BBC News*, 26 May 2006, http://news.bbc.co.uk/1/hi/world/europe/5019582.stm

Luo, Lei, Jiaxiao Wang, Jitao Lv, Zhengang Liu, Tianran Sun, Yi Yang, and Yong-Guan Zhu. "Carbon Sequestration Strategies in Soil Using Biochar: Advances, Challenges, and Opportunities," *Environmental Science & Technology* 57, no. 31 (2023): 11357–11372.

Luo, Changhong, et al. "Subsidy, Tax or Green Awareness: Government Policy Selection for Promoting Initial Shore Power Usage and Sustaining Long-Run Use," *Journal of Cleaner Production* 442 (2024): 140946.

Luoma, Colin. "Reckoning with Conservation Violence on Indigenous Territories: Possibilities and Limitations of a Transitional Justice Response," *International Journal of Transitional Justice* 17, no. 1 (2023): 89–106.

Lupien, Pascal. "Indigenous Movements, Collective Action, and Social Media: New Opportunities or New Threats?," *Social Media and Societ*y 6, no. 2 (2020): 1–11.

Lyons, Andrea L., et al. "Climate Change, Wildfire, and Past Forest Management Challenge Conservation of Canada Lynx in Washington, USA," *The Journal of Wildlife Management* 87, no. 5 (2023): e22410.

Lyster, Rosemary. *Climate Justice and Disaster Law* (Cambridge: Cambridge University Press, 2015).

M

Mackey, Brendan, et al. "Assessing the Cumulative Impacts of Forest Management on Forest Age Structure Development and Woodland Caribou Habitat in Boreal Landscapes: A Case Study from Two Canadian Provinces." *Land* 13, no. 1 (2023): 1–17.

MacNeil, Robert. *Thirty Years of Failure: Understanding Canadian Climate Policy* (Winnipeg: Fernwood Publishing, 2019).

MacNeil, Robert. "On the Immorality and Futility of Canadian Climate Policy under Trudeau," *Capitalism Nature Socialism* 32, no. 4 (2021): 111–129.

Mac-Seing, Muriel. "The Climate Crisis: Unpacking the Intersectional Implications of 'Leaving No one Behind'," *Global Health Promotion* 30, no. 3 (2023): 3–5.

Madhanagopal, Devendraraj, et al. eds. "Environment, Climate, and Social Justice: Perspectives and Practices from the Global South" (Singapore: Springer Nature Singapore, 2022), 1–387.

Madondo, Martha and Hunadi Mapula Nkwana. "Exploring South Africa's National Climate Change Response White Paper, 2011: Challenges and Prospects for Public Policy," *Administratio Publica* 29, no. 2 (2021): 187–207.

Magill, Bobby. "Biden Raises Oil Drilling Royalties and Fees on Federal Land (1)," *Bloomberg Law*, 12 April (2024). https://news.bloomberglaw.com/environment-and-energy/biden-raises-oil-drilling-royalties-and-fees-on-federal-land

Magnan, Alexandre K., et al. "Strengthen Climate Adaptation Research Globally," *Science* 376, no. 6600 (2022): 1398–1400.

Mahon, M.B., et al. "Global Change Drivers and the Risk of Infectious Disease," *bioRxiv* (2022): 2022–07.

Maier, Raphaela, Alfred Posch, Christoph Proß, Simon Plakolb, and Karl W. Steininger. "Cutting Social Costs by Decarbonizing Passenger Transport," *Transportation Research Part D: Transport and Environment* 122 (2023): 1–23.

Makleff, Shelly, et al. "Mental Health and Community Resilience Among Vulnerable Populations Affected by Natural Hazards: Protocol for Scoping Reviews," *Methods and Protocols* 5, no. 88 (2022): 1–7.

Makungo, Rachel, and Tinyiko R. Nkuna. "Prioritising Climate Coping and Adaptation Strategies for Smallholder Farmers in Luvuvhu River Catchment, South Africa," *Climate Services* 30 (2023): 100361–100373.

Malakar, Kousik Das, et al. "Climate Change, Justice, and Rights," in *Climate Change and Socio-Ecological Transformation: Vulnerability and Sustainability*, eds. Kousik Das Malakar, et al. (Singapore: Springer Nature Singapore, 2023), 375–437.

Malakoff, David. "Trump Picks Prominent Climate Skeptic as EPA Chief," Science Insider, 7 December (2016). https://www.science.org/content/article/trump-picks-prominent-climate-skeptic-epa-chief

Maldonado, Julie Koppel, Benedict Colombi, and Rajul Pandya, eds., *Climate Change and Indigenous Peoples in the United States: Impacts, Experiences and Actions* (Berlin: Springer, 2016), 49–59.

Malik, Ishfaq Hussain, and James D. Ford. "Addressing the Climate Change Adaptation Gap: Key Themes and Future Directions," *Climate* 12, no. 2 (2024): 12–24.
Malik, J., T. Hong, M. Wei, and S. Rotmann. "Prioritize Energy Sufficiency to Decarbonize Our Buildings," *Nature Human Behaviour* 8, no. 3 (2024): 406–410.
Malin, Stephanie A., and Stacia S. Ryder. "Developing Deeply Intersectional Environmental Justice Scholarship," *Environmental Sociology* 4, no. 1 (2018): 1–7.
Malley, Christopher S., et al. "A Roadmap to Achieve the Global Methane Pledge," *Environmental Research: Climate* 2, no. 1 (2023): 1–16.
Maloy, J.S. "Beyond Crisis and Emergency: Climate Change as a Political Epic," *Ethics & International Affairs* 38, no. 1 (2024): 103–125.
Mapfumo, Emmanuel, et al. "Historic Climate Change Trends and Impacts on Crop Yields in Key Agricultural Areas of the Prairie Provinces in Canada: A Literature Review," *Canadian Journal of Plant Science* 103, no. 3 (2023): 243–258.
Manikandan, Sivasubramanian, Sundaram Vickram, Ramasamy Subbaiya, Natchimuthu Karmegam, Soon Woong Chang, Balasubramani Ravindran, and Mukesh Kumar Awasthi. "Comprehensive Review on Recent Production Trends and Applications of Biochar for Greener Environment," *Bioresource Technology* (2023): 129725–129737.
Mann, Samuel, Tara McKay, and Gilbert Gonzales. "Climate Change-Related Disasters & the Health of LGBTQ+ Populations," *The Journal of Climate Change and Health* (2024): 100304.
Markowitz, Steven. "Asbestos-Related Lung Cancer and Malignant Mesothelioma of the Pleura: Selected Current Issues," in *Seminars in Respiratory and Critical Care Medicine*, 36, no. 3 (Thieme Medical Publishers, 2015), 334–346.
Martín, Carlos Eduardo. "The Evolution of Climate Action in the Environmental Justice Movement, 2010–2020," *Environmental Justice* 15, no. 3 (2022): 170–178.
Martin, Jennifer. "Climate Change and Heatwaves: Understanding the Experiences and Implications for Unsheltered Homeless Populations. A Scoping Review," *Public Health Institute Journal* (2023): 31–31.
Martinez, Brandon P., and Alan A. Aja. "How Race Counts for Latinx Homeownership," *Critical Sociology* 47, no. 6 (2021): 993–1011.
Martinez-Alier, Joan. "Scale, Environmental Justice, and Unsustainable Cities," *Capitalism Nature Socialism* 14, no. 4 (2003a): 43–63.
Martinez-Alier, Joan. *The Environmentalism of the Poor: A Study of Ecological Conflicts and Valuation* (Cheltenham, UK: Edward Elgar Publishing, 2003b).
Martinez-Alier, Joan, et al. "Is There a Global Environmental Justice Movement?" *Journal of Peasant Studies* 43, no. 3 (2016): 731–755.
Maru, Vivek. "A Pathway to Climate and Environmental Justice," *American Journal of Law and Equality* 3 (2023): 103–149.
Marya, Rupa, and Raj Patel. *Inflamed: Deep Medicine and the Anatomy of Injustice* (London: Penguin UK, 2021).
Masakure, Oliver. "Structural Inequalities, Exclusion, and Minorities in Africa," in *Expanding Perspectives on Human Rights in Africa*, eds. M. Raymond Izarali, Oliver Masakure, and Bonny Ibhawoh (New York: Routledge, 2019), 65–85.
Masondo, Zama Nonkululeko, and Ovett Nwosimiri. "Environmental Pollution and Climate Change: An Ethical Evaluation of the Carbon Tax Policy in South Africa," *Journal of Humanities* 31, no. 1 (2023): 113–133.
Mastrucci, Alessio, Leila Niamir, Benigna Boza-Kiss, Nuno Bento, Dominik Wiedenhofer, Jan Streeck, Shonali Pachauri et al. "Modeling Low Energy Demand

Futures for Buildings: Current State and Research Needs," *Annual Review of Environment and Resources* 48, no. 1 (2023): 761–792.

Mathias, James A., Kimberly M. Juenger, and Jennifer J. Horton. "Advances in the Energy Efficiency of Residential Appliances in the US: A Review," *Energy Efficiency* 16, no. 5 (2023): 1015–1034.

Matola, Joseph Upile, and Kendra Connock. "South Africa's Post-COVID Climate Response and the Path to its NDC Goals," *The COVID-19 Macroeconomic Policy Response in Africa (CoMPRA)* (2023). https://idl-bnc-idrc.dspacedirect.org/server/api/core/bitstreams/989eae89-b8ed-4cf8-a1c6-df2387d13f9e/content

Matsler, A. Marissa, Sara Meerow, Ian C. Mell, and Mitchell A. Pavao-Zuckerman. "A 'Green' Chameleon: Exploring the Many Disciplinary Definitions, Goals, and Forms of 'Green Infrastructure'," *Landscape and Urban Planning* 214 (2021): 104145–104156.

Matsumoto, Sarah. "Protecting Water, Sustaining Communities: Transforming Groundwater Management Entities into Sources of Power During and After Environmental Crises," *University of Missouri-Kansas City Law Review* 92, no. 4 (2024): 825–757.

Mattar, Sennan D., Tahseen Jafry, Patrick Schröder, and Zarina Ahmad. "Climate Justice: Priorities for Equitable Recovery from the Pandemic," *Climate Policy* 21, no. 10 (2021): 1307–1317.

McCauley, Darren, and Raphael Heffron. "Just Transition: Integrating Climate, Energy and Environmental Justice," *Energy Policy* 119 (2018): 1–7.

McClean, Jessica. "From Dispossession to Compensation: A Political Ecology of the Ord Final Agreement as a Partial Success Story for Indigenous Traditional Owners," *Australian Geographer* 43, no. 4 (2012): 339–355.

McClean, Jordan. "The Role of Social Protection Policies in Enhancing Climate Resilience," *Policy Briefing* 296, no. 2 (South African Institute of International Affairs, 2024): 1–9.

McDonald, Matt. "Geoengineering, Climate Change and Ecological Security," *Environmental Politics* 32, no. 4 (2023): 565–585.

McDougall, Gay. "Minorities, Poverty and the Millennium Development Goals: Assessing Global Issues," *International Journal on Minority and Group Rights* 14, no. 2/3 (2007): 333–355.

McDougall, Gay. *The First United Nations Mandate on Minority Issues* (Leiden: Brill, 2016).

McGraw, George, and Radhika Fox. "Closing the Water Access Gap in the United States: A National Action Plan," US Water Alliance, 6 November 2019. https://uswateralliance.org/sites/uswateralliance.org/files/Closing%20the%20Water%20Access%20Gap%20in%20the%20United%20States_DIGITAL.pdf

McIntyre-Mills, Janet J. "Regenerative Education and Safe Habitats for Diverse Species: Caterpillar Dreaming Butterfly Being," in *Transformative Education for Regeneration and Wellbeing: A Critical Systemic Approach to Support Multispecies Relationships and Pathways to Sustainable Environments*, eds. Janet J. McIntyre-Mills and Yvonne Corcoran-Nantes (Singapore: Springer Nature Singapore, 2022), 3–29.

McKibben, Bill. "Putin's War Shows Autocracies and Fossil Fuels Go Hand in Hand. Here's How to Tackle Both," *The Guardian*, 11 April (2022). https://www.theguardian.com/environment/2022/apr/11/putin-autocracies-fossil-fuels-climate-action

McKibben, Bill, et al. "If Trump Wins the Election, It Will Doom Our Efforts to Slow Climate Disaster," *The Guardian*, 6 September (2024). https://www.theguardian.com/us-news/ng-interactive/2024/sep/06/presidential-election-climate-crisis-project-2025-trump

McMahon, Gary, and Susana Moreira. "The Contribution of the Mining Sector to Socioeconomic and Human Development," *Extractive Industries for Development Series* 30 (2014). https://documents1.worldbank.org/curated/ru/713161468184136844/pdf/872980NWP0Mini00Box385186B00PUBLIC0.pdf

Meadowcroft, James. "Climate Change Governance," World Bank Policy Research Working Paper No. 4941 (2016). https://papers.ssrn.com/sol3/papers.cfm?abstract_id=1407959

Meadows, Robin. "Climate Change, Wildfires, and Health in Canada," in *Climate Change and Human Health Scenarios: International Case Studies*, ed. Rais Akhtar (Cham, Switzerland: Springer Nature Switzerland, 2024), 385–397.

Means, Taneisha N., Andrew Eslich, and Kaitlin Prado. "Judicial Diversity in the United States Federal Judiciary," in *Research Handbook on Law and Courts*, eds. Susan M. Sterett and Lee Demetrius Walker (Cheltenham, UK: Edward Elgar Publishing, 2019), 231–245.

Meattle, Chavi, et al. "The South African Climate Finance Landscape 2023," Presidential Climate Commission (2023). https://pccommissionflo.imgix.net/uploads/images/PCC-SA-CLIMATE-FINANCE-LANDSCAPE.pdf

Meerow, Sara, and Carrie L. Mitchell. "Weathering the Storm: The Politics of Urban Climate Change Adaptation Planning," *Environment and Planning A: Economy and Space* 49, no. 11 (2017): 2619–2627.

Mehmet, Ozay, and Vedat Yorucu. "Human Damage to the Global Ecosystem," in *From Land Disputes to Sustainable Environmental Development: A Near East Perspective*, eds. Ozay Mehmet and Vedat Yorucu (Cham, Switzerland: Springer International Publishing, 2024), 151–165.

Mehta, Geeta, and Dhriti Jade. "Nine Steps Towards Rehabilitation and Restoration of Degraded Urban Rivers," in *Managing Urban Rivers*, eds. Victor R. Shinde, et al. (Amsterdam: Elsevier, 2024), 241–259.

Mehta, Lyla, et al. *Water for Food Security, Nutrition and Social Justice* (New York: Routledge, 2019).

Meighan, Paul John. "'What is Language For Us?' The Role of Relational Technology, Strength-Based Language Education, and Community-Led Language Planning and Policy Research to Support Indigenous Language Revitalization and Cultural Reclamation Processes." (PhD dissertation, McGill University, Canada, 2023).

Mele, Christopher. "'Using' Race in the Politics of Siting Environmentally Hazardous Facilities," *Environmental Sociology* 2, no. 2 (2016): 166–179.

Mels, Tom. "Mainstreaming Environmental Justice? Right to the Landscape in Northern Sweden," in *Arctic Justice*, eds. Corine Wood-Donnelly and Johanna Ohlsson (Bristol: Bristol University Press, 2023), 81–95.

Melton, Ada Pecos. "Indigenous Justice Systems and Tribal Society," *Judicature* 79 (1995): 126–133.

Memmott, Trevor, et al. "Sociodemographic Disparities in Energy Insecurity Among Low-income Households Before and During the COVID-19 Pandemic," *Nature Energy* 6, no. 2 (2021): 186–193.

Méndez, Michael. *Climate Change from the Streets: How Conflict and Collaboration Strengthen the Environmental Justice Movement* (New Haven: Yale University Press, 2020).

Meng, Kyle C., and Ashwin Rode. "The Social Cost of Lobbying Over Climate Policy," *Nature Climate Change* 9, no. 6 (2019): 472–476.

Menton, Mary, et al. "The COVID-19 Pandemic Intensified Resource Conflicts and Indigenous Resistance in Brazil," *World Development* 138 (2021): 105222.

Merenlender, Adina, and Brendan Buhler. *Climate Stewardship: Taking Collective Action to Protect California* (Oakland, California: University of California Press, 2021).

Merner, L. Delta, et al. "Comparative Analysis of Legal Mechanisms to Net-Zero: Lessons from Germany, the United States, Brazil, and China," *Carbon Management* 15, no. 1 (2024): 1–16.

Merven, Bruno, et al. *Climate Mitigation in South Africa.* Working Paper 174. SA-TIED, May (2016). https://sa-tied.wider.unu.edu/sites/default/files/SA-TIED-WP174.Pdf

Metallic, Naiomi Walqwan. "Five Linguistic Methods for Revitalizing Indigenous Laws," *McGill Law Journal* 68, no. 1 (2023): 47–87.

Mi, Chenxi, Tom Shatwell, Jun Ma, Yaqian Xu, Fangli Su, and Karsten Rinke. "Ensemble Warming Projections in Germany's Largest Drinking Water Reservoir and Potential Adaptation Strategies," *Science of the Total Environment* 748 (2020): 141366–141377.

Mikulewicz, Michael, et al. "Intersectionality & Climate Justice: A Call for Synergy in Climate Change Scholarship," *Environmental Politics* 32, no. 7 (2023): 1275–1286.

Mills-Novoa, Megan, and Diana M. Liverman. "Nationally Determined Contributions: Material Climate Commitments and Discursive Positioning in the NDCs," *Wiley Interdisciplinary Reviews: Climate Change* 10, no. 5 (2019): 1–15. https://wires.onlinelibrary.wiley.com/doi/pdfdirect/10.1002/wcc.589

Minguet, A. "Environmental Justice Movements and Restorative Justice," *The International Journal of Restorative Justice* 4, no. 1 (2021): 58–78.

Minority Rights Group International. "Annual Report & Audited Accounts," Minority Rights Group International (2021). https://minorityrights.org/wp-content/uploads/2022/10/1.0-MRGAnnualReportAuditedAccounts2021Signed.pdf

Mir, Wasey Ali Yadullahi, et al. "Asbestos-Related Diseases and Its Impact on Health: An Updated Review Article," *Current Pulmonology Reports* 12, no. 4 (2023): 244–255.

Miriti, Maria N., et al. "Overcoming Blind Spots to Promote Environmental Justice Research," *Trends in Ecology & Evolution* 36, no. 4 (2021): 269–273.

Mirza, Faisal Mehmood, Avik Sinha, Javeria Rehman Khan, Olga A. Kalugina, and Muhammad Wasif Zafar. "Impact of Energy Efficiency on CO2 Emissions: Empirical Evidence from Developing Countries," *Gondwana Research* 106 (2022): 64–77.

Mirzabaev, Alisher, et al. "Severe Climate Change Risks to Food Security and Nutrition," *Climate Risk Management* 39 (2023): 1–10.

Mirzania, Pegah, et al. "Barriers to Powering Past Coal: Implications for a Just Energy Transition in South Africa," *Energy Research & Social Science* 101 (2023): 1–24.

Missirian, David E. "Native Nations' Land Ownership and Our Disservice to Their People and Culture a Proposed Legislative Solution and a Lesson to Be Learned," *American Indian Law Journal* 9, no. 2 (2021): 380–401.

Mitchell, Bruce, and Juan Franco. "HOLC 'Redlining' Maps: The Persistent Structure of Segregation and Economic Inequality," National Community Reinvestment Coalition (2018). https://ncrc.org/holc/

Mizutani, Julia. "In the Backyard of Segregated Neighborhoods: An Environmental Justice Case Study of Louisiana," *Georgetown Environmental Law Review* 31 (2018): 363–390.

Moatti, Claudia, and Christel Müller. "The People in Ancient Times and the Rise of 'Popularism'," in *Routledge Handbook of Illiberalism*, eds. András Sajó, Renáta Uitz, and Stephen Holmes (New York: Routledge, 2021), 203–217.

Moayedi, Masoumeh, and Dariush Hayati. "Identifying Strategies for Adaptation of Rural Women to Climate Variability in Water Scarce Area," *Frontiers in Water* 5 (2023): 1177684–1177696.

Mohai, Paul, and Bunyan Bryant. "Environmental Racism: Reviewing the Evidence," in *Race and the Incidence of Environmental Hazards*, eds. Bunyan Bryant and Paul Mohai (New York: Routledge, 2019), 163–176.

Mohai, Paul, David Pellow, and J. Timmons Roberts. "Environmental Justice," *Annual Review of Environment and Resources* 34, no. 1 (2009): 405–430.

Mohanty, Mohit Prakash, and Slobodan P. Simonovic. "Changes in Floodplain Regimes Over Canada Due to Climate Change Impacts: Observations from CMIP6 models," *Science of the Total Environment* 792 (2021): 1–15.

Molero, Fernando, et al. "Consequences of Perceived Personal and Group Discrimination Against People with Physical Disabilities," *Rehabilitation Psychology* 64, no. 2 (2019): 212–220.

Momeni, Saeed, Farhad Kooban, Saman Alipouri Niaz, Sahel Niyafard, and Alireza Soleimani. "Waste Heat Recovery, Efficient Lighting, and Proper Insulation: A Comprehensive Study of Energy Consumption and Savings in the Residential Sector," *Asian Journal of Civil Engineering* 25, no. 3 (2024): 2499–2508.

Monarrez, Tomas, and Carina Chien. "Dividing Lines: Racially Unequal School Boundaries in US Public School Systems. Research Report," Urban Institute, 1 September (2021). https://www.urban.org/research/publication/dividing-lines-racially-unequal-school-boundaries-us-public-school-systems

Moore, Brendan, et al. "Transformations for Climate Change Mitigation: A Systematic Review of Terminology, Concepts, and Characteristics," *Wiley Interdisciplinary Reviews: Climate Change* 12, no. 6 (2021): 1–25.

Moreda, Tsegaye. "Large-Scale Land Acquisitions, State Authority and Indigenous Local communities: Insights from Ethiopia," *Third World Quarterly* 38, no. 3 (2017): 698–716.

Morelli, John. "Environmental Sustainability: A Definition for Environmental Professionals," *Journal of Environmental Sustainability* 1, no. 1 (2011): 19–27.

Morello-Frosch, Rachel, and Osagie K. Obasogie. "The Climate Gap and the Color Line – Racial Health Inequities and Climate Change," *New England Journal of Medicine* 388, no. 10 (2023): 943–949.

Moriarty, Patrick, and Damon Honnery. "Renewable Energy and Energy Reductions or Solar Geoengineering for Climate Change Mitigation?" *Energies* 15, no. 19 (2022): 7315.

Morrison-Saunders, Angus et al. "Strengthening Impact Assessment: A Call for Integration and Focus," *Impact Assessment and Project Appraisal* 32, no.1 (2014): 1–8.

Morse, Bradford W. "Tsilhqot'in Nation v. British Columbia: Is It a Game Changer in Canadian Aboriginal Title Law and Crown-Indigenous Relations?" *Lakehead Law Journal* 2, no. 2 (2017): 65–88.

Mosby, Yessie. "Here is Why We are Taking the Australian Government to the UN Over its Inaction on Climate Change," *The Guardian*, 14 August (2020). https://

www.theguardian.com/commentisfree/2020/aug/14/here-is-why-we-are-taking-the-australian-government-to-the-un-over-its-inaction-on-climate-change

Mosoetsa, Sarah, and David Francis. "Framing Poverty and Inequality Studies in South Africa," *Transformation: Critical Perspectives on Southern Africa* 101, no. 1 (2019): 1–10.

Moss, Astoneia O. "Climate, Clarity, Controversy: A Constitutional, Statutory, and Policy Analysis of the SEC's Proposed Climate Disclosure Rules," *Emory Corporate Governance and Accountability Review* 11, no. 1 (2024): 47–82.

Motairek, Z. Chen, M.H.E. Makhlouf, S. Rajagopalan, and S. Al-Kindi. "Historical Neighbourhood Redlining and Contemporary Environmental Racism," *Local Environment* 80 (2022): 171–175.

Motairek, Issam, Sanjay Rajagopalan, and Sadeer Al-Kindi. "The 'Heart' of Environmental Justice," *American Journal of Cardiology* 189 (2023): 148–149.

Motalebi, Maedeh, Ali Rashidi, and Mohammad Mahdi Nasiri. "Optimization and BIM-Based Lifecycle Assessment Integration for Energy Efficiency Retrofit of Buildings," *Journal of Building Engineering* 49 (2022): 1–24.

Motswaledi, Thabang Richard, Phemelo Olifile Marumo, and O. Adeyemi Oluwatobi. "Eskom's Lack of Energy Provision and Human Insecurity in South Africa: A Critical Analysis," *African Renaissance* 2023, no. si1 (2023): 71–93.

Moyer, Laura, Allison P. Harris, and Rorie Spill Solberg. "'Better Too Much Than Not Enough': The Nomination of Women of Color to the Federal Bench," *Journal of Women, Politics & Policy* (2022): 1–13.

Msimango, Nonhlanhla, Carla Orffer, and Roula Inglesi-Lotz. "South Africa's Energy Policy," *Energy Policy* 183 (2023): 1–41.

Mthethwa, Sandile, and Edilegnaw Wale. "Household Vulnerability to Climate Change in South Africa: A Multilevel Regression Model," *Development Southern Africa* 40, no. 2 (2023): 466–481.

Mudzingiri, Calvin, et al. "The Effect of Local Climatic Conditions on Household Consumption: A Case Study of South Africa," *Humanities and Social Sciences Communications* 11, no. 1 (2024): 1–16.

Muller, Nicholas Z., Robert Mendelsohn, and William Nordhaus. "Environmental Accounting for Pollution in the United States Economy," *American Economic Review* 101, no. 5 (2011): 1649–1675.

Müller, Jörg, et al. "Weather Explains the Decline and Rise of Insect Biomass over 34 Years," *Nature* 628, no. 8007 (2024): 349–354.

Munguia, Nora, Javier Esquer, Hector Guzman, Janim Herrera, Jesus Gutierrez-Ruelas, and Luis Velazquez. "Energy Efficiency in Public Buildings: A Step Toward the UN 2030 Agenda for Sustainable Development," *Sustainability* 12, no. 3 (2020): 1212–1229.

Myers, Garth. "Sustainable Development and Environmental Justice in African Cities," *Geography Compass* 2, no. 3 (2008): 695–708.

N

Nagle, Mary Kathryn. "Environmental Justice and Tribal Sovereignty: Lessons from Standing Rock," *The Yale Law Journal Forum* 127 (2017): 667–684.

Nagy, Rosemary. "Transformative Justice in a Settler Colonial Transition: Implementing the UN Declaration on the Rights of Indigenous Peoples in Canada," *The International Journal of Human Rights* 26, no. 2 (2022): 191–216.

Napieralski, Jacob, Atreyi Guin, and Catherine Sulich. "Buried But Not Dead: The Impact of Stream and Wetland Loss on Flood Risk in Redlined Neighborhoods," *City and Environment Interactions* 21 (2024): 1–8.

Nasir, Mohammad Izzamil Mohd, and Asyirah Abdul Rahim. "Climate Adaptation Strategy via Urban Green Space: A Bibliometric Analysis," *Journal of Asian Geography* 2, no. 1 (2023): 31–36.

Natarajan, Usha. "Climate Justice," in *The Routledge Handbook of Law and Society*, eds. Mariana Valverde, et al. (New York: Routledge, 2021), 102–106.

Nath, Ishan B., Valerie A. Ramey, and Peter J. Klenow. "How Much Will Global Warming Cool Global Growth?" *NBER Working Paper Series* (2023).

National Academies of Sciences, Engineering, and Medicine. "Communities, Climate Change, and Health Equity – State-Level Implementation: Proceedings of a Workshop—in Brief," (2022). https://www.nationalacademies.org/our-work/communities-climate-change-and-health-equity-state-level-implementation

National Institute of Environmental Health Sciences. "Perfluoroalkyl and Polyfluoroalkyl Substances (PFAS)," (2024). https://www.niehs.nih.gov/health/topics/agents/pfc#:%7E:text=PFAS%20are%20a%20group%20of,the%20U.S.%20Environmental%20Protection%20Agency

National Research Council. *Nature's Numbers: Expanding the National Economic Accounts to Include the Environment* (Washington, DC: National Academies Press, 1999), 19.

Natural Justice. "Submission Climate Change Bill (South Africa)," Natural Justice (2022). https://naturaljustice.org/publication/submission-climate-change-bill-south-africa/

Neethirajan, Suresh. "Innovative Strategies for Sustainable Dairy Farming in Canada Amidst Climate Change," *Sustainability* 16, no. 265 (2023): 1–35.

Nel, Etienne, Lochner Marais, and Zolile Mqotyana. "The Regional Implications of Just Transition in the World's Most Coal-Dependent Economy: The Case of Mpumalanga, South Africa," *Frontiers in Sustainable Cities* 4 (2023): 1–16.

Nemavhidi, Muano, and Ademola Oluborode Jegede. "Carbon Tax as a Climate Intervention in South Africa: A Potential Aid or Hindrance to Human Rights?" *Environmental Law Review* 25, no. 1 (2023): 11–27.

Nesmith, Ande A., et al. *The Intersection of Environmental Justice, Climate Change, Community, and the Ecology of Life* (Berlin: Springer, 2021).

Newell, Peter. "Climate Justice," *The Journal of Peasant Studies* 49, no. 5 (2022): 915–923.

Newell, Peter, et al. "Toward Transformative Climate Justice: An Emerging Research Agenda," *Wiley Interdisciplinary Reviews: Climate Change* 12, no. 6 (2021): 1–22.

New York City Mayor's Office. "JNYC: A Study of Environmental Justice in New York City," Mayor's Office of Environmental and Climate Justice April (2024). https://climate.cityofnewyork.us/ejnyc-report/

Ngcamu, Bethuel Sibongiseni. "Climate Change Effects on Vulnerable Populations in the Global South: A Systematic Review," *Natural Hazards* 118, no. 2 (2023): 977–991.

Nhamo, Godwell, and Lazarus Chapungu. "Prospects for a Sustainable and Climate-Resilient African Economy Post-COVID-19," *Global Environmental Change* 86 (2024): 1–14.

Niehaus, Manuela, and Kirsten Davies. "Voices for the Voiceless: Climate Protection from the Streets to the Courts," *Journal of Human Rights and the Environment* 12, no. 2 (2021): 228–253.

Nigam, Amulya, and Madhuri Parikh. "Comparative Analysis of Climate Justice Framework in USA, UK, Japan, China, Germany and France," *Journal of Humanities and Social Sciences Studies* 5, no. 9 (2023): 41–53.

Nikkhah, Hasan, et al. "A Comprehensive Review on Atmospheric Water Harvesting Technologies: From Thermodynamic Concepts to Mechanism and Process Development," *Journal of Water Process Engineering* 53 (2023): 1–29.

Nikolakis, William, and Ngaio Hotte. "How Law Shapes Collaborative Forest Governance: A Focus on Indigenous Peoples in Canada and India," *Society & Natural Resources* 33, no. 1 (2020): 46–64.

Nikolakis, William, and Harry Nelson. "To Log Or Not to log? How Forestry Fits With the Goals of First Nations in British Columbia," *Canadian Journal of Forest Research* 45, no. 6 (2015): 639–646.

Nisbet, Euan G., et al. "Atmospheric Methane: Comparison Between Methane's Record in 2006–2022 and during Glacial Terminations," *Global Biogeochemical Cycles* 37, no. 8 (2023): e2023GB007875.

Nkrumah, Bright. "Courting Emissions: Climate Adjudication and South Africa's Youth," *Energy, Sustainability and Society* 11 (2021): 1–12.

NOAA Research. "No Sign of Greenhouse Gases Increases Slowing in 2023," National Oceanic and Atmospheric Administration (2024). https://research.noaa.gov/2024/04/05/no-sign-of-greenhouse-gases-increases-slowing-in-2023/?campaign_id=54&emc=edit_clim_20240409&instance_id=119740&nl=climate-forward&regi_id=73699248&segment_id=163030&te=1&user_id=3cc6620316c0a93e52f5fac3913b2a16

Nogueira, Leticia M., and K. Robin Yabroff. "Climate Change and Cancer: The Environmental Justice Perspective," *JNCI: Journal of the National Cancer Institute* 116, no. 1 (2024): 15–25.

Nordhaus, William D. "The 'DICE' Model: Background and Structure of a Dynamic Integrated Climate-Economy Model of the Economics of Global Warming," *Cowles Foundation Discussion Papers*, 1252 (1992): 1–134.

Norris, Pippa. "Measuring Populism Worldwide," Party Politics 26, no. 6 (2020): 697–717.

Norton, Lindsay. "Let's Talk Dirty: Revealing the United States Sanitation Crisis and Its Disproportionate Effect on Poor and Minority Communities," *Villanova Environmental Law Journal* 34, no. 1 (2023): 85, 87.

Ntombela, Sifiso., et al. "Greening the South Africa's Economy Could Benefit the Food Sector: Evidence from a Carbon Tax Policy Assessment," *Environment and Resource Economics* 74 (2019): 1–28. https://repository.up.ac.za/bitstream/handle/2263/70144/Ntombela_Greening_2019.pdf?sequence=1

Nwankwo, Onnnaedo. "A Watershed Moment: The Health and Economic Impact of Water Sustainability in the Navajo Nation Post Pandemic," *American Indian Law Journal* 10 (2022): 169–194.

O

Oberg, Michael Leroy. "Good Neighbors: The Onondagas and the Fort Schuyler Treaty of September 1788," *New York History* 88, no. 4 (2007): 390–417.

O'Connor, Maeve. "Paris Maligned II: Climate Alignment Assessments Reveal Oil and Gas Company Transition Risk Exposure," Carbon Tracker (2024). https://carbontracker.org/reports/paris-maligned-2/

Ochs, Sara. "The Role of Truth-Telling in Indigenous Justice," 11 *Journal of Race, Gender and Ethnicity* (2024): 117–134.

O'Donnell, Emily C., Simon N. Gosling, Noelwah R. Netusil, Faith Ka Shun Chan, and Nanco J. Dolman. "Perceptions of Blue-Green and Grey Infrastructure as Climate Change Adaptation Strategies for Urban Water Resilience," *Journal of the British Academy* 9, no. 9 (2021): 143–182.

Odimegwu, Frank. "Climate Change and Agriculture: Analysis and Implication on South Africa," *African Social Science and Humanities Journal* 3, no. 2 (2022): 1–19.

Offenheim, Alyssa. *The Impact Assessment Act, 2019, Climate Change, and Political Priorities in Canada* (York University, Master's Thesis, 2020).

Ogunbode, Charles A. "Climate Justice is Social Justice in the Global South," *Nature Human Behaviour* 6, no. 11 (2022): 1443–1443.

Ogunseitan, Oladele A. "The Environmental Justice Agenda for E-Waste Management," *Environment: Science and Policy for Sustainable Development* 65, no. 2 (2023): 15–25.

OHCHR. "Transforming Our World: Human Rights in the 2030 Agenda for Sustainable Development, Position Paper," United Nations (nd). http://www.ohchr.org/Documents/Issues/MDGs/Post2015/HRAndPost2015.pdf

OHCHR. "Born Free and Equal: Sexual Orientation and Gender Identity in International Human Rights Law," United Nations (2012). https://www.ohchr.org/sites/default/files/Documents/Publications/BornFreeAndEqualLowRes.pdf

OHCHR. Indigenous Peoples can lead us all through the turbulence and risks of our era, 17 July (2023). https://www.ohchr.org/en/statements-and-speeches/2023/07/indigenous-peoples-can-lead-us-all-through-turbulence-and-risks-our

OHCHR. Press Release of the United Nations Human Rights Office of the High Commissioner: Rethinking Business and Economic Paradigms for People and Planet to Survive: UN Expert 6 March (2024a). https://www.ohchr.org/en/press-releases/2024/03/rethinking-business-and-economic-paradigms-people-and-planet-survive-un

OHCHR. Rights of Persons Belonging to National or Ethnic, Religious and Linguistic Minorities, A/HRC/55/35 10 January (2024b). https://documents.un.org/doc/undoc/gen/g23/268/21/pdf/g2326821.pdf?token=lcqxthDfwWmJ993cZT&fe=true

Ohenhen, Leonard O., et al. "Disappearing Cities on US Coasts," *Nature* 627, no. 8002 (2024): 108–115.

Olabanji, Mary Funke, Thando Ndarana, Nerhene Davis, and Emma Archer. "Climate Change Impact on Water Availability in the Olifants Catchment (South Africa) with Potential Adaptation Strategies," *Physics and Chemistry of the Earth*, Parts A/B/C 120 (2020): 102939–102948.

Oladipo, Gloria. "Native American Communities Lashed by Covid, Worsening Chronic Inequities," *The Guardian*, 13 March (2021). https://www.theguardian.com/us-news/2021/dec/13/pandemic-challenges-native-american- communities

Olajiga, Oladiran Kayode, Emmanuel Chigozie Ani, Zamathula Queen Sikhakane, and Tosin Michael Olatunde. "A Comprehensive Review of Energy-Efficient Lighting Technologies and Trends," *Engineering Science & Technology Journal* 5, no. 3 (2024): 1097–1111.

O'Malley, Jeffrey, et al. "The Sustainable Development Goals: Sexual and Gender Minorities," United Nations Development Programme (2018). https://www.undp.org/sites/g/files/zskgke326/files/publications/SDGs_SexualAndGenderMinorities.pdf

O'Manique, Claire, et al. "Public Enemies: Assessing MDB and G20 international finance institutions' energy finance," OilChange International (2024). https://priceofoil.org/content/uploads/2024/04/G20-Public-Enemies-April-2024.pdf

Ong, David M. "Prospects for Transitional Environmental Justice in the Socio-Economic Reconstruction of Kosovo," *Tulane Environmental Law Journal* 30, no. 2 (2017): 217–272.

O'Rourke, Jennifer. "The Overlooked Communities of Forced Displacement in the United States: Humanizing the Relocation of Indigenous Tribes in the Face of Climate Change," *University of Cincinnati Law Review* 92, no. 3 (2024): 850–873.

Otten, Ainsley, Aamir Fazil, Anna Chemeris, Patrick Breadner, and Victoria Ng. "Prioritization of Vector-Borne Diseases in Canada Under Current Climate and Projected Climate Change," *Microbial Risk Analysis* 14 (2020): 1–14.

Outhwaite, Charlotte L., Peter McCann, and Tim Newbold. "Agriculture and Climate Change are Reshaping Insect Biodiversity Worldwide," *Nature* 605, no. 7908 (2022): 97–102.

Outka, Uma, and Elizabeth Kronk Warner. "Reversing Course on Environmental Justice under the Trump Administration," *Wake Forest Law Review* 54 (2019): 393–423.

Ouweneel, Birgitt and Nicholas P. Simpson. "Securing a Climate-resilient Pathway for South Africa," in *Climate Security in the Anthropocene: Exploring the Approaches of United Nations Security Council Member-States*, eds. Judith Nora Hardt, et al. (Cham, Switzerland: Springer International Publishing, 2023), 231–257.

Owen, John R., et al. "Energy Transition Minerals and Their Intersection With Land-Connected Peoples," *Nature Sustainability* 6, no. 2 (2023a): 203–211.

Owen, John R., Deanna Kemp, Alex M. Lechner, Jill Harris, Ruilian Zhang, and Éléonore Lèbre. "Energy Transition Minerals and Their Intersection with Land-Connected Peoples," *Nature Sustainability* 6, no. 2 (2023b): 203–211.

Oxfam, Climate Equality: A planet for the 99%, November (2023): 67. https://policy-practice.oxfam.org/resources/climate-equality-a-planet-for-the-99-621551/

P

Pachauri, Rajendra K., et al. "Climate Change 2014: Synthesis Report. Contribution of Working Groups I, II and III to the Fifth Assessment Report of the Intergovernmental Panel on Climate Change," The Intergovernmental Panel on Climate Change (2014). https://www.ipcc.ch/report/ar5/syr/

Palinkas, Lawrence A., Jessenia De Leon, Kexin Yu, Erika Salinas, Cecilia Fernandez, Jill Johnston, Md Mostafijur Rahman et al. "Adaptation Resources and Responses to Wildfire Smoke and Other Forms of Air Pollution in Low-Income Urban Settings: A Mixed-Methods Study," *International Journal of Environmental Research and Public Health* 20, no. 7 (2023): 1–16.

Palmer, Michael. "For Comparative Legal Studies," in *A Research Agenda for Comparative Law*, ed. Jaakko Husa (Cheltenham, UK: Edward Elgar Publishing, 2024), 129–155.

Panão, Marta J.N. Oliveira. "Energy Ratings as Drivers of Energy Sufficiency in Residential Buildings: A Comprehensive Review and Future Directions," *Energy and Buildings* (2024): 1–21.

Pang, Yiyuan, et al. "Bibliometric Analysis of Trends in Smart Irrigation for Smart Agriculture," *Sustainability* 15, no. 23 (2023): 1–23.

Papacostantis, Helen. "South Africa's Journey to Climate Change Regulation: *Earthlife Africa Johannesburg v Minister of Environmental Affairs* 2017 2 All SA 519 (GP)," *Potchefstroom Electronic Law Journal/Potchefstroomse Elektroniese Regsblad* 24, no. 1 (2021): 1–25.

Papies, Esther K., Tess Davis, Stephanie Farrar, Maddie Sinclair, and Lara H. Wehbe. "How (Not) to Talk About Plant-Based Foods: Using Language to Support the Transition to Sustainable Diets," *Proceedings of the Nutrition Society* 83, no. 3 (2024): 142–150.

Papke, David Ray, and Mary Elise Papke. "Rights & Remedies: Rental Housing for Low-Income Households in the United States," *Marquette Benefits & Social Welfare Law Review*, 25 (2023): 1–31.

Paquin, Caroline. "The Impact Assessment Act of 2019: Meaningful Participation of Canada's Indigenous Peoples, or Surface Level Engagement?" (PhD dissertation, Dalhousie University College of Sustainability, 2023).

Parkinson, Stuart, and Linsey Cottrell. *Estimating the Military's Global Greenhouse Gas Emissions* (Lancaster, UK: Scientists for Global Responsibility (SGR) and the Conflict and Environment Observatory (2022)). https://ceobs.org/wp-content/uploads/2022/11/SGRCEOBS-Estimating_Global_MIlitary_GHG_Emissions_Nov22_rev.pdf

Parry, Jo-Ellen, Sarah O'Keefe, and Anne Hammill. "Adaptation Governance in Canada." International Institute for Sustainable Development (2022), 2. https://www.iisd.org/system/files/2023-05/adaptation-options-canada.pdf

Parsloe, Sarah M. and Rashaunna C. Campbell. "'Folks Don't Understand What It's Like to Be a Native Woman': Framing Trauma via MMIW," *Howard Journal of Communications* 32, no. 3 (2021): 197–212.

Pastor, Manuel, J. Mijin Cha, Michael Méndez, and Rachel Morello-Frosch. "California Dreaming: Why Environmental Justice is Integral to the Success of Climate Change Policy," *Proceedings of the National Academy of Sciences* 121, no. 32 (2024): e2310073121.

Patel, Zarina. "South Africa's Three Waves of Environmental Policy: (Mis)aligning the Goals of Sustainable Development, Environmental Justice and Climate Change," *Geography Compass* 8, no. 3 (2014): 169–181.

Patnaik, Aneesh, et al. "Racial Disparities and Climate Change," Princeton Student Climate Initiative (2020). https://psci.princeton.edu/tips/2020/8/15/racial-disparities-and-climate-change

Paulina Smolinski. "Biden to Set 10-Year Deadline to Remove All Lead Pipes Nationwide," CBS News, 08 October (2024). https://www.cbsnews.com/news/biden-lead-pipes-infrastructure/

Pearce, Fred. "Nations Are Undercounting Emissions, Putting UN Goals at Risk," YaleEnvironment360 (2024). https://e360.yale.edu/features/undercounted-emissions-un-climate-change

Pearson, Christopher H. "Exploitation: A Missing Element to Our Understanding of Environmental Justice," *Ethics, Policy & Environment* 26, no. 3 (2022): 1–13.

Pecharroman, Lidia Cano, Melissa O. Tier, and Elke U. Weber. "Feature Importance of Climate Vulnerability Indicators with Gradient Boosting Across Five Global Cities," *arXiv* preprint arXiv:2411.10628 (2024).

Pelch, Katherine E., et al. "PFAS Health Effects Database: Protocol for a Systematic Evidence Map," *Environment International* 130 (2019): 1–8.

Perry, Andre, Jonathan Rothwell, and David Harshbarger. "The Devaluation of Assets in Black Neighborhoods," Brookings Institute, 27 November (2018). https://www.brookings.edu/articles/devaluation-of-assets-in-black-neighborhoods/

Peters, Michael A. "Science, Truth and Conspiracy in the Age of Trump," *Educational Philosophy and Theory* 55, no. 14 (2023): 1647–1652.

Petersen, Nancy, et al. "Respectful Tribal Partnership: What Philanthropy Can Learn from the Navajo Nation's Collaborative Response to the COVID-19 Crisis," *The Foundation Review* 14, no 2 (2022): 93–103.

Petersen, Brian and Hélène B. Ducros, eds. *Justice in Climate Action Planning* (Cham, Switzerland: Springer, 2022).

Petzold, Jan., et al. "Indigenous Knowledge on Climate Change Adaptation: A Global Evidence Map of Academic Literature," *Environmental Research Letters* 15, no. 11 (2020): 1–18.

Pezzullo, Phaedra C. "Environmental Justice and Climate Justice," in *The Routledge Handbook of Environmental Movements*, eds. Maria Grasso and Marco Giugni (New York: Routledge, 2022), 229–244.

Philibert, A., et al. "Past Mercury Exposure and Current Symptoms of Nervous System Dysfunction in Adults of a First Nation Community (Canada)," *Environmental Health*, 21, no. 1 (2022): 1–17.

Piccard, Ann. "Death by Boarding School: The Last Acceptable Racism and the United States' Genocide of Native Americans," *Gonzaga Law Review* 49 (2013): 137–185.

Pili, Olwethu, and Bongani Ncube. "Smallholder Farmer Coping and Adaptation Strategies for Agricultural Water Use During Drought Periods in the Overberg and West Coast Districts, Western Cape, South Africa," *Water SA* 48, no. 1 (2022): 97–109.

Plumer, Brad, and Lisa Friedman. "Biden, Headed to the Exit, Sets an Aggressive Climate Goal for the U.S.," *The New York Times*, 19 December (2024). https://www.nytimes.com/2024/12/19/climate/biden-new-us-climate-target.html?smid=nytcore-ios-share&referringSource=articleShare

Plourde, André. "What's a Country Like Canada to Do? Addressing the Challenge of Global Climate Change," in *Canada and Great Power Competition: Canada Among Nations*, eds. D. Carment, Laura Macdonald, and Jeremy T. Paltiel (Cham, Switzerland: Springer International Publishing, 2022), 209–232.

Plumer, Brad. "U.S. Efforts to Cut Emissions Stalled in 2024 as Power Demand Surged," *The New York Times*, 9 January (2025). https://www.nytimes.com/2025/01/09/climate/us-emissions-solar-wind-power.html

Pongratz, Julia, et al. "Land Use Effects on Climate: Current State, Recent Progress, and Emerging Topics," *Current Climate Change Reports* (2021): 1–22.

Pooe, Katlego. "An Analysis of the South African Carbon Tax Act," University of Johannesburg (South Africa) (2022). https://ujcontent.uj.ac.za/esploro/outputs/9913355907691

Popovich, Nadja, Livia Albeck-Ripka, and Kendra Pierre-Louis. "The Trump Administration is Reversing 100 Environmental Rules. Here's the Full List," *The New York Times*, 20 January (2021). https://www.nytimes.com/interactive/2020/climate/trump-environment-rollbacks-list.html

Porter, Jody. "Children of the Poisoned River," *CBC News* (2017). https://www.cbc.ca/news2/interactives/children-of-the-poisoned-river-mercury-poisoning-grassy-narrows-first-nation/

Porter, Libby, et al. "Climate Justice in a Climate Changed World," *Planning Theory & Practice* 21, no. 2 (2020): 293–321.

Pouw, Nicky, and Joyeeta Gupta. "Inclusive Development: A Multi-Disciplinary Approach," *Current Opinion in Environmental Sustainability* 24 (2017): 104–108.

Prüss-Üstün, et al. "Preventing Disease Through Healthy Environments: A Global Assessment of the Burden of Disease from Environmental Risks," World Health Organization (2016). https://iris.who.int/bitstream/handle/10665/204585/9789241565196_eng.pdf?sequence=1

Pulido, Laura. "Conversations in Environmental Justice: An Interview with David Pellow," *Capitalism Nature Socialism* 28, no. 2 (2017): 43–53.

Purohit, Anuja. "Law Reform through Deep Ecology: A Critical Review of the Canadian Environmental Protection Act, 1999." 2023 (Masters thesis University of Toronto). https://tspace.library.utoronto.ca/bitstream/1807/130224/3/Purohit_Anuja_202311_MSc_thesis.pdf

Q

Qian, Emily. "The Staying Power of Asbestos Hazards: A Critical Environmental Justice Case Analysis," *Sociology Between the Gaps: Forgotten and Neglected Topics* 9, no. 1 (2023): 1–7.

R

Rabson, Mia. "The Federal Carbon Price Goes Up Today. Here's What It Will Mean For You," *National Observer*, 1 April (2024). https://www.nationalobserver.com/2024/04/01/news/federal-carbon-price-rises

Racehorse, Vanessa. "Achieving Climate Justice Through Land Back: An Overview of Tribal Dispossession, Land Return Efforts, and Practical Mechanisms for #LandBack," *Colorado Environmental Law Journal* 34, no. 175 (2023): 176–192.

Racehorse, Vanessa and Anna Hohag. "Achieving Climate Justice Through Land Back: An Overview of Tribal Dispossession, Land Return Efforts, and Practical Mechanisms for #LandBack," *Colorado Natural Resources Energy & Environmental Law Review* 34 (2023): 175–212.

Radcliffe, Sarah A. "Gendered Frontiers of Land Control: Indigenous Territory, Women and Contests over Land in Ecuador," *Gender, Place & Culture* 21, no. 7 (2014): 854–871.

Raihan, Asif. "A Review of the Global Climate Change Impacts, Adaptation Strategies, and Mitigation Options in the Socio-Economic and Environmental Sectors," *Journal of Environmental Science and Economics* 2, no. 3 (2023): 36–58.

Räikkä, Juha. "Climate Change Conspiracy Theories," in *Handbook of the Philosophy of Climate Change*, eds. Gianfranco Pellegrino and Marcello Di Paola (Cham, Switzerland: Springer International Publishing, 2023), 1161–1177.

Rajapakse, Jay, Brian Hudson, and Charlotte Brown. "Poor Water Quality and Related Health Issues in Remote Indigenous Populations of Some of the World's Wealthiest Nations," in *Safe Water and Sanitation for a Healthier World: A Global View of Progress Towards SDG 6*, ed. Jay Rajapakse (Cham, Switzerland: Springer International Publishing, 2022), 195–214.

Rapozo, Pedro. "Necropolitics, State of Exception, and Violence Against Indigenous People in the Amazon Region During the Bolsonaro Administration," *Brazilian Political Science Review* 15 (2021): e0002.

Reed, Graeme, et al. "Indigenising Climate Policy in Canada: A Critical Examination of the Pan-Canadian Framework and the ZéN Roadmap," *Frontiers in Sustainable Cities* 3 (2021): 1–12.
Reed, Graeme, et al. "Toward Indigenous Visions of Nature-Based Solutions: An Exploration Into Canadian Federal Climate Policy," *Climate Policy* 22, no. 4 (2022): 514–533.
Refsgaard, Jens Christian, Karsten Arnbjerg-Nielsen, Martin Drews, Kirsten Halsnæs, Erik Jeppesen, Henrik Madsen, Anil Markandya, Jørgen E. Olesen, John R. Porter, and Jens H. Christensen. "The Role of Uncertainty in Climate Change Adaptation Strategies—A Danish Water Management Example," *Mitigation and Adaptation Strategies for Global Change* 18 (2013): 337–359.
Regier, Kristen, and Tenley M. Conway. "Where to Expand Green Infrastructure to Support Equitable Climate Change Adaptation in the City of Toronto?" *Cities and the Environment (CATE)* 15, no. 1 (2022): 1–13.
Renglet, Charlotte. "The Recognition of the Special Relationship of Indigenous Peoples with Their Environment under International Law: A Potential Advantage in Climate Litigation?" *International Journal on Minority and Group Rights* 29, no. 4 (2022): 720–746.
Renner, Heather M., et al. "Catastrophic and Persistent Loss of Common Murres After a Marine Heatwave," *Science* 386, no. 6727 (2024): 1272–1276.
Rentschler, Jun, and Nadezda Leonova. "Global Air Pollution Exposure and Poverty," *Nature Communications* 14, no. 1 (2023): 4432.
Republican Party. "Republican Party Platform," (2024). https://s3.documentcloud.org/documents/24795758/read-the-2024-republican-party-platform.pdf
Revesz, Richard L. "Air Pollution and Environmental Justice," *Ecology Law Quarterly* 49 (2022): 187–252.
Reyes-García, Victoria, et al. "Introduction: Understanding Climate Change Impacts on Indigenous Peoples and Local Communities: A Global Perspective from Local Studies," in *Routledge Handbook of Climate Change Impacts on Indigenous Peoples and Local Communities*, ed. Victoria Reyes-García et al. (New York: Routledge, 2023), 1–18.
Reza, Mohammad Selim, and Gabriela Sabau. "Impact of Climate Change on Crop Production and Food Security in Newfoundland and Labrador, Canada," *Journal of Agriculture and Food Research* 10 (2022): 100405.
Rhodium Group. "Preliminary US Greenhouse Gas Estimates for 2024," 9 January (2025). https://rhg.com/research/preliminary-us-greenhouse-gas-estimates-for-2024/
Ricalde, Iñigo, Sebastián Vicuña, Oscar Melo, James E. Tomlinson, Julien J. Harou, and Greg Characklis. "Assessing Tradeoffs in the Design of Climate Change Adaptation Strategies for Water Utilities in Chile," *Journal of Environmental Management* 302 (2022): 114035–114047.
Rice, Jennifer L., Joshua Long, and Anthony Levenda. "Against Climate Apartheid: Confronting the Persistent Legacies of Expendability for Climate Justice," *Environment and Planning E: Nature and Space* 5, no. 2 (2022): 625–645.
Richards, Thomas B., et al. "Lung Cancer Survival in the United States by Race and Stage (2001–2009): Findings from the CONCORD-2 Study," *Cancer* 123 (2017): 5079–5099.
Richardson, Jonathan L. "Increasing Rat Numbers in Cities Are Linked to Climate Warming, Urbanization, and Human Population," *Science Advances* 11, no. 5 (2025): 1–9.

Ricke, Katharine, and Anthony Harding. "How May Solar Geoengineering Impact Global Prospects for Climate Change Mitigation?," *Oxford Review of Economic Policy* 39, no. 4 (2023): 828–841.
Riley, Angela R., and Sarah Glenn Thompson. "Mapping Dual Sovereignty and Double Jeopardy in Indian Country Crimes," *Columbia Law Review* 122, no. 7 (2022): 1899–1956.
Rioux-Gobeil, Fabienne, and Annick Thomassin. "A Just Energy Transition for Indigenous Peoples: From Ideal Deliberation to Fairness in Canada and Australia," *Energy Research & Social Science* 114 (2024): 1–9.
Rivadeneira, Natalia Rubiano, and Wim Carton. "(In) justice in Modelled Climate Futures: A Review of Integrated Assessment Modelling Critiques Through a Justice Lens," *Energy Research & Social Science* 92 (2022): 1–11.
Robert, Amanda. "A Fresh Approach," *ABA Journal* 108, no. 4 (2022): 65–66.
Roberts, Brett G. "Returning the Land: Native Americans and National Parks," *Ave Maria Law Review* 21 (2023): 148–204.
Robinson, Mary. *Climate Justice: Hope, Resilience, and the Fight for a Sustainable* Future (London: Bloomsbury Publishing, 2018).
Rockström, Johan, et al. "Safe and Just Earth System Boundaries," *Nature* 619, no. 7968 (2023): 101–111.
Rodon, Thierry. "Institutional Development and Resource Development: The Case of Canada's Indigenous Peoples." *Canadian Journal of Development Studies/Revue canadienne d'études du développement* 39, no. 1 (2018): 119–136.
Rodriguez, Mayra, Guangtao Fu, David Butler, Zhiguo Yuan, and Lauren Cook. "The Effect of Green Infrastructure on Resilience Performance in Combined Sewer Systems under Climate Change," *Journal of Environmental Management* 353 (2024): 120229–120240.
Rodríguez-Garavito, César, ed., *Litigating the Climate Emergency: How Human Rights, Courts, and Legal Mobilization Can Bolster Climate Action* (Cambridge: Cambridge University Press, 2022).
Rodríguez-Pose, Andrés, Neil Lee, and Cornelius Lipp. "Golfing with Trump. Social Capital, Decline, Inequality, and the Rise of Populism in the US," *Cambridge Journal of Regions, Economy and Society* 14, no. 3 (2021): 457–481.
Roemer, Neoshia R. "The Indian Child Welfare Act as Reproductive Justice," *Boston University Law Review* 103 (2023): 55–116.
Romero-Lankao, et al. "A Framework to Centre Justice in Energy Transition Innovations," *Nature Energy* 8, no. 11 (2023): 1192–1198.
Rosen, Yereth. "Caribou Declines Causing Angst for Alaska Hunters are Part of Wider North America Trend," Alaska Beacon, 14 March (2024). https://alaskabeacon.com/2024/03/14/caribou-declines-causing-angst-for-alaska-hunters-are-part-of-wider-north-america-trend/?emci=41d8aefd-a8e1-ee11-85f9-002248223848&emdi=fdcf6621-0be2-ee11-aaf0-002248223197&ceid=145810
Roth, Sammy. "Why Communities Fighting for Fair Policing Also Demand Environmental Justice," *Los Angeles Times*, 4 June (2020). https://www.latimes.com/environment/newsletter/2020-06-04/why-communities-fighting-for-fair-policing-also-demand-environmental-justice-boiling-point
Rothstein, Richard. *The Color of Law: A Forgotten History of How our Government Segregated America* (New York: Liveright Publishing, 2017).
Rothwell, Jonathan, and Andre M. Perry. "How Racial Bias in Appraisals Affects the Devaluation of Homes in Majority-Black Neighborhoods," Brookings Institute, 5

December (2022). https://www.brookings.edu/articles/how-racial-bias-in-appraisals-affects-the-devaluation-of-homes-in-majority-black-neighborhoods/

Roy, B., and K. Hanaček. "From the Environmentalism of the Poor and the Indigenous Toward Decolonial Environmental Justice," in *The Barcelona School of Ecological Economics and Political Ecology: A Companion in Honour of Joan Martinez-Alier*, eds. Sergio Villamayor-Tomas and Roldan Muradian (Cham, Switzerland: Springer International Publishing, 2023), 305–315.

Ruckstuhl, Katharina, et al. eds. *The Routledge Handbook of Indigenous Development* (Milton Park, UK: Taylor & Francis, 2022).

Rulli, Megan. "When Fast-Tracking Slows You Down: Reconsidering Nationwide Permit 12 Use for Large-Scale Oil Pipelines," *Dickinson Law Review* 126 (2021): 273–303.

Rumble, Olivia. "Climate Change Legislative Development on the African Continent," in *Law | Environment | Africa*, eds. Patricia Kameri-Mbote, et al. (Baden-Baden: Nomos Verlagsgesellschaft, 2019), 31–60.

Ruppel, Oliver C. "Overview of International Soil Law," *Soil Security* 6 (2022a): 1–6.

Ruppel, Oliver C. "South Africa: Climate Change, Responsibility and Liability – The Legal System, Public and Private Law Considerations," in *Climate Change, Responsibility and Liability*, eds. Eva Schulev-Steindl, et al. (Baden-Baden: Nomos Verlagsgesellschaft, 2022b), 201–278.

Rusco, Michael. "Oklahoma v. Castro-Huerta, Jurisdictional Overlap, Competitive Sovereign Erosion, and the Fundamental Freedom of Native Nations," *Marquette Law Review* 106 (2022): 889–947.

S

Sabatello, Maya, Scott D. Landes, and Katherine E. McDonald. "People with Disabilities in COVID-19: Fixing our Priorities," *The American Journal of Bioethics* 20, no. 7 (2020): 187–190.

Sabti, Resha. "Liquified Natural Gas (LNG), Is it Part of Canada's Energy Resilience? How Realistic?." (PhD dissertation, University of British Columbia, 2024).

Sabzevar, Morteza Salmani, Amirreza Rezaei, and Bagher Khaleghi. "Incremental Adaptation Strategies for Agricultural Water Management Under Water Scarcity Condition in Northeast Iran," *Regional Sustainability* 2, no. 3 (2021): 224–238.

Safotso, Gilbert Tagne, and Nguafac Julius. "Addressing Global Challenges from a Linguistic Perspective," *International Journal of Linguistics, Literature and Translation* 6, no. 4 (2023): 141–146.

Saito, Kohei. *Marx in the Anthropocene: Towards the Idea of Degrowth Communism* (Cambridge: Cambridge University Press, 2023).

Salter, Raya. "Tracing the Roots of Energy Justice in Action: Environmental Justice, Climate Justice, and the New York Climate Leadership and Community Protection Act," in *Handbook on Energy Justice*, eds. Stefan Bouzarovski, Sara Fuller, and Tony G. Reames (Cheltenham, UK: Edward Elgar Publishing, 2023), 131–143.

Samuels, M.I., et al. "Climate Vulnerability and Risks to an Indigenous Community in the Arid Zone of South Africa," *Journal of Arid Environments* 199 (2022): 104718.

Sant'Anna, André Albuquerque. "Not So Natural: Unequal Effects of Public Policies on the Occurrence of Disasters," *Ecological Economics* 152 (2018): 273–281.

Santam, nd. "Santam Insurance Barometer 2022/23," Santam (nd). https://www.santam.co.za/about-us/santam-insurance-barometer-202223/

Santos, Sofia Ferreira. "Amazon Rainforest: Deforestation Rate Halved in 2023," *BBC News*, 12 January (2024). https://www.bbc.com/news/world-latin-america-67962297

Sapignoli, Maria, and Robert K. Hitchcock. "Indigenous Peoples in Southern Africa," *The Round Table* 102, no. 4 (2013): 355–365.

Sarathchandra, Dilshani, and Kristin Haltinner. "How Believing Climate Change is a 'Hoax' Shapes Climate Skepticism in the United States," *Environmental Sociology* 7, no. 3 (2021): 225–238.

Sarathchandra, Dilshani, and Kristin Haltinner. "How Media, Information Sources, and Trust Shape Climate Change Denial or Doubt," *Social Currents* 10, no. 6 (2023): 583–602.

Sardo, Michael Christopher. "Responsibility for Climate Justice: Political not Moral," *European Journal of Political Theory* 22, no. 1 (2023): 26–50.

Sareen, Siddharth, et al. "Solidaric Solarities: Governance Principles for Transforming Solar Power Relations," *Progress in Environmental Geography* 2, no. 3 (2023): 143–165.

Sargeant, Malcolm. *Discrimination and the Law* (New York: Routledge, 2017).

Sarkin, Jeremy. "The Trials and Tribulations of South Africa's Truth and Reconciliation Commission," *South African Journal on Human Rights* 12, no. 4 (1996): 617–640.

Sarkin, Jeremy. "The Development of a Human Rights Culture in South Africa," *Human Rights Quarterly* 20, no. 3 (1998): 628–665.

Sarkin, Jeremy. "The Common Law in South Africa: Pro-Apartheid or Pro-Democracy," *Hastings International & Comparative Law Review* 23, no. 1 (1999a): 1–27.

Sarkin, Jeremy. "The Drafting of South Africa's Final Constitution from a Human-Rights Perspective," *The American Journal of Comparative Law* 47, no. 1 (1999b): 67–87.

Sarkin, Jeremy. "A Review of Health and Human Rights After Five Years of Democracy in South Africa," *Medicine and Law* 19 (2000): 287–307.

Sarkin, Jeremy. "The Reshaping of the Co-Operatives Legal Architecture as a Result of the 2013 Amendments to the 2005 Co-Operatives Act: Promoting Democratic Governance and Economic Sustainability or Control and Overregulation?," *South Africa Mercantile Law Journal* 27, no. 2 (2015): 275–307.

Sarkin, Jeremy. "South Africa's Reconciliation Process: Tools, Resources and Obstacles in the Journey to Deal With its Atrocious Past," in *Reconciling with the Past: Resources and Obstacles in a Global Perspective*, eds. Annika Frieberg and C.K. Martin Chung (London: Routledge, 2017a).

Sarkin, Jeremy. "The Interrelationship of Nation Building, Reconciliation and Guarantees of Non-Recurrence (or Non-Repetition) of Human Rights Violations Using the South African Transition as an Illustration: How Victim Rights and Needs Can Be Better Achieved," *Revista de Victimología* 5 (2017b): 9–32.

Sarkin, Jeremy. "Colonial and Apartheid Victimisation: Towards Finding Solutions to Deal with an Atrocious Past" in *Victimology in Africa*, ed. R. Peacock (Pretoria: Van Schaik, 2019), 345–361.

Sarkin, Jeremy Julian. "Towards a Greater Understanding of Guarantees of Non-Repetition (GNR) or Non-Recurrence of Human Rights Violations: How GNR Intersects Transitional Justice with Processes of State (Re) Building, The Rule of Law, Democratic Governance, Reconciliation, Nation Building, Social Cohesion and Human Rights Protection." *Stanford Journal of International Law* 57 (2021): 191–231.

Sastry, Parinitha, Emil Verner, and David Marques-Ibanez. "Business as Usual: Bank Climate Commitments, Lending, and Engagement," Working Paper Series, European Central Bank (2024). https://www.ecb.europa.eu/pub/pdf/scpwps/ecb.wp2921~603e225101.de.pdf?f3854e151126bea0371149d197b37353

Satterthwaite, David, Diane Archer, Sarah Colenbrander, David Dodman, Jorgelina Hardoy, Diana Mitlin, and Sheela Patel. "Building Resilience to Climate Change in Informal Settlements," *One Earth* 2, no. 2 (2020): 143–156.

Schapper, Andrea. "Climate Justice and Human Rights," *International Relations* 32, no. 3 (2018): 275–295.

Scherer, Laura, Oscar Rueda, and Sergiy Smetana. "Environmental Impacts of Meat and Meat Replacements," in *Meat and Meat Replacements*, eds. Herbert L. Meiselman and José Manuel Lorenzo (Sawston, UK: Woodhead Publishing, 2023), 365–397.

Schindler, D.W. "Past, Present and Future Challenges to Management of Freshwater Quality in Canada." *Water Quality in the Americas* 128–147 (Mexica: The Inter-American Network of Academies of Sciences IANAS 2019). https://www.researchgate.net/profile/Sylvia-Bonilla-2/publication/331687977_Water_Quality_in_Uruguay_Current_status_and_challenges/links/5c88114892851c1df93d4323/Water-Quality-in-Uruguay-Current-status-and-challenges.pdf#page=129

Schleussner, Carl-Friedrich, et al. "Overconfidence in Climate Overshoot," *Nature* 634, no. 8033 (2024): 366–373.

Schlosberg, David. "Theorising Environmental Justice: The Expanding Sphere of a Discourse," *Environmental Politics* 22, no. 1 (2013): 37–55.

Schlosberg, D., and L.B. Collins. "From Environmental to Climate Justice: Climate Change and the Discourse of Environmental Justice," *Wiley Interdisciplinary Review of Climate Change*, 5 no. 3 (2014): 359–374.

Schneider, Susan A. "Climate Change, Food Security, and the Myth of Unlimited Abundance," *Journal of Food Law & Policy* 19, no. 1 (2023): 30–57.

Schnell, William F. "The Benefit of the Doubt: How Big Oil Makes Us Think," in *The Routledge Handbook of Language and Mind Engineering*, eds. Chris Shei and James Schnell (London: Routledge, 2024), 91–104.

Scholten, Jordan, et al. "Mining, Climate Change and Indigenous Peoples in Ontario, Canada: Intersecting Impacts and the Role of Corporate Social Responsibility," in *Local Communities and the Mining Industry*, eds. N. Brunet and S. Longboat (London: Routledge, 2023), 158–176.

Scholz, John T., and Cheng-Lung Wang. "Cooptation or Transformation? Local Policy Networks and Federal Regulatory Enforcement," *American Journal of Political Science* 50, no. 1 (2006): 81–97.

Schrock, Greg, Ellen M. Bassett, and Jamaal Green. "Pursuing Equity and Justice in a Changing Climate: Assessing Equity in Local Climate and Sustainability Plans in US Cities," *Journal of Planning Education and Research* 35, no. 3 (2015): 282–295.

Schubert, Julia. "The Politics of Climate Engineering Research," in *Climate, Science and Society*, eds. Zeke Baker, et al. (Abingdon, Oxon: Routledge, 2023), 253–260.

Schulz, Bailey, and Jessica Guynn. "As Homeowner's Insurance Prices Climb, More Americans Ask: Is it worth it?" *USA Today*, 23 June (2024). https://eu.usatoday.com/story/money/2024/06/23/americans-not-buying-homeowners-insurance/74144566007/

Schwartz, M.L. "International Legal Protection for Victims of Environmental Abuse," *Yale Journal of International Law*, 18 (1993): 355–375.

Scoones, Ian. "Livestock, Methane, and Climate Change: The Politics of Global Assessments," *Wiley Interdisciplinary Reviews: Climate Change* 14, no. 1 (2023): 1–8.

Scott, Dayna Nadine. "Environmental Justice and the Hesitant Embrace of Human Rights," in *Elgar Encyclopedia of Environmental Law*, ed. Michael Faure (Cheltenham, UK: Edward Elgar Publishing, 2016), 447–459.

Sedlak, David. *Water for All: Global Solutions for a Changing Climate* (London: Yale University Press, 2024).

Segone, Marco, and Florence Tateossian. "No One Left Behind. A Focus on Gender and Social Equity," *Evaluation for Agenda 2030* (2017): 23–24.

Seiyefa, Ebimboere, and Felix Idongesit Oyosoro. "Green Crime in West Africa: Uncovering the Threats to Human Security and Ecosystem Integrity in Cote d'Ivoire, Nigeria, and Senegal," *GeoJournal* 89, no. 62 (2024): 1–19.

Selje, Tom, Lena Anna Schmid, and Boris Heinz. "Community-Based Adaptation to Climate Change: Core Issues and Implications for Practical Implementations," *Climate* 12, no. 10 (2024): 155–172.

Senadheera, Sachini Supunsala, Souradeep Gupta, Harn Wei Kua, Deyi Hou, Sumin Kim, Daniel CW Tsang, and Yong Sik Ok. "Application of Biochar in Concrete – A Review," *Cement and Concrete Composites* (2023): 1–16.

Senathirajah, Kala, et al. "Estimation of the Mass of Microplastics Ingested – A Pivotal First Step Towards Human Health Risk Assessment," *Journal of Hazardous Materials* 404 (2021): 124004.

Sequeira, Miguel Macias, João Pedro Gouveia, and João Joanaz de Melo. "(Dis) Comfortably Numb in Energy Transitions: Gauging Residential Hard-to-Reach Energy Users in the European Union," *Energy Research & Social Science* 115 (2024): 103612–103631.

Sered, Susan, and Rushika Fernandopulle. *Uninsured in America, Updated: Life and Death in the Land of Opportunity* (Oakland, California: University of California Press, 2023).

Seyranian, Viviane, Hazel Atuel, and William D. Crano. "Dimensions of Majority and Minority Groups," *Group Processes & Intergroup Relations* 11, no. 1 (2008): 21–37.

Shafiee, Mojtaba, Pardis Keshavarz, Ginny Lane, Punam Pahwa, Michael Szafron, Derek Jennings, and Hassan Vatanparast. "Food Security Status of Indigenous Peoples in Canada According to the 4 Pillars of Food Security: A Scoping Review," *Advances in Nutrition* 13, no. 6 (2022): 2537–2558.

Shah, Shipra, and Jahangeer A. Bhat. "Ethnomedicinal Knowledge of Indigenous Communities and Pharmaceutical Potential of Rainforest Ecosystems in Fiji Islands," *Journal of Integrative Medicine* 17, no. 4 (2019): 244–249.

Shaheed, Ahmed. "Freedom of Religion or Belief," United Nations (2020). https://digitallibrary.un.org/record/3888724?ln=en.&v=pdf

Shapiro, Michelle. "Flushed and Forgotten: Sanitation and Wastewater in Rural Communities in the United States," Alabama Center for Rural Enterprise, The Columbia Law School Human Rights Clinic, and The Institute for the Study of Human Rights at Columbia University (2019). https://www.humanrightscolumbia.org/sites/default/files/Flushed%20and%20Forgotten%20-%20FINAL%20%281%29.pdf

Shelley, Nathanael Paul. "The Concept of Ethnicity in Early Antiquity: Ethno-Symbolic Identities in Ancient Greek, Biblical Hebrew, and Middle Babylonian Texts" (PhD dissertation, Columbia University, 2016).

Shelton, Dinah. "Human Rights and the Environment: What Specific Environmental Rights Have Been Recognized?" *Denver Journal of International Law and Policy* 35, no. 1 (2006): 129–171.

Sherwin, Evan D., et al. "US Oil and Gas System Emissions from Nearly One Million Aerial Site Measurements," *Nature* 627, no. 8003 (2024): 328–334.

Shi, Linda. "From Progressive Cities to Resilient Cities: Lessons from History for New Debates in Equitable Adaptation to Climate Change," *Urban Affairs Review* 57, no. 5 (2021): 1442–1479.

Shi, Linda, and Susanne Moser. "Transformative Climate Adaptation in the United States: Trends and Prospects," *Science* 372, no. 6549 (2021): eabc8054, 2.

Shindell, Drew, et al. "The Methane Imperative," *Frontiers in Science* 2 (2024): 1–28.

Shirazi, Reshaad. "It's High Time to Dump the High-Crime Area Factor," *Berkeley Journal of Criminal Law* 21 (2016): 76–110.

Shook, Phillip, and Jun-Ki Choi. "Predicting the Impact of Utility Lighting Rebate Programs on Promoting Industrial Energy Efficiency: A Machine Learning Approach," *Environments* 9, no. 8 (2022): 100–113.

Shrader-Frechette, Kristin. "Health Risks from Supposedly Remediated US Hazardous-Waste Sites: An Early-Warning Signal," *Advances in Environmental and Engineering Research* 3, no. 3 (2022): 1–37.

Sibanda, Arorisoe, and Sadhana Manik. "Reflecting on Climate Change Education (CCE) Initiatives for Mitigation and Adaptation in South Africa," *Environmental Education Research* 29, no. 12 (2023): 1814–1831.

Sibiya, Nomfundo, et al. "Empowering the Voiceless: Securing the Participation of Marginalised Groups in Climate Change Governance in South Africa," *Sustainability* 14, no. 12 (2022): 7111–7131.

Sibiya, Nomfundo Patricia, et al. "Overcoming Bureaucratic Resistance: An Analysis of Barriers to Climate Change Adaptation in South Africa," *Climate* 11, no. 7 (2023): 145–164.

Siegfried, Gabriella A. "Environmental and Climate Justice," in *Elgar Encyclopedia of Climate Policy*, eds. Daniel J. Fiorino, Todd A. Eisenstadt, and Manjyot Kaur (Cheltenham, UK: Edward Elgar Publishing, 2024), 86–90.

Simmons, B. Alexander, Christoph Nolte, and Jennifer McGowan. "Delivering on Biden's 2030 Conservation Commitment," *BioRxiv* (2021): 2021–2022.

Singh, Julie Arrighi, et al. "Climate Change Exacerbated Rainfall Causing Devastating Flooding in Eastern South Africa," *World Weather Attribution KZN Floods Scientific Report* (2022) 1–31.

Singh, Nishtha. "Climate Justice in the Global South: Understanding the Environmental Legacy of Colonialism," in *Global Climate Justice: Theory and Practice,* eds. Fausto Corvino, and Tiziana Andina (Bristol: E-International Relations Publishing, 2023).

Sinha, Anita. "A Lineage of Family Separation," *Brooklyn Law Review* 87 (2021): 445–500.

Sisco, Matthew R., Sarah O'Brien, Sara M. Constantino, Lara Aleluia Reis, and Valentina Bosetti. "Humans are Dangerously Unaware of and Unresponsive to Daily Air Pollution a Multi-national Study on Human Worry, Awareness, and Adaptation to Air Pollution," *Awareness, and Adaptation to Air Pollution* (2022), SSRN 1-52. https://papers.ssrn.com/sol3/papers.cfm?abstract_id=4144380

Slaff, Grace. "The Administration of Injustice: The Conflict Between Federal and Tribal Criminal Jurisdiction," *American Indian Law Review* 47, no. 2 (2023): 261–299.

Sloan, Lacey M., and Cathryne L. Schmitz. "Environmental Degradation: Communities Forging a Path Forward," *Journal of Transdisciplinary Peace Praxis* 1, no. 1 (2019): 13–38.

Smead, Richard G. "West Virginia v. Environmental Protection Agency: The Case and What It Means," *Climate and Energy* 39, no. 2 (2022): 29–32.

Smetana, Sergiy, et al. "Meat Substitutes: Resource Demands and Environmental Footprints," *Resources, Conservation and Recycling* 190 (2023): 1–12.

Smith, Callum, J.C.A. Baker, and D.V. Spracklen. "Tropical Deforestation Causes Large Reductions in Observed Precipitation," *Nature* 615, no. 7951 (2023): 270–275.

Smith, Genee S., et al. "Climate Change, Environmental Disasters, and Health Inequities: The Underlying Role of Structural Inequalities," *Current Environmental Health Reports* 9, no. 1 (2022): 80–89.

Smith, Kirk R., et al. "Human Health: Impacts, Adaptation, and Co-Benefits," in *Climate Change 2014 Impacts, Adaptation and Vulnerability: Part A: Global and Sectoral Aspects*, (Stockholm: Stockholm Environment Institute, 2015), 709–754. https://www.sei.org/mediamanager/documents/Presentations/SEI-Klein-IPCC-AR5-WG2-presentation.pdf

Smith, M.A. Peggy. "A Reflection on First Nations in Their Boreal Homelands in Ontario: Between a Rock and a Caribou," *Conservation and Society* 13, no. 1 (2015): 23–38.

Smith, Shelley. "Climate Change and South Africa: A Critical Analysis of the National Climate Change Response White Paper and the Push for Tangible Practices and Media-Driven Initiatives," *Global Media Journal: Africa Edition* 7, no.1 (2013): 47–66.

Soeherman, Jimmy K., Andrew J. Jones, and Paul J. Dauenhauer. "Overcoming the Entropy Penalty of Direct Air Capture for Efficient Gigatonne Removal of Carbon Dioxide," *ACS Engineering Au* 3, no. 2 (2023): 114–127.

Sokol, Karen C. "Forging Global Rule of Law Through Climate Litigation against the United States and the Fossil Fuel Industry," *Yearbook of International Disaster Law Online* 4, no. 1 (2023a): 237–262.

Soloway, Mollie. "Measuring Environmental Justice: Analysis of Progress Under Presidents Bush, Obama, and Trump," *Environmental Law Repertory.* 51 (2021): 10038–10053.

Soltani, Lamia, and Thuraya Mellah. "Exploring Farmers' Adaptation Strategies to Water Shortage Under Climate Change in the Tunisian Semi-Arid Region," *Environmental Management* 71, no. 1 (2023): 74–86.

South Africa. The Constitution of the Republic of South Africa, 1996.

South Africa. "Energy Act 2008" Act 34 of 2008 (2008). https://www.gov.za/sites/default/files/gcis_document/201409/316381263.pdf

South Africa. *National Climate Change Response White Paper* (2011). https://www.gov.za/sites/default/files/gcis_document/201409/nationalclimatechangeresponsewhitepaper0.pdf

South Africa. "Climate Change Act, 2024," *Government Gazette* Vol. 709, No. 50966, 23 July (2024).

South Africa Department of Forestry, Fisheries and the Environment. "Second National Action Programme for South Africa to Combat Desertification, Land Degradation and the effects of Drought (2018–2030)," Government of South Africa (2018). https://www.dffe.gov.za/sites/default/files/docs/nap_desertification_land_degradation_droughteffects.pdf

South Africa Department of Forestry, Fisheries and the Environment. "South Africa's Updated Nationally Determined Contribution (NDC)," September (2021). https://www.dffe.gov.za/sites/default/files/reports/draftnationalydeterminedcontributions_2021updated.pdf

South Africa Department of Forestry, Fisheries, and the Environment. "Government Gazette Staatskoerant," Government of South Africa, 24 April (2024). https://www.dffe.gov.za/sites/default/files/legislations/policyposition_rhinoelephantleopardlion.conservation_g50541gon4750.pdf

South Africa Ministry of Environment, Forestry and Fisheries. *The High-Level Panel of Experts for the Review of Policies, Legislation and Practices on Matters of Elephant, Lion, Leopard and Rhinoceros Management, Breeding, Hunting, Trade and Handling*, 15 December (2020). https://www.dffe.gov.za/sites/default/files/reports/2020-12-22_high-levelpanel_report.pdf

South Africa National Treasury. "Media Statement: Publication of the 2019 Carbon Tax Act," Government of South Africa (2019). https://www.treasury.gov.za/comm_media/press/2019/2019052701%20Media%20statement%20-%20Carbon%20Tax%20Act.pdf

South Africa Presidential Climate Commission. *The State of Climate Action in South Africa: Priorities for Action for the Government of National Unity* (RSA Presidential Climate Commission, 2024). https://pccommissionflo.imgix.net/uploads/images/24_REP_State-of-Climate-Action-in-South-Africa_v4.pdf

South Africa Secretariat of the Presidential Climate Commission. "Laying the foundation for a just transition framework for South Africa." *Pretoria: Presidential Climate Commission* (2021). https://www.nstf.org.za/wp-content/uploads/2022/03/ReportJustTransition.pdf

Sovacool, Benjamin K. "Information and the Extractive Industries Transparency Initiative," in *Energy & Ethics*, ed. Benjamin K. Sovacool (New York: Palgrave Macmillan, 2013), 90–111.

Sovacool, Benjamin K. "Is Sunshine the Best Disinfectant? Evaluating the Global Effectiveness of the Extractive Industries Transparency Initiative (EITI)," *The Extractive Industries and Society* 7, no. 4 (2020): 1451–1471.

Sovacool, Benjamin K. "Reckless or Righteous? Reviewing the Sociotechnical Benefits and Risks of Climate Change Geoengineering," *Energy Strategy Reviews* 35 (2021): 100656.

Sparks, Matt S., Isaiah Farahbakhsh, Madhur Anand, Chris T. Bauch, Kathryn C. Conlon, James D. East, Tianyuan Li et al. "Health and Equity Implications of Individual Adaptation to Air Pollution in a Changing Climate," *Proceedings of the National Academy of Sciences* 121, no. 5 (2024): e2215685121.

Spears, Ellen Griffith. "The 'Tar Wars' and Climate Justice Activism in North America: A Transboundary Movement Linking the US and Canada," in *The Routledge Handbook of Environmental Movements*, eds. Maria Grasso and Marco Guigni (London: Routledge, 2022), 45–62.

Spiller, Elisheba, et al. "Mortality Risk from PM 2.5: A Comparison of Modeling Approaches to Identify Disparities Across Racial/Ethnic Groups in Policy Outcomes," *Environmental Health Perspectives* 129, no. 12 (2021): 127004.

Sprainer, Helen. "Air Quality Equity: Why the Clean Air Act Failed to Protect Low-Income Communities and Communities of Color from COVID-19," *New York University Environmental Law Journal* 30 (2022): 123–155.

Srivathsan, Amrita, et al. "Convergence of Dominance and Neglect in Flying Insect Diversity," *Nature Ecology & Evolution* 7, no. 7 (2023): 1012–1021.

Stadler, Leigh, Songo Anele Benya, Gina Ziervogel, and Petra Holden. "Integrating Intersectionality into Climate Risk Assessments: Review of Gendered Vulnerability in South Africa," *Authorea Preprints* (2024): 1–24.

Stark, Kekek Jason, and Monte Mills. "Re-Indigenizing Yellowstone," *Wyoming Law Review* 22 (2022): 397–487.

Stavenhagen, Rodolfo. "Indigenous Peoples' Rights to Education," *European Journal of Education* 50, no. 3 (2015): 254–57.

Stein, Michael Ashley, et al. eds. *Mental Health, Legal Capacity, and Human Rights* (Cambridge: Cambridge University Press, 2021).

Stein, Penelope J.S., et al. "The Role of the Scientific Community in Strengthening Disability-Inclusive Climate Resilience," *Nature Climate Change* 13, no. 2 (2023): 108–109.

Sterl, Sebastian, et al. "Leapfrogging the Darkness: Contextualising the Meaning of Energy Transitions for African Countries," *Progress in Energy* 6, no. 2 (2024): 1–11.

Sternfels, Bob. et al. "Our Future Lives and Livelihoods: Sustainable and Inclusive and Growing," McKinsey & Company (2022). https://www.mckinsey.com/featured-insights/sustainable-inclusive-growth/our-future-lives-and-livelihoods-sustainable-and-inclusive-and-growing

St-Laurent, Martin-Hugues, et al. "Lowering the Rate of Timber Harvesting to Mitigate Impacts of Climate Change on Boreal Caribou Habitat Quality in Eastern Canada," *Science of the Total Environment* 838 (2022): 1–14.

Stoett, Peter, et al. "The Transnationalization of Hazardous Waste," in *Spheres of Transnational Ecoviolence: Environmental Crime, Human Security, and Justice*, eds. Peter Stoett and Delon Alain Omrow (Cham, Switzerland: Palgrave Macmillan, 2021), 73–101.

Stolberg, Sheryl Gay, et al. "Trump Transition Updates: Trump Wants Robert F. Kennedy Jr. To Be Health Secretary," *The New York Times*, 14 November (2024). https://www.nytimes.com/live/2024/11/14/us/trump-gaetz-news?smid=nytcore-ios-share&referringSource=articleShare

Strakosch, Elizabeth. "Violence as Care: Indigenous Policy and Settler Colonialism," in *Handbook of Indigenous Public Policy*, eds. Sheryl Lightfoot and Sarah Maddison (Cheltenham, UK: Edward Elgar Publishing, 2024), 18–34.

Sturman, Kathryn et al. *Mission Critical: Strengthening Governance of Mineral Value Chains for the Energy Transition* (Oslo: Extractive Industries Transparency Initiative [EITI], 2022). https://espace.library.uq.edu.au/view/UQ:9792a9c

Suedi, Yusra, and Marie Fall. "Climate Change Litigation before the African Human Rights System: Prospects and Pitfalls," *Journal of Human Rights Practice* 16, no. 1 (2024): 146–159.

Suiseeya, Kimberly R. Marion. "Negotiating the Nagoya Protocol: Indigenous Demands for Justice," *Global Environmental Politics* 14, no. 3 (2014): 102–124.

Sulla, Victor, and Precious Zikhali. "Overcoming Poverty and Inequality in South Africa: An Assessment of Drivers, Constraints and Opportunities," World Bank, 4 January (2022). https://documents.worldbank.org/en/publication/documents-reports/documentdetail/530481521735906534/overcoming-poverty-and-inequality-in-south-africa-an-assessment-of-drivers-constraints-and-opportunities

Sultana, Farhana. "The Unbearable Heaviness of Climate Coloniality," *Political Geography* 99 (2022a): 1–16.

Sultana, Farhana. "Critical Climate Justice," *The Geographical Journal* 188, no. 1 (2022b): 118–124.

Sunilra, J.N.V., and R. Binu. "Social Mind in Disability – A Decadal Changes in the Concepts, Theories and Models of Disability," *Theories and Models of Disability* (2021): 1–12.

Sunstein, Cass R. "Chevron as Law," *Georgetown Law Journal* 107 (2018): 1613–1683.

Suradi, Suradi, et al. "Indigenous Peoples, Poverty and the Role of Social Workers," *Asian Social Work Journal* 4, no. 1 (2019): 11–19.

Sustainable Development Solutions Network. "Chapter 1: Getting to Know the Sustainable Development Goals." (2015). https://sustainabledevelopment.un.org/content/documents/2217Getting%20started.pdf

Swarnakar, Pradip, and Rajshri Shukla. "International Climate Financing and Just Energy Transition," in *The Role of Coal in a Sustainable Energy Mix for India*, eds. M. Mohanty and Runa Sarkar (New York: Routledge, 2024), 343.

Sweet, Justine. "A Climate Change Bill That Needs Changing?" *LexisNexis Polity*, 6 August (2018). https://www.polity.org.za/article/a-climate-change-bill-that-needs-changing-2018-08-06

Szablowski, David, and Bonnie Campbell. "Struggles over Extractive Governance: Power, Discourse, Violence, and Legality," *The Extractive Industries and Society* 6, no. 3 (2019): 635–641.

Szapudi, István. "Solar Radiation Management with a Tethered Sun shield," *Proceedings of the National Academy of Sciences* 120, no. 32 (2023): 1–3.

Sze, Jocelyne S., et al. "Reduced Deforestation and Degradation in Indigenous Lands Pan-Tropically," *Nature Sustainability* 5, no. 2 (2022): 123–130.

T

Tachovsky, Michael. "Environmental Dead Zones: The Socioeconomic Impacts of Contaminated Neighborhoods" (PhD dissertation, Fielding Graduate University, 2021).

Tadaki, Marc, et al. "Four Propositions about How Valuation Intervenes in Local Environmental Politics," *People and Nature* 3, no. 1 (2021): 190–203.

Tahir, Isabel. "Addressing the United States Climate Crisis and Climate Displacement: A Transition from the 'Otherization' of Climate Change to a Focus on Domestic Solutions," *California. Law Review* 110 (2022): 601–637.

Tam, Kim-Pong, and Hoi-Wing Chan. "Conspiracy Theories and Climate Change: A Systematic Review," *Journal of Environmental Psychology* (2023): 102129.

Tavakoli, Sadi, et al. "Exploring the Technical Feasibility of Carbon Capture Onboard Ships," *Journal of Cleaner Production* 452 (2024): 1–15.

Taylor, Jessica E., et al. "Climate Change Adaptation Trends Among Indigenous Peoples: A Systematic Review of the Empirical Research Focus over the Past 2 decades," *Mitigation and Adaptation Strategies for Global Change* 28, no. 6 (2023): 1–29.

Temper, Leah, and Joan Martinez-Alier. "The God of the Mountain and Godavarman: Net Present Value, Indigenous Territorial Rights and Sacredness in a Bauxite Mining Conflict in India," *Ecological Economics* 96 (2013): 79–87.

Terrell, Kimberly Ann et al. "Toxic Air Pollution and Concentrated Social Deprivation Are Associated with Low Birthweight and Preterm Birth in Louisiana," *Environmental Research: Health* 2, no. 2 (2024): 021002.

Tessum, Christopher W., et al. "PM2. 5 Polluters Disproportionately and Systemically Affect People of Color in the United States," *Science Advances* 7, no. 18 (2021): eabf4491.

Thakrar, Sumil K., Srinidhi Balasubramanian, Peter J. Adams, Inês ML Azevedo, Nicholas Z. Muller, Spyros N. Pandis, Stephen Polasky et al. "Reducing Mortality From Air Pollution in the United States By Targeting Specific Emission Sources," *Environmental Science & Technology Letters* 7, no. 9 (2020): 639–645.

The Decolonizing Climate Policy Project, Decolonizing Climate Policy in Canada: Report from Phase One, March (2021). https://static1.squarespace.com/static/5e8e4b5ae8628564ab4bc44c/t/6061cb5926611066ba64a953/1617021791071/pcf_critique_FINAL.pdf

The Elders. "Failure to Respect Rule of Law Risks Collapse of Global Stability," (2024a). https://theelders.org/news/failure-respect-rule-law-risks-collapse-global-stability

The Elders. "Climate and Nature Crisis Policy Position Paper," (2024b). https://theelders.org/sites/default/files/2024-09/The%20Elders_Climate%20%26%20Nature%20Policy%20Position%20Paper_2024.pdf

The White House. "Executive Order on Tackling the Climate Crisis at Home and Abroad," 27 January (2021a). https://www.whitehouse.gov/briefing-room/presidential-actions/2021/01/27/executive-order-on-tackling-the-climate-crisis-at-home-and-abroad/

The White House. "Executive Order 13990: Protecting Public Health and the Environmental and Restoring Science to Tackle the Climate Crisis," 20 January (2021b). https://www.whitehouse.gov/briefing-room/presidential-actions/2021/01/20/executive-order-protecting-public-health-and-environment-and-restoring-science-to-tackle-climate-crisis

The White House. "Executive Order on Guaranteeing an Education Environment Free from Discrimination on the Basis of Sex, Including Sexual Orientation or Gender Identity," 8 March (2021c). www.whitehouse.gov/briefing-room/presidential-actions/2021/03/08/executive-order-on-guaranteeing-an-educational-environment-free-from-discrimination-on-the-basis-of-sex-including-sexual-orientation-or-gender-identity

The White House. "White House Environmental Justice Advisory Council," 30 March (2021d). https://www.whitehouse.gov/environmentaljustice/white-house-environmental-justice-advisory-council/

The White House. "Memorandum on Restoring Trust in Government through Scientific Integrity and Evidence-Based Policymaking | The White House," 27 January (2021e). https://www.whitehouse.gov/briefing-room/presidential-actions/2021/01/27/memorandum-on-restoring-trust-in-government-through-scientific-integrity-and-evidence-based-policymaking/

The White House. "President Biden's Executive Order on Revitalizing our Nation's Commitment to Environmental Justice for All," 21 April (2023a). https://www.whitehouse.gov/briefing-room/presidential-actions/2023/04/21/executive-order-on-revitalizing-our-nations-commitment-to-environmental-justice-for-all/

The White House. "FACT SHEET: Biden-Harris Administration Launches American Climate Corps to Train Young People in Clean Energy, Conservation, and Climate Resilience Skills, Create Good-Paying Jobs and Tackle the Climate Crisis," 20 September (2023b). https://www.whitehouse.gov/briefing-room/statements-releases/2023/09/20/fact-sheet-biden-harris-administration-launches-american-climate

-corps-to-train-young-people-in-clean-energy-conservation-and-climate-resilience-skills-create-good-paying-jobs-and-tackle-the-clima/
The White House. "Remarks by President Biden at the White House Conservation in Action Summit," 21 March (2023c). https://www.whitehouse.gov/briefing-room/speeches-remarks/2023/03/21/remarks-by-president-biden-at-the-white-house-conservation-in-action-summit/
The White House. "Remarks by President Biden on the Biden-Harris Administration's Record of Delivering for Tribal Communities, Including Keeping His Promise to Make This Historic Visit to Indian Country | Laveen Village, AZ," The White House, 26 October (2024a). https://www.whitehouse.gov/briefing-room/speeches-remarks/2024/10/25/remarks-by-president-biden-on-the-biden-harris-administrations-record-of-delivering-for-tribal-communities-including-keeping-his-promise-to-make-this-historic-visit-to-indian-country-lavee/
The White House. "Fact Sheet: Biden-Harris Administration Kicks Off Infrastructure Week by Highlighting Historic Results Spurred by President Biden's Investing in America Agenda," 13 May (2024b). https://www.whitehouse.gov/briefing-room/statements-releases/2024/05/13/fact-sheet-biden-harris-administration-kicks-off-infrastructure-week-by-highlighting-historic-results-spurred-by-president-bidens-investing-in-america-agenda/
The White House. "Remarks by President Biden Honoring Our Nation's Labor History and Establishing the Frances Perkins National Monument | the White House." 16 December (2024c). https://www.whitehouse.gov/briefing-room/speeches-remarks/2024/12/16/remarks-by-president-biden-honoring-our-nations-labor-history-and-the-biden-harris-administrations-work-to-strengthen-americas-workforce/?utm_source=substack&utm_medium=email
The White House. "FACT SHEET: President Biden Recognizes Fuller Story of Pacific Islands Heritage | the White House," The White House, 2 January (2025). https://www.whitehouse.gov/briefing-room/statements-releases/2025/01/02/fact-sheet-president-biden-recognizes-fuller-story-of-pacific-islands-heritage/
The World Bank. "Sharp, Long-Lasting Slowdown to Hit Developing Countries Hard," The World Bank (2023). https://www.worldbank.org/en/news/press-release/2023/01/10/global-economic-prospects
Thind, Maninder P.S., et al. "Fine Particulate Air Pollution from Electricity Generation in the US: Health Impacts by Race, Income, and Geography," *Environmental Science & Technology* 53, no. 23 (2019): 14010–14019.
Throsby, David, and Ekaterina Petetskaya. "Sustainability Concepts in Indigenous and Non-Indigenous Cultures," *International Journal of Cultural Property* 23, no. 2 (2016): 119–140.
ThummBorst, Zachary J. "Water Justice: The Ninth Circuit Examines the Fair Housing Act in the Context of Water Services in Southwest Fair Housing Council, Inc., v. Maricopa Domestic Water Improvement District," *Villanova Environmental Law Journal* 34 (2023): 273–299.
Tigre, Maria Antonia, and Joana Setzer. "Human Rights and Climate Change for Climate Litigation in Brazil and Beyond: An Analysis of the Climate Fund Decision," *Georgetown Journal of International Law* 54, no. 4 (2023): 1–53.
Tiseo, Ian. "Carbon Dioxide Emissions from Fossil Fuel and Industrial Purposes in South Africa from 1970 to 2022," *Statista*, 19 September (2023). https://www.statista.com/statistics/486073/co2-emissions-south-africa-fossil-fuel-and-industrial-purposes/

Tladi, Bonolo, Njabulo Kambule, and Lee-Ann Modley. "Assessing the Social and Environmental Impacts of the Just Energy Reansition in Komati, Mpumalanga Province, South Africa," *Energy Research & Social Science* 111 (2024): 1–10.

Toki, Valmaine. *Indigenous Rights, Climate Change and Governance* (Cheltenham, UK: Edward Elgar Publishing, 2024).

Toledo, Victor M. "Indigenous Peoples and Biodiversity," *Encyclopaedia of Biodiversity* 3 (2001): 451–463.

Tollefson, Jeff. "How a Huge US Energy Bill Could Revive Biden's Climate Agenda," *Nature* 608, no. 7923 (2022): 458–459.

Tollefson, Jeff. "The Plan to 'Trump-Proof' U.S. Science against Meddling," *Nature* 613, no. 7945 (2023): 621–622.

Tormos-Aponte, Fernando. "The Influence of Indigenous Peoples in Global Climate Governance," *Current Opinion in Environmental Sustainability* 52 (2021): 125–131.

Totry-Jubran, Manal. "Towards Critical Environmental Justice Approach in Multicultural Divided States: Lessons from the Israeli Regulative System," *Fordham Environmental Law Review* 35 (2023): 60–70.

Tretheway, Mike, et al. "Northern and Arctic Air Connectivity in Canada," International Transport Forum Discussion Paper (2021), 14. https://www.econstor.eu/bitstream/10419/245857/1/1747443212.pdf

Trott, Carlie D., Stephanie Lam, Jessica Roncker, Emmanuel-Sathya Gray, R. Hayden Courtney, and Trevor L. Even. "Justice in Climate Change Education: A Systematic Review," *Environmental Education Research* 29, no. 11 (2023): 1535–1572.

Tschakert, Petra, et al. "Multispecies justice: Climate-Just Futures with, for and beyond Humans," *Wiley Interdisciplinary Reviews: Climate Change* 12, no. 2 (2021): 1–27.

Tsuji, Stephen R.J. "Canada's Impact Assessment Act, 2019: Indigenous Peoples, Cultural Sustainability, and Environmental Justice," *Sustainability* 14, no. 6 (2022): 1–52.

Tucker, Claire, and Marga Jordaan. "South Africa: President Assents to the Climate Change Bill, but No Date Set for Commencement" (Johannesburg: Bowmans, 2024), 1–7. https://bowmanslaw.com/insights/south-africa-president-assents-to-the-climate-change-bill-but-no-date-set-for-commencement/

Tyagi, Aman, and Kathleen M. Carley. "Climate Change Conspiracy Theories on Social Media." *arXiv preprint arXiv:2107.03318* (2021): 1–10.

Tyler, Emily, and Kathryn Hochstetler. "Institutionalising Decarbonisation in South Africa: Navigating Climate Mitigation and Socio-Economic Transformation," *Environmental Politics* 30, no. supl (2021): 5184–5205.

Tyler, Seth P., and Thomas Webler. "The Challenge of Community Acceptance of Small Nuclear Reactors," *Energy Research & Social Science* 118 (2024): 103831.

U

Uddin, M.N., Shahnawaz Anwer, Hsi-Hsien Wei, Hung-Lin Chi, Meng Ni, and N. Tamanna. "Energy Efficient Behavioural Trends in Residential Sectors for Low-Income Cultural Background: A Case-Study of Slums in Chittagong, Bangladesh," in *Thirty-Seventh Annual Conference* (2021): 774.

Udo, Fidelis, and Maheshvari Naidu. "Assessing Local Government's Response to Black Women's Vulnerability and Adaptation to the Impacts of Floods in the Context of Intersectionality: The Case of eThekwini Metropolitan Municipality, South Africa," *Environmental Policy and Governance* 34, no. 1 (2024): 31–41.

UK Climate Change Committee. "Independent Assessment of UK Climate Risk Advice to Government for the UK's third Climate Change Risk Assessment (CCRA3)," UK Climate Change Committee (2021), 11. https://www.theccc.org.uk/publication/independent-assessment-of-uk-climate-risk/

UN Innocenti. *Child Poverty in the Midst of Wealth: Innocenti Report Card 18.* Florença: UNICEF Office of Research-Innocenti (2023).

UN News. "Global Temperatures Set to Break Records During Next 5 Years: WMO," United Nations (2023). https://news.un.org/en/story/2023/05/1136732

UN News. "New UN Initiative Aims to Counter Climate Disinformation," 19 November (2024). https://news.un.org/en/story/2024/11/1157191

UNICEF. "Child Food Poverty: Nutrition Deprivation in Early Childhood. Child Nutrition Report 2024," United Nations (2004). https://www.unicef.org/media/157661/file/Child-food-poverty-2024.pdf

United Nations. "Sociocultural Biases and Discrimination," *Report on the World Social Situation*, 21 November (2003). https://www.un.org/esa/socdev/rwss/docs/2003/fullreport.pdf

United Nations. "Disability and Development Report: Realizing the Sustainable Development Goals by, for and with Persons with Disabilities." (2019). https://digitallibrary.un.org/record/3829277/files/aDisability2019.pdf?ln=en

United Nations Convention to Combat Desertification (UNCCD). *Drought in Numbers*, UNCCD (2022). https://www.unccd.int/resources/publications/drought-numbers

United Nations Department of Economic and Social Affairs. "UNPFII Mandated Areas – Economic and Social Development," United Nations (nd). https://www.un.org/development/desa/indigenouspeoples/mandated-areas1/economic-and-social-development.html

United Nations Development Programme (UNDP). "Microplastics on Human Health: How Much Do They Harm Us?" United Nations Development Programme (2023). https://www.undp.org/kosovo/blog/microplastics-human-health-how-much-do-they-harm-us

United Nations Development Programme (UNDP). "80 Percent of People Globally Want Stronger Climate Action by Governments according to UN Development Programme Survey," 20 June (2024a). https://www.undp.org/press-releases/80-percent-people-globally-want-stronger-climate-action-governments-according-un-development-programme-survey

United Nations Development Programme (UNDP). "The Peoples' Climate Vote 2024," United Nations Development Programme and the University of Oxford (2024b). https://peoplesclimate.vote/.

United Nations Environmental Programme (UNEP). "Indigenous Peoples and the Nature They Protect," United Nations Environmental Project (2020). https://www.unep.org/news-and-stories/story/indigenous-peoples-and-nature-they-protect

United Nations Environmental Programme (UNEP). "Five Threats to the Water that Sustains Our Farms," United Nations (2022). https://www.unep.org/news-and-stories/story/five-threats-water-sustains-our-farms

United Nations Environmental Programme (UNEP). "Novel Meat and Dairy Alternatives Could Help Curb Climate-Harming Emissions – UN," United Nations (2023). https://www.unep.org/news-and-stories/press-release/novel-meat-and-dairy-alternatives-could-help-curb-climate-harming

United Nations Environment Programme (UNEP). "Rise in Nitrous Oxide Emissions Endangers Pathway to 1.5°C, the Ozone Layer, and Human Health," (2024a).

https://www.unep.org/news-and-stories/press-release/rise-nitrous-oxide-emissions-endangers-pathway-15degc-ozone-layer
United Nations Environment Programme (UNEP). "Food Waste Index Report 2024. Think Eat Save: Tracking Progress to Halve Global Food Waste," United Nations (2024b). https://www.unep.org/resources/publication/food-waste-index-report-2024
United Nations Environment Programme (UNEP). *Emissions Gap Report 2024*, United Nations (2024c). https://www.unep.org/resources/emissions-gap-report-2024
United Nations Food and Agriculture Organization. "Achieving SDG 2 Without Breaching the 1.5°C Threshold: A Global Roadmap, Part 1 – How Agrifood Systems Transformation Through Accelerated Climate Actions Will Help Achieving Food Security and Nutrition Today and Tomorrow," United Nations (2023a). https://openknowledge.fao.org/items/b86cd543-e8ca-4e95-b5d2-06ea6af23842
United Nations Food and Agricultural Organization et al. "The State of Food Security and Nutrition in the World 2023. Urbanization, Agrifood Systems, Transformation, and Healthy Diets Across the Rural-Urban Continuum," United Nations (2023b). https://www.fao.org/documents/card/en/c/cc3017en
United Nations Framework Convention on Climate Change (UNFCCC). The Paris Agreement. 2015. UNFCCC, Preamble (2015). https://unfccc.int/sites/default/files/english_paris_agreement.pdf
United Nations Framework Convention on Climate Change (UNFCCC). "Updated Nationally Determined Contribution (NDC) of South Africa," June (2022). https://unfccc.int/sites/default/files/NDC/2022-06/South%20Africa.pdf
United Nations Framework Convention on Climate Change (UNFCCC) (Paris Agreement). "Technical dialogue of the first global stocktake Synthesis report by the co-facilitators on the technical dialogue" for the Fifty-ninth session in United Arab Emirates, 30 November to 6 December 2023, FCCC/SB/2023/9 September (2023). https://unfccc.int/sites/default/files/resource/sb2023_09_adv.pdf
United Nations General Assembly, Report of the Special Rapporteur on the implications for human rights of the environmentally sound management and disposal of hazardous substances and wastes. "The toxic impacts of some proposed climate change solutions" A/HRC/54/25 13 July (2023). https://documents-dds-ny.un.org/doc/UNDOC/GEN/G23/132/58/PDF/G2313258.pdf?OpenElement
United Nations Human Rights Council Special Procedures. "International Independent Expert Mechanism to Advance Racial Justice and Equality in the Context of Law Enforcement, Visit to the United States of America," A/HRC/54/CRP.7, 26 September (2023), paras 78 and 80. https://www.ohchr.org/sites/default/files/documents/hrbodies/hrcouncil/sessions-regular/session54/A_HRC_54_CRP.7.docx
United Nations Secretary General Antonio Guterres to the UN General Assembly, 13 February (2023). https://news.un.org/en/story/2023/02/1133452
United Nations Special Rapporteur Water and Sanitation, End of Mission Statement by the Special Rapporteur on the human rights to safe drinking water and sanitation, Mr Pedro Arrojo-Agudo at the conclusion of the country visit to Canada Ottawa, 19 April (2024). https://www.ohchr.org/sites/default/files/documents/issues/water/statements/2024-04-19-eom-sr-water-canada-en.pdf
United States Bureau of Land Management. "National Petroleum Reserve in Alaska," (2022). https://www.blm.gov/programs/energy-and-minerals/oil-and-gas/about/alaska/NPR-A

United States Bureau of Land Management. "National Petroleum Reserve in Alaska |Bureau of Land Management," (2024). https://www.blm.gov/programs/energy-and-minerals/oil-and-gas/about/alaska/NPR-A
United States Department of Agriculture (USDA). "Food Loss and Waste," US Department of Agriculture (nd). https://www.usda.gov/foodlossandwaste
United States Department of Housing and Urban Development (HUD) / HUD.gov. "HUD Releases January 2024 Point-In-Time Count Report," 27 December (2024). https://www.hud.gov/press/press_releases_media_advisories/HUD_No_24_327
United States Department of the Interior. "Federal Indian Boarding School Initiative | U.S. Department of the Interior," 11 May (2022). https://www.doi.gov/priorities/strengthening-indian-country/federal-indian-boarding-school-initiative
United States Department of the Interior. "Interior Department Substantially Reduces Scope of Willow Project," US Department of the Interior, 13 March (2023a). https://www.doi.gov/pressreleases/interior-department-substantially-reduces-scope-willow-project
United States Department of the Interior. "Biden-Harris Administration Announces Sweeping Protections for up to 16 Million Acres of Land and Water in Alaska," US Department of the Interior, 12 March (2023b). https://www.doi.gov/pressreleases/biden-harris-administration-announces-sweeping-protections-16-million-acres-land-and
United States Department of the Interior. "Biden-Harris Administration Finalizes Strategy to Guide Balanced Management and Conservation." Comunicado de imprensa (2024). https://www.doi.gov/pressreleases/biden-harris-administration-finalizes-strategy-guide-balanced-management-conservation
United States Environment Programme Federal Actions to Address Environmental Justice in Minority Populations and Low-Income Populations, Executive Order No. 12898, 59 Fed. Reg. 7,629, 16 February (1994).
United States Environmental Protection Agency. "Summary of Executive Order 12898 – Federal Actions to Address Environmental Justice in Minority Populations and Low-Income Populations," United States Environmental Protection Agency (1994). https://www.epa.gov/laws-regulations/summary-executive-order-12898-federal-actions-address-environmental-justice#:~:text=59%20FR%207629%3B%20February%2016%2C%201994&text=Its%20purpose%20is%20to%20focus,E.O
United States Environmental Protection Agency. "EPA Settles Alleged Asbestos Violations with Companies in Georgia | US EPA," 30 September 2024 (2024a). https://www.epa.gov/newsreleases/epa-settles-alleged-asbestos-violations-companies-georgia
United States Environmental Protection Agency. "Biden-Harris Finalizes Suite of Standards to Reduce Pollution from Fossil Fuel-Fired Power Plants," United States Environmental Protection Agency, 25 April (2024b). https://www.epa.gov/newsreleases/biden-harris-administration-announces-national-strategy-reduce-food-loss-and-waste-and
United States Environmental Protection Agency. "Understanding the Impact of EPA's Proposed Rules for Chemical Plants: EPA's Community Risk Assessment and Risk-Based Demographic Assessment," USEPA, 6 April (2024c). https://www.epa.gov/system/files/documents/2023-04/HON%20P%26R.%20Demographic%20Analysis.Fact%20Sheet.%204.6.23.pdf
United States Environmental Protection Agency. "Biden-Harris Administration Announces National Strategy to Reduce Food Loss and Waste and Recycle

Organics," United States Environmental Protection Agency, 12 June (2024d). https://www.epa.gov/newsreleases/biden-harris-administration-announces-national-strategy-reduce-food-loss-and-waste-and

United States Environmental Protection Agency. "Biden-Harris Administration Issues Final Rule Requiring Replacement of Lead Pipes within 10 Years, Announces Funding to Provide Clean Water to Schools and Homes I US EPA," 7 October (2024e). https://www.epa.gov/newsreleases/biden-harris-administration-issues-final-rule-requiring-replacement-lead-pipes-within

United States Environmental Protection Agency. "Fact Sheet, 'What the EPA is doing to address ethylene oxide (EtO) and to learn more about the Chemical," USEPA 9 April (2024f). https://www.epa.gov/hazardous-air-pollutants-ethylene-oxide/what-epa-doing-address-ethylene-oxide-eto-and-learn-more#:~:text=Final%20EPA%20Actions%20to%20Reduce%20EtO%2DRelated%20Risk,-The%20air%20toxics&text=On%20March%2014%2C%202024%2C%20EPA%20announced%20finalized%20requirements%20for%20nearly,reduce%20risk%20in%20nearby%20communities

United States Environmental Protection Agency. "Biden-Harris Administration Finalizes Stronger Clean Air Standards for Chemical Plants, Lowering Cancer Risk and Advancing Environmental Justice," 9 April (2024g). https://www.epa.gov/newsreleases/biden-harris-administration-finalizes-stronger-clean-air-standards-chemical-plants

University of Cape Town and the African Climate & Development Initiative. "Comments on the Climate Change Bill," May 2022, accessed 2 August (2024). https://acdi.uct.ac.za/sites/default/files/content_migration/acdi_uct_ac_za/1205/files/Climate%2520Change%2520Bill%2520Comments_Final.pdf

Urbancic, Nusa. "How Granting Indigenous Peoples' Land Titles Contributes to Forest Conservation in Latin America," *Birkbeck Law Review* 7 (2020): 26–57.

US Global Change Research Program, The Fifth National Climate Assessment (NCA5) (2023).

USA Office of Science and Technology Policy Office of Management and Budget, Department of Commerce. "National Strategy to Develop Statistics for Environmental-Economic Decisions: A U.S. System of Natural Capital Accounting and Associated Environmental – Economic Statistics," January (2023). https://www.aeaweb.org/forum/3447/national-strategy-statistics-environmental-economic-decisions

V

Valeggia, Claudia R., and J. Josh Snodgrass. "Health of Indigenous Peoples," *Annual Review of Anthropology* 44 (2015): 117–35.

VanderZwaag, David L., et al. "Canada and Ocean Climate Adaptation: Tracking Law and Policy Responses, Charting Future Directions," *Frontiers in Marine Science* 10 (2023): 1–16.

Vecellio, Daniel J., et al. "Evaluating the 35 C Wet-Bulb Temperature Adaptability Threshold for Young, Healthy Subjects (PSU HEAT Project)," *Journal of Applied Physiology* 132, no. 2 (2022): 340–345.

Venkataraman, Maya, Stefan Grzybowski, Darlene Sanderson, Jordie Fischer, and Arlin Cherian. "Environmental Racism in Canada," *Canadian Family Physician* 68, no. 8 (2022): 567–573.

Venner, Kayin, Melissa García-Lamarca, and Marta Olazabal. "The Multi-Scalar Inequities of Climate Adaptation Finance: A Critical Review," *Current Climate Change Reports* 10, no. 3 (2024): 46–59.

Verkuijl, C., Dutkiewicz, J., et al. "FAO's 1.5 °C Roadmap for Food Systems Falls Short," *National Library of Medicine*, no. 4 (2024): 264–266.

Vermeulen, Sonja J., Dhanush Dinesh, S. Mark Howden, Laura Cramer, and Philip K. Thornton. "Transformation in Practice: A Review of Empirical Cases of Transformational Adaptation in Agriculture Under Climate Change," *Frontiers in Sustainable Food Systems* 2 (2018): 65–81.

Verschuuren, Jonathan, ed. *Research Handbook on Climate Change Adaptation Law* (Cheltenham, UK: Edward Elgar Publishing, 2022).

Vieira, Marina A.R. de Mattos, and Lieselotte Viaene. "Indigenous Peoples' Rights at the United Nations Human Rights Council: Colliding (Mis) Understandings?" *Journal of Human Rights Practice* (2024): 1–21.

Vijayan, Dhanasingh Sivalinga, et al. "Advancements in Solar Panel Technology in Civil Engineering for Revolutionizing Renewable Energy Solutions – A Review," *Energies* 16, no. 18 (2023): 1–33.

Vilá, Olivia, Gavin Smith, Bethany Cutts, Samata Gyawali, and Samiksha Bhattarai. "Equity in FEMA Hazard Mitigation Assistance Programs: The Role of State Hazard Mitigation Officers," *Environmental Science & Policy* 136 (2022): 632–641.

Vinco, Elena, et al. "Climate Policy and Canadian Crop Production: A Qualitative Study of Farmers' Attitudes and Perceptions Towards Nitrous Oxide Reductions," *Journal of Cleaner Production* 418 (2023): 1–11.

Viseh, Hiva, and David N. Bristow. "How Climate Change Could Affect Different Cities in Canada and What that Means for the Risks to the Built-Environment Functions," *Urban Climate* 51 (2023): 1–23.

Vogel, Brennan, and Ryan C.L. Bullock. "Institutions, Indigenous Peoples, and Climate Change Adaptation in the Canadian Arctic," *GeoJournal* 86, no. 6 (2021): 2555–2572.

Vogel, Brennan, et al. "Indigenous-Led Nature-based Solutions for the Climate Crisis: Insights from Canada," *Sustainability* 14, no. 11 (2022a): 1–19.

Vogel, Coleen et al. "'Empowering Youth as Change Agents for Climate Change in South Africa': Challenges, Caveats and Course Corrections," *Journal of Youth Studies* 25, no. 6 (2022b): 812–832.

Von Hedemann, Nicolena, et al. "Climate Adaptation Research Priorities and Funding: A Review of US Federal Departments' Climate Action Plans," *Climate Policy* 23, no. 10 (2023): 1288–1301.

Vorobjovas-Pinta, Oscar, and Anne Hardy. "Resisting Marginalisation and Reconstituting Space Through LGBTQI+ Events," *Journal of Sustainable Tourism* 29, no. 2–3 (2021): 448–466.

W

Waldron, Ingrid. "Pipelines and Protests: Legacies of Struggle and Resistance in the Fight against Environmental Racism in Canada," *Brown Journal of World Affairs* 28 (2021): 7–19.

Walker, Gordon. "Environmental Justice, Impact Assessment and the Politics of Knowledge: The Implications of Assessing the Social Distribution of Environmental Outcomes," *Environmental Impact Assessment Review* 30, no. 5 (2010): 312–318.

Walker, Gordon. *Environmental Justice: Concepts, Evidence and Politics* (London: Routledge, 2012).

Walker, S.E., E.A. Smith, N. Bennett, E. Bannister, A. Narayana, T. Nuckols, K. Pineda Velez, J. Wrigley, and K.M. Bailey. "Defining and Conceptualizing Equity and Justice in Climate Adaptation," *Global Environmental Change* 87 (2024): 1–12.

Wallace-Wells, David. "John Kerry: 'I Feel Deeply Frustrated'," *The New York Times*, 6 March (2024). https://www.nytimes.com/2024/03/06/opinion/john-kerry-climate-change-china.html/

Wallenhorst, Nathanaël. "Introduction to the Anthropocene," in *A Critical Theory for the Anthropocene*, ed. Nathanaël Wallenhorst (Cham, Switzerland: Springer International Publishing, 2023), 63–78.

Waller, Lisa, and Kerry McCallum. "Settler Colonial Representations of Indigenous Disadvantage," in *The Routledge Companion to Media and Poverty*, ed. Sandra L. Borden (New York: Routledge, 2021): 212–222.

Wang, Feiyue, and Robie W. Macdonald. "Socioeconomic, Political, and Legal Ramifications of Environmental and Biochemical Toxicology: The Complicated Story of Mercury," in *Environmental and Biochemical Toxicology: Concepts, Case Studies and Challenges*, eds. Jürgen Gailer and Raymond J. Turner (Berlin: Walter De Gruyter, 2022), 287–312.

Wang, Yuzhou, et al. "Air Quality Policy Should Quantify Effects on Disparities," *Science* 381, no. 6655 (2023): 272–274.

Wang, Guohao et al. "A Comprehensive Review of Building Lifecycle Carbon Emissions and Reduction Approaches," *City and Built Environment* 2, no. 1 (2024): 1–28.

Warren, K.B. and, J. Jackson. "Introduction: Studying Indigenous Activism in Latin America," in *Indigenous Movements, Self-Representation, and the State in Latin America*, eds. Kay B. Warren and Jean E. Jackson (Austin: University of Texas Press, 2001), 1–46.

Warren, Martin S., et al. "The Decline of Butterflies in Europe: Problems, Significance, and Possible Solutions," *Proceedings of the National Academy of Sciences* 118, no. 2 (2021): e2002551117.

Warren, Rachel, et al. "Risks Associated with Global Warming of 1.5 to 4° C Above Pre-Industrial Levels in Human and Natural Systems in Six Countries," *Climatic Change* 177, no. 3 (2024): 1–24.

Waters, Linda et al. "Under-Reported and Under-Served: Disparities in US Disaster Federal Aid-to-Damage Ratios After Hurricanes," *International Journal of Disaster Risk Reduction* 106, no. 41 (2024): 105–240.

Watts, Kim, Thierry Vansweevelt, and Britt Weyts. "Climate Change, Compensation and Unpredictable or Uninsurable Loss: A Possible Path Forward Using Compensation Funds," *Journal of European Tort Law* 14, no. 3 (2023): 316–351.

Webber, Jeremy. "Understanding Populism," *Social & Legal Studies* 32, no. 6 (2023): 849–876.

Webley, Lisa. "The Why and How to of Conducting a Socio-Legal Empirical Research Project," in *Routledge Handbook of Socio-Legal Theory and Methods*, eds. Naomi Creutzfeldt, Marc Mason, and Kirsten McConnachie (New York: Routledge, 2019), 58–69.

Webster, Edward, Imraan Valodia, and David Francis. "Towards a Southern Approach to Inequality: Inequality Studies in South Africa and the Global South," in *Inequality Studies from the Global South*, eds. David Francis, Imrann Valodia, and Edward Webster (New York: Routledge, 2020), 3–21.

Wen, Jiahong, et al. "Disaster Risk Reduction, Climate Change Adaptation and their Linkages with Sustainable Development over the Past 30 Years: A Review," *International Journal of Disaster Risk Science* 14, no. 1 (2023): 1–13.

West, BrieAnn. "Mediating Our Future: The Role of the Land Buy-Back Program in Rebuilding Confidence and Strengthening Trust Between Tribal Nations and the United States Government," *Journal of the National Association of Administrative Law* (2016): 482–524.

West, Thales A.P., Sven Wunder, Erin O. Sills, Jan Börner, Sami W. Rifai, Alexandra N. Neidermeier, Gabriel P. Frey, and Andreas Kontoleon. "Action Needed to Make Carbon Offsets from Forest Conservation Work for Climate Change Mitigation," *Science* 381, no. 6660 (2023): 873–877.

Westra, Laura. *Environmental Justice and the Rights of Indigenous Peoples: International and Domestic Legal Perspectives* (London: Routledge, 2012).

Whitehead, Ann, and Dzodzi Tsikata. "Policy Discourses on Women's Land Rights in Sub–Saharan Africa: The Implications of the Re–turn to the Customary," *Journal of Agrarian Change* 3, no. 1–2 (2003): 67–112.

Whyte, Kyle. "Critical Investigations of Resilience: A Brief Introduction to Indigenous Environmental Studies & Sciences," *Daedalus* 147, no. 2 (2018): 136–47.

Wiatros-Motyka, Malgorzata, et al. "Global Electricity Review 2024," Ember (2004). https://ember-climate.org/insights/research/global-electricity-review-2024/?_hsenc=p2ANqtz-_Ymfb5Oxd3AS6Md429J1WszcTqbBh3jJjccHtuLSarhG3r03HoH03Mv2TrHoJh9Tj5HK71Io3cFmqptp9aWifPBr-zPg&_hsmi=306147912&campaign_id=54&emc=edit_clim_20240509&instance_id=122871&nl=climate-forward&regi_id=73699248&segment_id=166157&te=1&user_id=3cc6620316c0a93e52f5fac3913b2a16&utm_campaign=heatmap_am&utm_content=306147912&utm_medium=email&utm_source=hs_email

William, Micheal A., Aly M. Elharidi, Ahmed A. Hanafy, Abdelhamid Attia, and Mohamed Elhelw. "Energy-Efficient Retrofitting Strategies for Healthcare Facilities in Hot-Humid Climate: Parametric and Economical Analysis," *Alexandria Engineering Journal* 59, no. 6 (2020): 4549–4562.

Williams, Jeremy. *Climate Change is Racist: Race, Privilege and the Struggle for Climate Justice* (Icon Books, 2021a).

Williams, Kirsten. "The Impact of Foresight: Reframing Discriminatory Intent to Properly Remedy Environmental Racism," *Houston Law Review* 59 (2021b): 1231–1260.

Williams, Terry Tempest. "For the Benefit and Enjoyment of All the People," in *For the Enjoyment of the People: The Creation of National Identity in American Public Lands*, ed. Mary E. Stuckey (Lawrence, Kansas: University Press of Kansas, 2023).

Williams, Anna, Joseph Dickman, and Regan Smurthwaite. "Advancing Evaluation and Learning on Transformational Change: Lessons from the Climate Investment Funds' Transformational Change Learning Partnership." *American Journal of Evaluation* 42, no. 1 (2021): 90–109.

Williams, Augusta, Larissa McDonogh-Wong, and John D. Spengler. "The Influence of Extreme Heat on Police and Fire Department Services in 23 US Cities," *GeoHealth* 4, no. 11 (2020): 1–13.

Williams, David Samuel, et al. "A Policy Content Analysis for Evaluating Urban Adaptation Justice in İstanbul," *Environmental Science & Policy* 136 (2022): 476–485.

Williams, Portia Adade, Olivier Crespo, and Mumuni Abu. "Adapting to Changing Climate Through Improving Adaptive Capacity at the Local Level – the Case of

Smallholder Horticultural Producers in Ghana," *Climate Risk Management* 23, (2019): 124–135, 129–131.
Wilson, John D., and Zach Zimmerman. "The Era of Flat Power Demand is Over," Grid Strategies, December (2023). https://gridstrategiesllc.com/wp-content/uploads/2023/12/National-Load-Growth-Report-2023.pdf
Winker, Raquel, et al. "Wildfires and Climate Justice: Future Wildfire Events Predicted to Disproportionally Impact Socioeconomically Vulnerable Communities in North Carolina," *Frontiers in Public Health* 12 (2024): 1–13.
Winkler, Inga T., and Catherine Coleman Flowers. "America's Dirty Secret: The Human Right to Sanitation in Alabama's Black Belt," *Columbia Human Rights Law Review* 49 (2017): 181–228.
Winkler, Inga T., and Margaret L. Satterthwaite. "Leaving No One Behind? Persistent Inequalities in the SDGs," *The International Journal of Human Rights* 21(8) (2017): 1073–1097.
Winters, Joseph, "After Decades of Missed Targets, Canada Finally has a Climate Roadmap," *The Grist*, 31 March (2022). https://grist.org/climate/after-decades-missed-targets-canada-climate-roadmap/
Witze, Alexandra. "Geologists reject the Anthropocene as Earth's New Epoch-After 15 years of Debate," *Nature* 627, no. 8003 (2024): 249–250.
Woodruff, Sierra C. "Coordinating Plans for Climate Adaptation," *Journal of Planning Education and Research* 42, no. 2 (2022): 218–230.
Woo-Durand, Catherine, et al. "Increasing Importance of Climate Change and the Threats to At-risk Species in Canada," *Environmental Reviews* 28, no. 4 (2020): 449–456.
World Bank. "In Southern Africa, Leveling the Playing Field at Birth Critical to Reducing Inequality, Intergenerational Poverty," Pretoria: World Bank Group, 14 March (2022). https://www.worldbank.org/en/region/afr/publication/in-southern-africa-leveling-the-playing-field-at-birth-critical-to-reducing-inequality-intergenerational-poverty
World Bank Group. "Indigenous Peoples," The World Bank (2023). https://www.worldbank.org/en/topic/indigenouspeoples#1
World Economic Forum. "What Is Environmental Racism and How Can We Fight It?" *World Economic Forum*, 1 September (2021). https://www.weforum.org/agenda/2020/07/what-is-environmental-racism-pollution-covid-systemic/
World Economic Forum, "In Spain, you can get solar power from your neighbours' roof," (2024), available at: https://www.weforum.org/videos/spain-solar-power/#:~:text=In%20Spain%2C%20you%20can%20get,neighbours'%20roof%20%7C%20World%20Economic%20Forum&text=Local%20buildings%20such%20as%20schools,becomes%20a%20'Solar%20Host'
World Health Organization (WHO). "Quantitative Risk Assessment of the Effects of Climate Change on Selected Causes of Death, 2030s and 2050s," World Health Organization (2014). https://www.who.int/publications/i/item/9789241507691#:~:text=Even%20under%20these%20conditions%2C%20it,000%20due%20to%20childhood%20undernutrition
World Health Organization (WHO). "Health, Environment and Climate Change. Draft WHO Global Strategy on Health, Environment, and Climate Change: The Transformation Needed to Improve Lives and Well-Being Sustainably Through Healthy Environments," A72/15 18 April (2019a). https://www.who.int/docs/default-source/climate-change/who-global-strategy-on-health-environment-and-climate-change-a72-15.pdf?sfvrsn=20e72548_2

World Health Organization (WHO). "Water, Sanitation, Hygiene, and Health: A Primer for Health Professionals," World Health Organization (2019b). https://iris.who.int/bitstream/handle/10665/330100/WHO-CED-PHE-WSH-19.149-eng.pdf?sequence=5

World Health Organization (WHO). "Dietary and Inhalation Exposure to Nano- and Microplastic Particles and Potential Implications for Human Health," World Health Organization (2022). https://iris.who.int/bitstream/handle/10665/362049/9789240054608-eng.pdf

World Meteorological Organization (WMO). "Climate Change Indicators Reached Record Levels in 2023: WMO," World Meteorological Organization (2024). https://wmo.int/news/media-centre/climate-change-indicators-reached-record-levels-2023-wmo

Wortzel, Jeremy D., et al. "Ascertainment Bias in a Historic Cohort Study of Residents in an Asbestos Manufacturing Community," *International Journal of Environmental Research and Public Health* 18, no. 5 (2021): 2211–2228.

Wright, David. "Canadian Net-Zero Emissions Accountability Act: A Bridge Over the Implementation Gap?" *University of New Brunswick Law Journal* 74 (2023): 59–81.

Wyatt, Stephen. "First Nations, Forest Lands, and "Aboriginal Forestry" in Canada: From Exclusion to Ecomanagement and Beyond," *Canadian Journal of Forest Research* 38, no. 2 (2008): 171–180.

Wyatt, Stephen, et al. "Indigenous Peoples' Expectations for Forestry in New Brunswick: are Rights Enough?" *Society & Natural Resources* 28, no. 6 (2015): 625–640.

Y

Yadin, Sharon. "Regulatory Shaming and the Problem of Corporate Climate Obstruction," *Harvard Journal on Legislation* 60 (2023): 337–382.

Yale Program on Climate Change Communication. "Climate Change in the American Mind: Politics & Policy, Spring 2024," 8 August (2024). https://climatecommunication.yale.edu/publications/climate-change-in-the-american-mind-politics-policy-spring-2024/

Yanosek, Kassia, and Victor, David G. "Capitalism's Green Revolution How Private Finance Can Help Decarbonize the Economy," *Foreign Affairs* 29 November 2023.

Ye, Jun, et al. "Novel Synergy Mechanism for Carbon Emissions Abatement in Shipping Decarbonization," *Transportation Research Part D: Transport and Environment* 127 (2024): 104059.

Yirenkyi, Edmund, and Kelly Vodden. "The Role of Philanthropy in Financing Climate Change Mitigation and Adaptation in Canada: A Systematic Literature Review," *Local Development & Society* (2024): 1–31.

Young, Stephen. *Indigenous Peoples, Consent and Rights: Troubling Subjects* (London: Routledge, 2019).

Yumagulova, Lilia, et al. "Indigenous Perspectives on Climate Mobility Justice and Displacement-Mobility-Immobility Continuum," *Climate and Development* (2023): 1–18.

Z

Zahar, Alexander. "The Paris Agreement's 'Nationally Determined Contribution': Law and Practice," in *Research Handbook on the Law of the Paris Agreement*, ed. Alexander Zahar (Edward Elgar Publishing, 2024): 143–166.

Zakharova, Olga, and Anna Glazkova. "Green Waste Practices as Climate Adaptation and Mitigation Actions: Grassroots Initiatives in Russia," *BRICS Law Journal* 11, no. 4 (2024): 145–167.

Zanjani, Leila V., Hugh Govan, Holly C. Jonas, Theodore Karfakis, Daniel M. Mwamidi, Jessica Stewart, Gretchen Walters, and Pablo Dominguez. "Territories of Life As Key to Global Environmental Sustainability," *Current Opinion in Environmental Sustainability* 63 (2023): 101298, 5.

Zhang, Boen et al. "Anthropogenic Climate Change Doubled the Frequency of Compound Drought and Heatwaves in Low-Income Regions," *Communications Earth & Environment* 5, no. 1 (2024): 1–8.

Zhong, Raymand. "Are We in the 'Anthropocene,' the Human Age? Nope, Scientists Say," *The New York Times,* 5 March (2024). https://www.nytimes.com/2024/03/05/climate/anthropocene-epoch-vote-rejected.html?smid=nytcore-ios-share&referringSource=articleShare

Zhou, Leocadia. "An Analysis of the Differences in Vulnerability to Climate Change: A Review of Rural and Urban Areas in South Africa," *Climate* 10, no. 118 (2022): 1–22.

Zhou, Zhenchao, et al. "Association Between Particulate Matter (PM) 2· 5 Air pollution and Clinical Antibiotic Resistance: A Global Analysis," *The Lancet Planetary Health* 7, no. 8 (2023): 1–11.

Zhu, He. "Geoengineering: Proactive Measures to Tackle Climate Change," *National Science Review* 11, no. 4 (2024): nwad271.

Zhu, Xufeng, and Xin Shang. "Positioning as Discursive Struggle for Equity: A Critical Discourse Analysis of the Nationally Determined Contributions (NDCs) of African Countries," *Critical Discourse Studies* 21, no. 2 (2024): 218–233.

Ziervogel, Gina, and Anna Taylor. "A Co-Produced National Climate Change Risk and Vulnerability Assessment Framework for South Africa," *Frontiers in Climate* 5 (2023): 1–15.

Ziervogel, Gina, et al. "Climate Change in South Africa: Risks and Opportunities for Climate-Resilient Development in the IPCC Sixth Assessment WGII Report," *South African Journal of Science* 118, no. 9–10 (2022): 1–5.

Zimm, Caroline, et al. "Justice Considerations in Climate Research," *Nature Climate Change* 14, no. 1 (2024): 22–30.

Zimmerman, Scott. "Drilling Deeper 2024: Global Oil & Gas Extraction Tracker," Global Energy Monitor, March (2024). https://globalenergymonitor.org/wp-content/uploads/2024/03/GEM-global-oil-and-gas-extraction-tracker-March-2024.pdf

Zoll, Deidre. "We Can't Address What We Don't Acknowledge: Confronting Racism in Adaptation Plans," in *Justice in Climate Action Planning*, eds. Brian Petersen and Hélène B. Ducros (New York: Springer, 2022), 3–23.

Zoll, Deidre, R. et al. "Intersectional Climate Perceptions: Understanding the Impacts of Race and Gender on Climate Experiences, Future Concerns, and Planning Efforts," *Urban Climate* 50 (2023): 1–18.

Zoomers, Annelies, and Mayke Kaag. *The Global Land Grab: Beyond the Hype* (London: Bloomsbury Publishing, 2014), 9–11.

Zraick, Karen, and Lisa Friedman. "Inside Trump's Renewed Effort to Undo a Major Climate Rule," *The New York Times*, 28 January (2025). https://www.nytimes.com/2025/01/28/climate/trump-endangerment-finding-epa.html

Zuni Cruz, Christine. "The Indigenous Decade in Review," *Southern Methodist University Law Review Forum* 73 (2020): 140–156.

Zweig Sophia A., et al. "Ensuring Rights While Protecting Health: Importance of Using a Human Rights Approach in Implementing Public Health Responses to COVID-19," *HHR Health and Human Rights Journal* 23, no. 2, December (2021): 173–186.

COURT CASES

United States

Adoptive Couple v. Baby Girl 570 U.S. 637 (2013).
Arizona et al. v. Navajo Nation et al. 26 F. 4th 794 (reversed).
Chevron U.S.A., Inc. v. Natural Resources Defense Council, Inc. 467 U.S. 837 (1984).
Corner Post, Inc. v. Board of Governors of the Federal Reserve System 603 U.S. ___ (2024).
Federal Power Commission v. the Tuscarora Indian Nation 362 U.S. 99 (1956).
Haaland, Secretary of the Interior, et al. v. Brackeen et al. 599 U.S. (2023).
Jeffrey Nuziard, et al. v. Minority Business Development Agency, et al. United States District Court for the Northern District of Texas Fort Worth Division No. 4:23-cv-00278-P (2024).
Juliana v. United States, 947 F.3d 1159, 1165 (9th Cir. 2020).
Liberty Energy v. SEC, No. 24-60109 (5th Cir. 2024).
Loper Bright Enterprises v. Raimondo 603 U.S. ___ (2024).
Ohio v. Environmental Protection Agency 603 U.S. ____ (2024).
Students for Fair Admissions Inc v. President and Fellow of Harvard College, 600 U.S. 181 (2023).
McGirt v. Oklahoma 591 U.S. __ (2020); 140 S.Ct. 2452, 2462 (2020).
Oklahoma v. Castro-Huerta 597 I.S. __ (2022); 142 S. Ct. 2486 (2022).
Sackett v. Environmental Protection Agency 598 U.S. 651 (2023).
United States v. Cooley 593 U.S. __ (2021); 141 S Ct 1638 (2021).
West Virginia v. Environmental Protection Agency, 597 U.S. ___ (2022).
Wynn v. Vilsack, 3:21-CV-514-MMH-LLL (M.D. Fla. Aug. 1, 2023).

Canada

Tsilhqot'in Nation v. British Columbia 2014 SCC 44.

Index